Management for Professionals

More information about this series at http://www.springer.com/series/10101

Peter Krüssel

Editor

Future Telco

Successful Positioning of Network Operators in the Digital Age

 Springer

Editor
Peter Krüssel
Detecon International GmbH
Cologne, Germany

ISSN 2192-8096   ISSN 2192-810X   (electronic)
Management for Professionals
ISBN 978-3-030-08521-6   ISBN 978-3-319-77724-5   (eBook)
https://doi.org/10.1007/978-3-319-77724-5

This Springer imprint is published by the registered company Springer Nature Switzerland AG.
The registered company address is: Gewerbestrasse 11, 6330 Cham, Switzerland

Foreword

The limits of our language are the limits of our world. This version of Ludwig Wittgenstein, the Austrian philosopher's proposition, has passed into our language as a kind of proverb. Telecommunications minimizes the limits of our world. We speak simultaneously about the past, present, and future—ignoring the real temporal and spatial distance between the conversationalists. The free, fast, direct, and unlimited access to information is a basic condition for knowledge and education, and both of these are in turn fundamental prerequisites for mastering the tasks and challenges confronting all civil societies in the twenty-first century.

The digital revolution currently affecting so many dimensions of our daily lives is shaping the future of communication in our culture. Resource management, industry, services, and high technology are parts of this culture. And all are facing the question of whether they will be the hammer or the anvil, whether they will themselves shape the course of events or simply be shaped by what happens.

Many players, both in the telco industry and outside of it, have in the meantime realized what opportunities are being offered by the predominant technological development of our young millennium to those who do not wait to see what their competitors are doing but become drivers and pioneers themselves. This—just as anything else of value—cannot be accomplished without commitment and anticipatory imagination. Demands are growing, the steady exponential rise in quantities must be mastered, and the quality of the related performance must be guaranteed. Speed, flexibility, and learning capability are playing an ever-greater role. New forms of procurement, distribution, and cooperation must be tried out, evaluated, and either rejected or adapted. This is equally true of hardware, software, and infrastructures, and it impacts employees. Persistence, innovation, creativity, curiosity, and the readiness—no, the eagerness—to accept and drive changes out of the understanding of their necessity: all of this is first and last dependent on the people who join us in our efforts to push against the walls limiting us and, one day, to tear them down. That is why we pay attention to one another.

Actively and positively shaping the future of the world with the aid of communication and telecommunications, that should and can and will be achieved! This is the key message that is conveyed in the many articles found in this editor's volume.

The authors whose works are compiled in this book deserve special praise for having illuminated all of the relevant aspects of digitalization in the various core functions of telecommunications companies from different perspectives. The primary challenges are analyzed, concrete solutions are briefly described, and efficacy of these solutions is demonstrated on the basis of numerous examples of success in daily practice. Profound experience, outstanding expertise, and valuable recommendations offer orientation to telco decision-makers as well as to representatives of regulatory authorities and the political establishment.

Deutsche Telekom AG   Timotheus Höttges
Bonn, Germany

Preface

Everyone is talking about digitalization—it is changing the way we live and do business. The pace at which this is happening is accelerating and encompassing more and more areas. Every company is affected; the market structures of all industries are changing profoundly. New, innovative companies are sprouting up, beginning life as start-ups, and then developing into unicorns; within only a short time, they dominate specific markets before unhesitatingly expanding their business fields into vertically and horizontally adjacent sectors and out onto the world market. Large, established corporations, supposedly immune to attack, may begin to tremble if they do not adapt their production systems and marketing models quickly enough. The effects can clearly be seen in the energy industry, the automotive sector, the media, the hotel business, the financial services, and, on a massive scale, in the telecommunications industry as well. Digitalization makes existing barriers to market entry to disappear, turns established business and remuneration models upside down, causes the merger of sectors, promotes the decoupling of added-value stages previously tightly meshed with one another, releases immense potential for efficiency and flexibilization in production, and accelerates innovation processes in a magnitude never seen before.

Network operators in particular are challenged by these developments. On the one hand, they have the infrastructure to serve as a key vehicle for the digitalization of other business branches and national economies, yet, on the other hand, they must master the challenges and seize the opportunities arising from digitalization for their own business purposes.

While the close coupling of network and services once meant that the network operators were the providers of both domains, they find themselves today being pushed into the simple role of connectivity providers in many cases because of digitalization and the IP-based technical separation in production of services and network operation. The traditional communications services of telecommunications operators (telcos) are being replaced more and more frequently by the services offered by OTTs. Successful innovative services appear to come primarily from such providers as well. Telcos—it seems—achieve only limited success in these fields. As a consequence, many carriers often focus on their genuine core asset—the further expansion of their networks.

Digitalization is evidently playing into the hands of the platform-based providers. They can exploit economies of scale and scope in equal measure. Operating from just a few central production points, they can make their services available world-wide at little cost. The laws of platform business inevitably lead to a situation in which only a few providers—and in extreme cases, only one—win the race: "the winner takes it all." Speed, the lack of market entry barriers, the non-remuneration of services, the fast gain of high market shares—a critical mass—and network effects determine the rules of the game. New services and business models are more easily reconciled with the conditions that both foster creativity and favor investments found in Silicon Valley than with the rigid structures of an established industry.

When it comes to questions about creating general conditions appropriate for fostering competition, politicians and regulatory authorities in Europe and Germany concentrate primarily on the network level. The service level, although just as important, withdraws into the background. Digitalization is understood first and foremost to be an infrastructure task: the provision of connectivity with the greatest possible bandwidths—based ideally on optic fiber—to all customers. A regulatory framework that secures a so-called *level playing field* among all providers in the field of services regardless of where they come from while taking into consideration as a totality all legal instruments such as competition law, tax law, media law, consumer protection, privacy, and sector-specific regulation does not appear to be achievable.

This book examines the question of how the network operators, faced with these general conditions, can successfully secure a sustainable position for themselves in the competitive environment. The target vision is the so-called *network-based service provider*, whose role is rather the one of a pacemaker of digitalization than the one of a mere transporter of data or of a bit pipe. Achieving such a position demands a major presence at the service level in addition to rigorous modernization of the network infrastructures. It also presumes the exploitation of opportunities and the mastering of the challenges arising from digitalization by the carriers. This task confronts carriers with enormous transformation needs. The contributions to this book offer recommendations for action aimed at the realization of this target vision.

The primary target audience of this book comprises managers and practitioners from the telecommunications industry but also extends to teachers and students from the sectors business, computer science, and communications technology as well as all those in society, business, and politics who concern themselves with the future of the telecommunications sector.

The first section of the book offers a prediction about the changes in the market structure that will be induced on the telecommunications market by digitalization. A model for the derivation of six fundamentally possible, future market scenarios is introduced. These market scenarios stand for various market structures on the provider side. One specific scenario, the so-called *heterogeneous power play*, is postulated as the most attractive scenario for network operators. In this scenario, network operators might have the chance to establish a successful position as *network-based service providers*. Telcos can drive this scenario by employing seven levers. These seven levers represent concrete fields of action for the network operators at the various added-value stages and provide the structural framework for the book. One part is devoted to each lever.

The first part, *Integrated Networks*, emphasizes the significance of integrated networks and offers suggestions for ways to address the important topics of network coverage and network capacity through the use of new technologies and methods.

The part *Modern Network Concepts*—the second lever—considers technologies that contribute significantly to enhanced efficiency through automation, standardization, and virtualization in production.

The third lever/part *Product Innovation* introduces a number of different examples to describe the opportunities that network operators can create for themselves by pushing their own innovations. The articles and interviews in this part present examples of promising product innovations that are based above all on the specific assets of telcos.

The part *Partnering* (the fourth lever) describes possibilities for expanding carriers' portfolios through partnerships with other companies, thereby strengthening the bonds of customer relationships.

The following part entitled *Regulation, Wholesale and Wholebuy*, the fifth lever, revolves thematically around the creation of appropriate regulatory framework conditions in the relationship between telcos and OTTs and offers examples of innovative and successful wholesale business models.

In the sixth part *Customer Centricity*—the sixth lever—questions about the optimal design of interfaces between customers and telcos are answered. In addition, it describes how telcos can obtain greater understanding for the interests of their customers through the use of new technologies.

These six levers are of fundamental, transformational significance for telecommunications companies. A transformation of this nature will be successful only if the seventh lever, *Internal Enabler*, is applied. Telcos are called upon to modify accordingly the traditional structures, cultures, roles, skills, and processes together with IT as the embodiment of the living processes in the technical system. This range of subjects is discussed in the concluding seventh part.

Finally, I would like to thank all the authors who contributed to the success of this book. My special thanks go to Detecon International GmbH, which played a decisive role in the realization of this publication.

Cologne, Germany   Peter Krüssel
2018

Contents

Anticipating the Future: Strategic Scenarios for Telco Markets and Initial Recommendations for Operators

Peter Krüssel and Friederike Göbel

1   Introduction

In recent years, many telecommunications companies (telcos) have been forced to adapt to profound changes and to complete a steep learning curve. The deregulation of the telecommunications markets by the end of the 1990s led to a strong fragmentation of the service provider side at national, regional and local levels as well as to a strong decrease in prices for access and services. The rapid technological developments and the market entrance of over-the-top players (OTTs) led to fantastic innovations as well as to increased competition for telcos at the service level. The structures of the telecommunications markets have changed dramatically in the last 10 years.

The article is motivated by the following questions: How does the spectrum of possible market structures for the next 3–5 years look like? Which future market structures are particularly advantageous for telcos? And how can telcos actively create these market structures that are beneficial to them?

In order to answer these questions we introduce a model, developed to derive strategic market scenarios (market scenario model) and apply it to the German telecommunications market. We evaluate the resulting scenarios in terms of their benefits for network operators and their likelihood of occurrence. Subsequently, we identify levers (i.e., areas of action) that telcos need to activate in order to achieve the most advantageous scenario for them.

P. Krüssel (✉) · F. Göbel
Detecon International GmbH, Cologne, Germany
e-mail: Peter.Kruessel@detecon.com; Friederike.Goebel@detecon.com

© Springer International Publishing AG, part of Springer Nature 2019
P. Krüssel (ed.), *Future Telco*, Management for Professionals,
https://doi.org/10.1007/978-3-319-77724-5_1

1

2   Key Elements of the Model

The key elements of the market scenario model are the player typology, the market structure matrix, and the market drivers (see Fig. 1). The *player typology* used in the model distinguishes four player groups:

- *Network operators*; fixed network operators such as cable television (CATV) operators or regional carriers such as NetCologne and EWE Tel in Germany, mobile network operators such as O2 in Germany, and integrated network operators such as Vodafone and Deutsche Telekom in Germany
- *Resellers* such as freenet in Germany operating independently of network operators
- *Over-the-top single-purpose providers* (OTT SPs) such as Deezer, Netflix or Spotify
- *Over-the-top ecosystem providers* (OTT EPs) such as Apple, Amazon or Facebook

The *market structure matrix* illustrates possible market scenarios by providing information on size, market activities and production base by player type. The comparison of matrices over time clearly indicates changes in the market structure. Trends like fixed-mobile convergence, consolidation, an increasing vertical integration by player

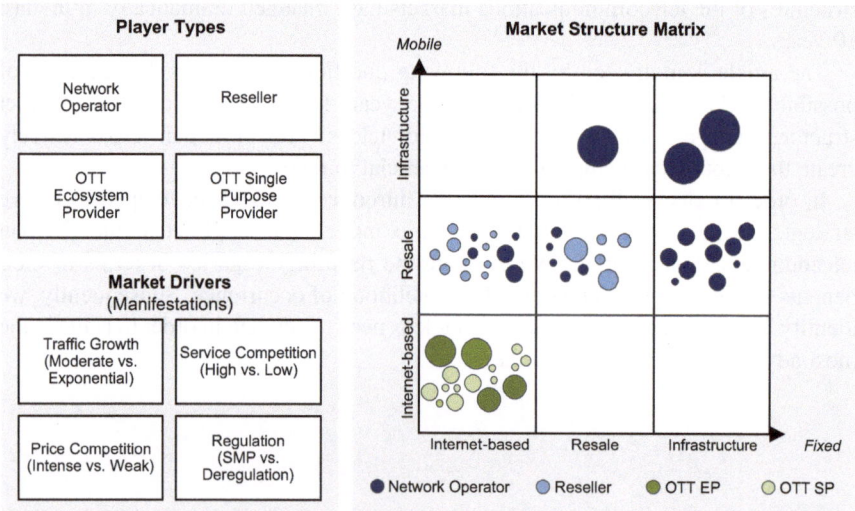

Fig. 1 Overview of key elements of the model

shifts among the player types can be easily depicted. The market structure matrix is organized as follows:

- The circle color indicates the player type.
- The circle size corresponds to the player size (in terms of revenue).
- The number of circles reflects roughly the number of players.
- Along both axes the three different options (own infrastructure, resale, internet based) of the production bases (fixed versus mobile networks) are displayed. This results in a division of the matrix into nine quadrants.
- The positioning of a circle within a quadrant provides information on the market activity of the player. Along the x-axis/y-axis the positioning indicates the extent of activities in the fixed/mobile segment.

Market drivers are the third key element of our model. We identify four key market drivers that affect the market structure and the player types. We assume that each market driver can show one of two opposing manifestations:

- *Traffic growth*: exponential versus moderate
- *Service competition*: high versus low
- *Price competition*: intense versus weak
- *Regulation*: significant market power (SMP) oriented versus deregulation for telcos

3   Market Drivers and Their Effects on Player Types

Each market driver has different effects on the player types and the market structure, depending on its manifestation. In this section we describe each market driver, its effects on the different player types and outline the corresponding changes in the market structure matrix.

3.1   Traffic Growth and Its Effects

Traffic growth refers to the increasing flow of data across the networks. It can either be exponential or moderate. In case of *exponential* traffic growth, current and planned investments in the expansion of infrastructure will not be sufficient to handle the increasing amount of data. Existing networks will not be able to deal with the increasing amount of data. Consequently, there will be an urgent need for the expansion of integrated networks. Providers that have only a weak infrastructure and limited investment power will be forced out of the market. Integrated heavy-asset players are clearly at an advantage, while mobile-only players are deadlocked. In case of *moderate* traffic growth, current and planned investments in the expansion of infrastructure will be sufficient. The use of conservative technologies (e.g., vectoring) results in an expansion of capacity of existing networks. This expansion will be enough for the gradually processing of emerging traffic.

Network Operators The stronger the traffic growth, the greater the need for additional investments in infrastructure, the worse the profitability, the smaller the budget available for investments in innovations on the services level and in new areas of business and the stronger the pressure to consolidate. Using the nomenclature of the market structure matrix: the stronger the traffic growth, the smaller the number of circles and the larger the size of the remaining circles. In case of moderate traffic growth, either the opposite effect occurs or the current situations remains unchanged.

Resellers In case of exponential traffic growth, the bargaining power of resellers towards network operators will become worse. This is due to the decreasing need of network operators to sell capacity to resellers. Moreover, the networks are operating at the limits to their capacities. The relevance of integrated seamless networks and services increases. Today most resellers' activities do not cover these services. Thus, their offering will turn into an USP for network operators. Prices for advance services are likely to increase. This increase in prices probably cannot be transferred to the end customer. Thus, the margins of the resellers will decrease. The pressure to cut costs increases. In order to gain bargaining power resellers must gain in size. Thus, the pressure to consolidate increases. In case of a moderate traffic growth, the opposite effects can be assumed. The stronger the traffic growth, the smaller the number of circles in the market structure matrix and the greater their size.

OTT EPs The stronger the traffic growth, the more scarce the availability of network resources, the smaller the chance to provide innovative services in a superior quality to the end customer, the greater the pressure to pay network operators for a quality assured transport or to invest in infrastructure, the worse the profitability. At country level, however, the impact on the large OTT EP is limited. This is because they operate globally and have immense financial resources. The number of circles as well as their size will therefore mostly remain the same. If they invest in infrastructure or push MVNO models, their position in the market structure matrix will move in the direction of network operators.

OTT SPs In general, the effects of exponential growth on OTT SPs are comparable to those on OTT EPs. As OTT SPs are smaller and larger in number than OTT EPs they have less bargaining and market power. Therefore, they will be pressured to consolidate. Alternatives like contributing to network costs in the form of cooperative ventures or investments in their own infrastructures are unlikely. Ultimately, exponential growth in traffic leads to consolidation or exits from the market, the number of circles declines or they even disappear completely owing to market exits or consolidation efforts by the network operators and, above all, the large OTT EPs.

3.2   Service Competition and Its Effects

The driver service competition refers to the activities of the different players on the service level. The formerly close connection of networks and services will dissolve. The provision of services increasingly shifts from telecommunication networks towards the internet (OTT services). This development results in an increased fragmentation of the value chain in the telecommunication industry. Service competition can either be high or low. In case of *high* service competition network operators would succeed in offering both, connectivity as well as network integrated services and platforms. They would build their own ecosystems consisting of diverse components der value added. These ecosystems could compete with those of OTTs on the same level. In case of *low* service competition network operators focus their activities on the infrastructure level (pure connectivity business). They would not participate in the production and selling of internet-based services at the services level. They would act as bit pipe provider with or without direct contact to the end customer.

Network Operators   The greater the level of service competition—in other words, the more intense competition between telcos and OTTs on the service side—the higher the short- to medium-term investment requirements in platforms, partnerships with other carriers and OTT providers, the lower the margins and profitability, the greater the consolidation pressure. At the same time, new revenue potential can arise for telcos in areas where they are or were not so strongly represented. As a consequence, the size of the network operator circles increases while at the same time reducing the number of connections. In the case of low service competition, telcos will sooner or later become bit pipes, in extreme cases they will act as wholesale providers for connectivity without direct customer contact. Their business model would be largely defined by efficiency in production and economies of scale with lower revenues and margins. The number of circles and their size would decrease accordingly.

Resellers   The lower the level of service competition and the stronger the OTTs dominate the internet-based production of services, the lower the chances of survival for resellers. Its core business (i.e., the provision of traditional communication services such as SMS and telephony) is threatened by OTTs' offerings based on new web technologies (e.g., WebRTC, eSIM, WhatsApp). The OTTs would invade the resellers' domain and destroy their basis for business. In the market structure matrix the circles representing the resellers would disappear. In case of high service competition, whether or not resellers have a chance of survival depends largely on the network operators.

OTT EPs and OTT SPs   The lower the level of service competition, the greater the success of OTT EPs and OTT SPs. Both player types profit. The number and size of circles in the market structure matrix tend to increase. Intensive service competition leads to the opposite effects. In particular, the OTT SPs could lose much of their importance in the course of consolidations.

3.3 Price Competition and Its Effects

The driver price competition relates to the development of prices end customers have to pay for telecommunication services like access, telephony and SMS (i.e., the traditional core products of network operators). Price competition can be either intense or weak. Intense price competition exists when prices decrease, while there is weak price competition, when prices stagnate or even increase slightly. The horizontal price competition between telcos has a major impact on profitability, on available funds for investments in infrastructure expansion and innovation, and on consolidation pressures. In case of *intense* price competition, we assume that investments in infrastructure and new business areas will be scaled back, as otherwise there would be significant margin pressure on network operators. Based on this, the number of secondary brands of network operators, no-frills and low-cost brands will increase. In addition, the consolidation pressure on carriers increases. Cost pressure and efficiency programs are at the forefront. In case of a *weak* price competition, we expect the opposite effect. Stable earnings allow carriers to invest in infrastructure and new business areas.

Network Operators The more intense the price competition, the lower the revenues, the lower the margin, the greater the pressure to enhance efficiency in production and the greater the pressure to consolidate among network operators. Due to consolidation the number of circles in the market structure matrix that represent network operators declines while their size increases. In the event of weak price competition, these trends are correspondingly weaker or non-existent.

Resellers Intense price competition also has a negative effect on the earnings and margins of resellers. In addition, network operators are likely to start the attempt to win the market segments addressed by resellers with their own secondary brands. Resellers' negotiating power vis-à-vis network operators is decreasing, so that we also expect resellers to consolidate. The number of circles representing resellers decreases while their size increases. In the case of weak price competition no change in the status quo is expected.

OTT EPs and OTT SPs Increasing horizontal price competition among network operators is advantageous for both OTT EPs and OTT SPs as they have a different approach of monetization for their services. They are financed via data-centric business models, primarily through the marketing of targeted advertising space and times or individual transactions and specific service revenues. The weakening of network operators and resellers strengthens their position. The obligation of network operators to cooperate with OTTs increases and enhances their business potential. The number and size of circles depicting two groups increases.

3.4   Regulation and Its Effects

The driver regulation is significant for the market scenarios because the regulatory framework affects the power relationships among the players by steering competition, access to resources, price, privacy, consumer protection etc. At this time, there are multiple initiatives under discussion in this field, including the *single market* initiative of the EU Commission. This initiative aims at creating a European telecommunications market with pan-European providers, including a limited but thorough access regulation combined with network neutrality regulation. At the same time the EU Commission took action against Google in the form of competition regulation initiatives. In our model we assume two opposing regulatory models: one that is SMP-oriented and one that foresees deregulation for telcos. In case of *SMP-oriented regulation,* the sector-specific regulation of the telecommunications market with a focus on access regulation at the network level is retained and strict network neutrality for the network operators is introduced. OTTs are not subject to regulation on either the sector-specific service side or the side of the competition, anti-trust laws or other frameworks. In the case of *deregulation* for telcos, there will be a regulatory cutback for telecommunications companies, including a reduction in the regulatory measures for access and charges, and a weakening of the strict network neutrality requirements. At the same time, the OTTs will be subjected to more extensive regulation, in particular by the regulation of competition.

Network Operators The greater the regulatory reduction for network operators and the greater the opportunities for differentiated traffic management and monetization, the greater the entrepreneurial freedom to exploit the opportunities for price differentiation and the development of new sources of revenue, the greater the incentives for investment in new infrastructures, services and business models. In the market structure matrix, the circles depicting network operators get bigger due to additional revenue opportunities. Their number remains essentially unchanged. In contrast, strict network neutrality and sector-specific regulation oriented to SMPs exclude these revenue opportunities and freedom for business operations. The circles depicting network operators become smaller and their number tends to decrease.

Resellers Resellers in Germany and Europe benefit from the current regulatory system. One good example can be seen in the requirements of the EU Commission related to mergers. When Telefónica took over E-Plus and Hutchison took over Orange in Austria, the companies were obligated to open their networks to MVNOs and resellers. In view of the consolidation on the market and the increasing market power of OTTs, regulatory measures of this kind are a key factor for the existence of resellers. To this extent, a loosening of the regulation of OTTs presents a major problem for resellers. Furthermore, the introduction of network neutrality in favor of the OTTs would represent a threat of substitution by the OTTs. In the case of strict network neutrality and less strict SMP regulation, the circles representing resellers become fewer in number and smaller in size. They might even disappear as a consequence of the substitution with the OTTs (e.g., Google Fi, Apple eSIM,

WhatsApp Voice) or because network operators are forced into second-brand activities.

OTT EPs Deciding on network neutrality and maintaining sector-specific telecommunications regulation would benefit OTT EPs. The resale and MVNO obligations, which were introduced in connection with mergers and acquisitions, especially in the mobile communications market, create further opportunities for OTT EPs. The obligations weaken the network operators and pave the way for OTT EPs connectivity to be integrated into their offers. The circles of the OTT EPs become larger, and some of them might develop in the direction of resale. Their number most likely remains unchanged. Possibly they are joined by additional OTT providers that have previously been among the smaller ones. In the event of heightened regulatory activities for platform-based Internet providers in the sectors data and consumer protection, tax laws, interconnection (opening of closed ecosystems to third parties), legal interception, and a loosening of strict network neutrality, we can assume rising costs for the OTT EPs, a correspondingly decline in profits, and increasing competition among themselves as well as from the network operators. In this case, we foresee stagnation or a slight shrinkage of the circles representing OTT EPs. The effects in comparison with their current size, however, are likely to be rather minor. Also their number is unaffected as this player type has a critical mass, a strong relationship to customers, and market power that is securely founded in numerous business models.

OTT SPs The effects of the driver regulation on OTT SPs are very similar to those on OTT EPs. Due to their smaller size and market power their impact, however, is much more noticeable. The greater the regulatory reduction in the direction of telcos and the regulatory build-up in the direction of OTT players, the greater the pressure on revenues and margins due to the costs of any upfront investments by telcos and the greater the consolidation pressure. This form of regulation tends to cause a decline in the number of circles for this group. The size presumably stagnates or even shrinks. In the extreme case, we can look for a significant containment of the OTT SPs because they are identified at an early stage as rewarding consolidation objects by either the telcos or the large OTT EPs. At its core, differentiated regulation dependent on market power is decisive here. In the event of strict network neutrality and strict regulation of the telcos, the reverse effects on the circles of the OTT SPs in the market structure matrix can be expected.

4   Application of the Model to the German Telecommunications Market

In this section we describe the development of the German telecommunications market in recent decades. Subsequently, we apply the market scenario model to today's market. On the basis of current market developments and observed case studies, we determine the manifestations that market drivers will have in the next 3–5 years.

4.1   The German Telecommunications Market: A Retrospective

Following the deregulation of the 1990s, which was based on pronounced SMP access regulation, the German market since the start of the new millennium has been marked by consolidation, falling prices, low levels of investment activities, and strong traffic growth. This was very much a result of the large number of providers entering the market at the turn of the millennium, leading to intense competition.

The effects on the market structure between 2006 and 2016 are illustrated in Fig. 2. In 2006, there was a large number of smaller, independent telecommunications providers operating locally and regionally and based primarily on fixed networks. There was also a large number of independent resellers. Fixed network and wireless network services were strictly separated on the provider side. There were significantly fewer OTTs, and OTT SPs was de facto non-existent. The market structure in 2016 is far different. The convergence of fixed and mobile networks can be clearly observed on the provider side, either on the basis of integrated networks or from the addition of resale services. Consolidation has advanced quite far. The number of independent local or regional providers has declined drastically, resellers have consolidated, and carriers' second brands have gained in importance. The OTT landscape has mushroomed, and its role has grown substantially in significance.

Set against the backdrop of the described model for the derivation of strategic market scenarios, these past developments can be traced back to intense price competition, immense growth in traffic, greater service competition, and strict SMP regulation.

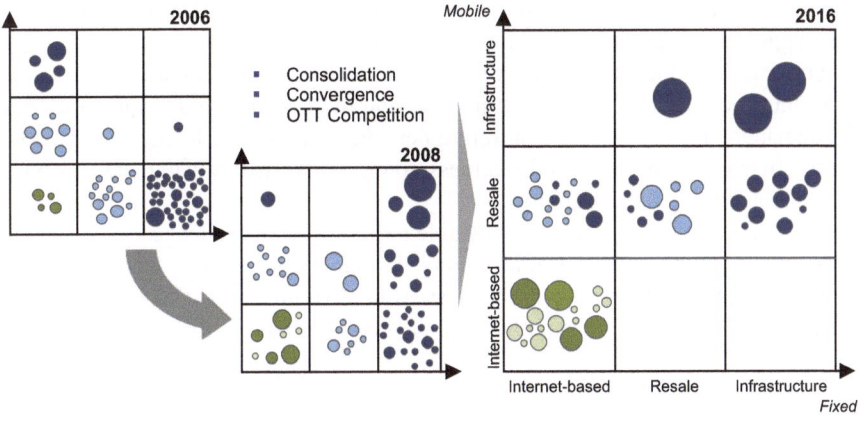

Fig. 2 Market structures on the German Telecommunications Market 2006–2016

4.2   Determination of Market Driver Manifestations

4.2.1   Traffic Growth

Traffic growth is and remains a key driver of changes in the market structures. It results from numerous technical developments like artificial intelligence, virtual and augmented reality, 360° videos, new high-resolution displays, new access technologies for fixed and mobile networks or edge computing.[1] Likewise, new services (e.g., 5G-oriented use cases, M2M, IoT, cloud-based services or mobile video streaming) as well as new business models (e.g., IaaS, PaaS, NaaS) contribute to an increasing amount of traffic. Another factor that accelerates traffic growth are the user habits. There is a growing number of personal, connected devices and a growing use of smartphones as personal switchboards for utilization of all possible services. Users have the desire for constant connectivity independent of their location.

In 2016 the data volume in fixed and mobile networks has once again increased rapidly in Germany. 22.5 billion GB of data has been transferred via fixed networks, which corresponds to an increase by 32% compared to the previous year. 918 million GB has been transferred via the mobile network. This corresponds to an increase by even 60% compared to 2015 (Bundesnetzagentur 2017).

Today already each mobile user in Germany uses 1 GB traffic per month. In 4 years this number will have increased fivefold. A main reason for this development will be the increased consumption of videos via the mobile network. This is not only due to the increasing popularity of Netflix and YouTube. The digitalization of the educational system and the working world will also strengthen the role of videos (Funkschau 2017).

The mobile data traffic in Germany will increase by an annual growth rate of 41% until 2021. By the end of 2021 smartphones will account for only 62% of the whole mobile traffic. Instead the proportion of M2M-connections and wearables will increase. The M2M data traffic will increase twelve-fold and therewith accounts for one fifth of the mobile traffic. In 2021 there will be 24 million wearable of which 2 million will have a mobile telephone connection. In 2021 each German citizen will have approximate 2.8 networked mobile devices (Funkschau 2017).

These developments have a direct impact on network operators in particular. They must competently handle the growing and more volatile traffic quantities and the demands that will be made on future networks with regard to capacity, (worldwide) coverage, integration and seamless connectivity, quality parameters, robustness, reliability, and affordability.

Not every network operator will be able to master these challenges, especially in regions or countries where traffic volume is rising significantly. Only integrated carriers (i.e., network operators who possess the two domains of fixed and mobile networks) will be able to accomplish this. In these countries, traffic growth will be a

---

[1]Edge computing is the centralization and virtualization respectively softwareization of products in networks accompanied by distributed intelligence and storage on the periphery of the networks.

major driver of consolidation among network operators. The goal will be to take care of the traffic quantity while maintaining the demanded quality and observing the constraints of economical operation. Numerous examples of mergers and acquisitions and divestment transactions[2] in Europe are evidence of this.

Based on these market observations we assume an exponential traffic growth in the German telecommunications market.

4.2.2 Service Competition

The service provision increasingly shifts from the telecommunication networks to the internet. This concerns not only the classic OTT-domain but also the core business of telcos (i.e., voice services and SMS). Thus, the competition on the service level does not only increase among telcos but also between OTTs and telcos. This trend is described in the annual report for the year 2016 of the Bundesnetzagentur. In the course of the proliferation of messenger services (e.g., WhatsApp) the number of SMS sent in Germany decreases drastically. While the number of SMS sent reached its peak in 2012 (60 billion SMS), there were only 16.6 billion SMS sent in 2015 and 12.7 billion SMS sent in 2016. Also the total volume of call minutes in the fixed network decreases (2016: 131.0 billion minutes; 2015: 139.9 billion minutes) while the total volume of call minutes in the mobile network stagnates (2015 and 2016: 115 billion minutes). By the end of 2016, almost 60% of all call minutes in fixed networks have been realized IP-based (Bundesnetzagentur 2017).

In some countries, connectivity services such as mobile contracts and fixed access are already part of the OTTs' portfolios and they compete with telcos along the entire value chain. They are relying in these cases on innovative access technologies of many different types or MVNO agreements with established mobile network operators.

Alphabet/Google has launched the *Google Fi* service for end customer in the USA as an MVNO or mobile virtual network provider, riding on the networks of the network operators as well as on public hotspots to obtain access. The primary interest could be the tight linking with one another of Google devices, the most widespread mobile operating system Google Android, the Google services universe (including search engine, browsing, ticketing services, various cloud services, smart home, TV, etc.), and the created Google Connectivity. Google could then present itself to customers as a provider with complete vertical integration. Since expenditures could be refinanced from users' data or from advertising, Google could if necessary offer the connectivity service as well as many of the services free of charge or at prices substantially lower than those of the incumbent mobile network providers.

---

[2]Recent examples are Telefónica's sale of EE to BT in Great Britain in 2015, Vodafone's acquisition of KDG in Germany and Ono in Spain in 2014, Orange's purchase of Jazztel in Spain in 2015, Telekom Austria's purchase of One in Macedonia and of Amis in Slovenia/Croatia in 2015, Numericable's purchase of SFR in France in 2014 and Telenet's purchase of Base in Belgium in 2016.

Alphabet is also engaged in the initiative *LinkNYC*; the campaign's goal is to provide free WiFi in New York through 7500 modern phone booths or Internet kiosks. The first Internet kiosk in futuristic look was unveiled at the beginning of 2016.

In addition, Alphabet is working with the *Google Loon* project and testing the deployment of gas-filled balloons that float in the stratosphere and operate with solar power as a means of providing Internet service to remote regions.

Another example of its determination to move toward access infrastructure is the investment made in the company *SpaceX* at the beginning of 2015 with the future goal of building up a network of near-earth satellites providing Internet access.

Facebook is driving these efforts as well and follows various initiatives related to the provision of access to the Internet. Cooperation with Eutelsat and the company Spacecom includes plans to position satellites in geostationary orbit; once in place, they can be used to provide Internet connectivity to the African continent. The *Aquila* project is investigating the use of giant solar-powered drones as a means of providing Internet connections in rural or remote areas.

The *Terragraph* and *Aries* projects keep their feet on the ground and are studying terrestrial access technologies that are also the subject of intense interest on the part of carriers and could be used in densely populated areas or cities. Terragraph relies on low-cost small cells that broadcast in the 60-GHz band and, as envisioned by Facebook, could be placed on lampposts at intervals of 200–250 meters. Aries, on the other hand, depends on improvement in spectral efficiency and energy efficiency by using massive Multiple Input Multiple Output (Facebook 2016a).

Another project Facebook is running is the *Telecom Infra Project* (TIP). The project is a cooperation between Facebook, network operators, system integrators, and technology companies and is expected to prepare and share best practices for the planning, construction, and operation of future-proof and efficient telecommunications networks (Facebook 2016b). A model that works analogously is the one of the *Open Compute Project* (opencompute 2016). One of the major objectives of TIP will certainly be to lower the costs for telecommunications network components in the areas access, backhaul, core, and management and to put pressure on established network outfitters. The latter could in turn help telcos in the future to construct and operate networks more efficiently, which would subsequently be of benefit to customers in the form of lower prices. At the end of the benefit chain, Facebook would possibly record increased use of its own services by consumers. The possibility that Facebook may be participating in these various connectivity initiatives as a step towards the construction and operation of its own networks remains in the realm of speculation.

The activities of the Facebook-led projects *internet.org* and *Free Basics* and their claim *Connecting the World* (internet.org 2016) indicate that this step will probably become reality in the remote regions of the world where the two-thirds of the world's population without Internet access live. The objectives in regions that already have a good infrastructure in place are probably more related to the goals of TIP: to reduce costs and prices for network operators or consumers as well as to drive the use of services, which would also benefit Facebook.

Amazon continues to expand steadily its portfolio of services. Launched back in the day strictly as a mail order retailer concentrating on single transactions, the company is seeking to generate strong, long-lasting customer loyalty by offering an increasingly broad range of products and services embedded in a system of links.

At its core, the aim is to establish relationships of continuing obligation with customers, similar to the business models of telecommunications companies or energy utilities, to supplement the single transactions.

Among the methods employed by Amazon to realize this goal are cloud-based services (Amazon Web Services), bundled products and a sophisticated loyalty program like Amazon Prime, and a determined expansion of services from the trade transactions strictly related to specific products for any imaginable product worlds to the offering of products and services in the sectors TV (including production of own content), mobile devices, ebooks and tablets, and the provision of household-related services and virtual personal household assistants (Amazon Alexa).

In March 2017 Amazon announced a cooperation with the MVNO Drillisch in the German market. Amazon is planning to play the role of a reseller of mobile connectivity contracts. Similar to the other OTTs, Amazon would be able to bundle connectivity into their other service offerings and compiling a very attractive package consisting of connectivity, video, audio, devices, virtual personal assistants or simply integrate the connectivity in the Amazon Prime offering.

Another future pathway of the OTTs for mapping connectivity in mobile networks for customers comes from technological developments.

A first example is the *embedded SIM* (eSIM). The eSIM is a freely programmable SIM card or chip that has been permanently installed or integrated in the end device. They can be programmed over-the-air via the cellphone, smartphone, or tablet with profiles and rate plans of various providers. The eSIM has an essential role to play for all M2M-based applications, for the Internet of Things, and for connected devices such as sensors or vehicles. Its widespread use in tablets and smartphones is still in the future. However, various sources assume that it will happen soon. Mobile companies have an ambivalent relationship to this innovation. Today, the SIM card is an important control point for telecommunications companies in their customer relationships. In the future, it may be possible to install on the eSIM different contract profiles and tariff plans from different mobile providers, allowing customers to choose between them according to their needs. A second alternative for OTTs to provide connectivity in the future will come from the *network slicing method*. This method has become the subject of intense discussion because of the technological innovations (e.g., SDN, NFV, 5G) and the extremely diversified 5G use cases. Network slices are discrete logical networks that can be provided by network operators for specific customers and their needs and business models and that are based on a common physical infrastructure (e.g., Network as a Service). Multiple virtual E2E networks can be operated on one common platform. These dedicated networks can be made available, for example, to large business customers according to their highly specific demands. Energy or automotive companies (as well as others) could be interested in operating these logical networks on their own in accordance with their various use cases featuring differing quality requirements and functions. This type of option could of course be of value to the OTTs.

The telcos (e.g., Deutsche Telekom, Vodafone in Germany) attempt to push back in the service competition by offering *zero-rating* for mobile services and collaborating closely with content suppliers. Moreover, national as well as regional carriers try to score points with customers with product offers addressing all of the members of a household or complete households or communities and bundling fixed and mobile networks to create login effects. Beyond this, adjacent services such as excellent customer service, TV services featuring own content production, and exclusive content partnerships or smart home solutions are offered. In addition, the services or products are enhanced by special data protection and security promises. Deutsche Telekom recently announced the product innovation *Magenta*. Magenta takes direct aim at the voice assistants of Amazon, Apple and Google.

Putting it pithily, the network operators, based on their control over the networks, are attempting to realize paid premium services in an open ecosystem while the OTT providers frequently place their bets on closed ecosystems (e.g., Apple), freemium models, and a best-effort provision of services.

Based on these market developments and the described examples we expect the service competition to be high in the German telecommunications market.

4.2.3 Price Competition

The outlook shows that the driver price competition will not go untouched. The entry of OTTs into the connectivity business will result in decreasing prices for put downward pressure on the prices for internet access and mobile rate plans of the carriers, particularly since OTTs have the opportunity to operate on a growth market new for them while using the appropriate bundles and lateral subsidies from their previous core business. Rob Nail, the co-founder of Singularity University in the USA, assumes that these activities will lead to free Internet access in 10 years or less (Welt 2016).

But even without this more long-term impact of new competitors, the prices in Germany remain under pressure because of the stiff competition among established providers. The price index for telecommunication services has decreased per year on average by approximately 1.85% over the last 10 years. The price index for telecommunication services has decreased by on average 1.04% per year over the last 5 years (destatis 2017).

Indicators of a continued decline in prices are the exploitation of allotment models by Vodafone and Telefónica, that want to refinance their upfront payments as quickly as possible, as well as the competition for bandwidths between CATV and local access providers.

In the mobile sector, data volumes and transmission speeds as part of the bundled packages are constantly becoming larger and faster or are even substituted by zero-rating offers. Moreover, LTE technology has been opened up for the wholesale market, making a long-term price premium for LTE more unlikely.

On the German market, Rewheel's Digital Fuel Monitor Comprehensive Report from 4 January 2016 determined a very high price for 4G-based data packages among the 28 EU countries and attributed this to the rigid oligopoly-like structures on the market (Rewheel 2016). There is obviously a need to catch up in the European comparison.

The weighting of the three rough segments defined essentially according to willingness to pay—premium shopper, smart shopper, and discount shopper—is shifting toward the discount shopper. The premium segment appears to be largely stable for the moment, but the segment of the smart shoppers is shrinking. There is a host of low-cost brands or second brands of carriers, particularly in the growing discount segment, that seek to make use of very sharp, often temporary offers of services and products structured similarly to campaigns to lure highly specific, price-sensitive customer groups. The most recent examples are the activities of the Drillisch sub-brands simply and WinSIM in Germany, which have galvanized the competition with campaigns of aggressive pricing. Drillisch especially, thanks to its MVNO status on the Telefónica network, has enormous entrepreneurial freedom with respect to the terms and conditions of its products. The price gap between these products and those of the established network operators is in some cases enormous. This will undoubtedly exert a gradual downward pull on the premium segment as well.

According to these observations, we assume an intense price competition in the German telecommunications market.

4.2.4   Regulation

Historically speaking, there are two regulation models worldwide that contrast with each other. The European regulatory framework follows an asymmetrical approach[3] focusing on networks including the opening of such, regulation charges, and transparency and non-discrimination obligations.

The other model is found in the US where the regulation focuses on services and their provisioning. With its adoption of the *Protection and Promoting the Open Internet Rule* of 13 April 2015 (Federal Register 2015) the American regulatory authority, the FCC, set new standards for the regulation of broadband and Internet services. The FCC has now put these principles into effect:

- *Blocking* of lawful content, applications, services, and devices that do not represent a threat to the network (*non-harmful devices*) is prohibited;
- *No throttling*: Network operators are forbidden to slow down specific services or applications on the network;
- *No paid prioritization*: Network operators are forbidden to demand payment for preferred provision of any services and applications.

---

[3]In contrast to a system of symmetrical regulation, which applies equally to all market players, in a system of asymmetrical regulation only certain providers (i.e. as a rule, the so-called incumbents, formerly government-owned monopolies) are bound by regulatory constraints.

In the USA, this regime of network neutrality functions largely without any regulation of charges and access obligations. Described in simplified form, the aim in the USA tends more to the regulation of the conduct of the network operators—similar to the regulation of competition—while in Europe (in the past) the networks and their opening have been regulated for specific sectors. The EU, however, has launched a change of course with its network neutrality initiative and supplemented the *traditional* sector-specific regulation with the Network Neutrality Regulation. This obligates Internet providers to make an Open Internet available to every Internet user by prohibiting any discrimination in the form of paid prioritization, throttling and blocked content (with the exception of unlawful content). Along with this Open Internet, however, providers may offer higher-quality special services *on top*, provided that this does not have any adverse impact on the Open Internet.

All in all, the new Network Neutrality Regulation represents further regulation of the telecommunications providers with consequences for market events and the relationship between network operators and OTTs: The negotiating position of telecommunications providers with respect to OTTs has worsened because consumers are now in a position to obtain all of the applications and services of the OTTs from any and every Internet connection. The limitations on the opportunities to prioritize traffic or to mold the Open Internet in any way desired mean that telecommunications providers must first invest in the Open Internet before special services can be implemented. This will not only increase the required investments, but also limits the market potential of services and applications with prioritized traffic because they will always be offered in competition with the Open Internet.

With regard to the actual discussions and initiatives concerning the regulation in Europe and Germany we cannot make a valid assumption on which regulation form is more likely. Thus, we consider the market driver regulation as still open.

Summing up, we determine the following manifestations for the four market drivers in the German telecommunications market:

- Traffic growth is exponential instead of moderate,
- Price competition is intense instead of weak
- Service competition is high instead of low
- Regulation is still open.

4.3   Options for Action of the Player Types

This set of market driver manifestations affects the options for action of our four player types and thus will influence the position of the players in the market structure matrix. Table 1 gives an overview of the options for actions by player type.

According to the chosen option, the positioning of the players in the market structure changes. Figure 3 illustrates the options for action by player type.

Table 1 Options for action by player type

Player type	Option for action	Description
Network operator	Expansion	The large network operators appear as integrated full-service providers on the market and attempt to display greater presence on the service side for the customer in addition to the classic infrastructure and connectivity business. They attempt to take back the ground lost to the OTT ecosystem providers and to cement their relationship to consumers
	Focus	Network operators focus on efficient network operation and appear as bit pipe providers or wholesale-only providers without any contact to end customers
Reseller	Exit	This player type loses the foundation of its business and disappears from the market as OTT ecosystem provider activities expand into its business model (bundling of connectivity, mobile contracts, devices for a complete ecosystem) or through the carriers within the scope of the expansion of multi-brand strategy or the consolidation by one of the groups
OTT EP	Integration	The ecosystem providers delve more deeply into the added-value chain, enter the business field on the infrastructure side, initially concentrating on mobile service (MVNO, hotspots, eSIM, network slicing, own infrastructure), and continue to force competition on the side of services, content, ecosystems through bundled products
	Focus	OTT EPs focus on the previously established OTT products on the German market and expand them further
OTT SP	Exit	The player type disappears from the market over the course of consolidations or cooperation efforts of the OTT ecosystem providers or carriers
	Focus	OTT SPs keep their current focus; the dynamics of the internet economy continually produce new innovations, business models, and players

4.4   Spectrum of Future Market Scenarios

The combination of all options for actions of the respective player type results in eight market scenarios. However, of these eight scenarios there are two scenarios that we assume to be unrealistic right away and thus they are not substitute of our further considerations. These two scenarios are the ones in which both player types, network operators as well as OTT EPs, opt *Focus*. As this combination of options would result in a market gap we consider these two options as mutually exclusive. Consequently, there are six possible market scenarios left (see Table 2).

- Scenario 1 (Heterogeneous Power Play) is characterized by consolidated and integrated network operators, strong OTT ecosystem providers, and a large number of specialized OTT single-purpose providers. In the long run, traditional resellers will disappear from the market and be replaced either by the network operators as they continue to expand their second brands or by OTTs exploiting the opportunities provided by new technologies.

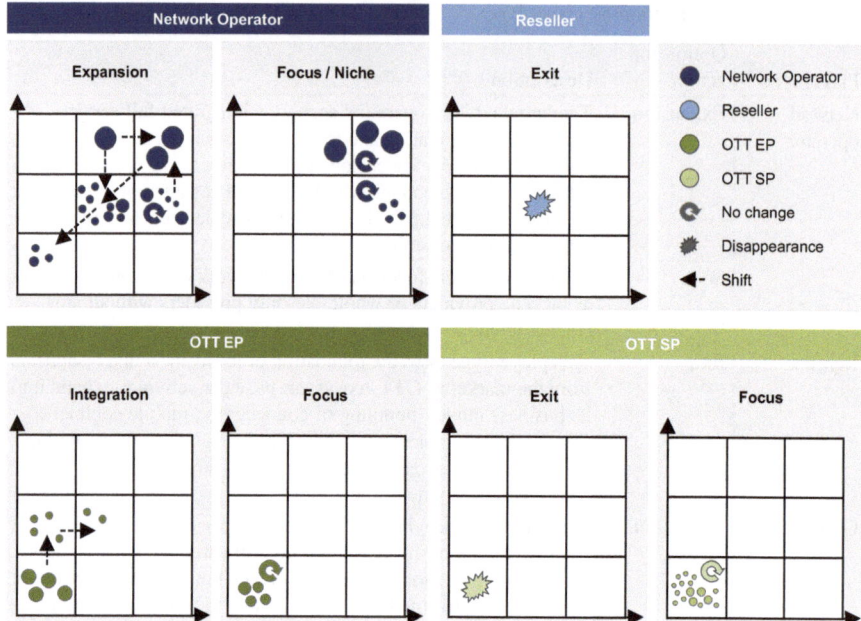

Fig. 3 Market structure matrices—options for action by player type

Table 2 Combinations of options for actions by player type and resulting market scenarios

		Network operators	Network operators
		Expansion	*Focus*
		Reseller Exit	
OTT EP *Integration*	OTT SP *Focus*	(1) Heterogeneous power play	(2) OTT diversity play
	OTT SP *Exit*	(3) Big player clash	(4) OTT big player dominance
OTT EP *Focus*	OTT SP *Focus*	(5) Co-existence	
	OTT SP *Exit*	(6) Telco play	

- Scenario 3 (Big Player Clash) is similar, but differs in that the OTT single purpose providers will exit the market or will be taken over by the ecosystem providers or network operators. The result will be a market structure of large OTTs and large network operators in confrontation with one another.
- Scenario 2 (OTT Diversity Play) and Scenario 4 (OTT Big Player Dominance) are characterized by the failed consolidation efforts of the network operators. In these scenarios, the OTTs are dominant in both the OTT segment and the traditional resale segment. The difference between Scenarios 2 and 4 is the

question of whether the small OTT SPs secure their relevance for the market or disappear.

- Scenario 5 (Co-Existence) and Scenario 6 (Telco Play) are the scenarios most favorable for the network operators. While the network operators consolidate and play a role in all of the market and service segments, the OTTs focus on Internet-based services. In terms of the relationship between OTTs and network operators, these two scenarios represent the circumstances that already existed about 10 years ago. However, we believe that a *backward somersault* of this nature is among the less probable scenarios.

Figure 4 illustrates the six market scenarios in the market structure matrix. Of these six scenarios, Scenario 1 (Heterogeneous Power Play) and Scenario 4 (OTT Big Player Dominance) are diametrically opposed and describe the full range of possibilities. In our opinion, both scenarios are highly probable. The market scenario with the highest expected value for network operators is Scenario 1.

4.5 Heterogeneous Power Play as the Best Market Scenario for Network Operators

In the scenario *Heterogeneous Power Play*, telcos continue to compete with OTTs to acquire consumers at the service level. It will be based on existing assets, including—besides the integrated network with its attributes such as worldwide, seamless connectivity differentiated by quality—the trust customers have in the brand and in data protection, respect for the private sphere, the local presence, and the service. The special sustainable value of these assets is a product of the numerous, differing, and (in part) highly sophisticated 5G use cases, above all from industry and the business customer sector. These use cases emphasize the quality parameters, reliability, and the security of the network. Best effort principles are not enough here; a "serious web" is demanded. Telecommunications companies continue to play a prominent role in the eyes of the consumer in this scenario. They offer additional services alongside integrated connectivity, they maintain or even extend the consumer relationship, and they occupy niches and micro-segments on the market by pursuing multi-brand strategies. In this way, they will hasten the disappearance of the player type of the reseller independent of network operators from the market in the long run.

The OTTs will in turn make use of the various opportunities described briefly above to complete their ecosystem by adding the key component of connectivity and then to compete with the telcos along the full length of the added-value chain. Focus will initially be on the mobile side; the fixed network and products integrated with it will stay with the telecommunications companies as a differentiation factor which must be exploited within the window of opportunity that remains. Just like the telcos, the OTT EPs will have a part to play in taking away the basis for the business of the classic resellers independent of network operators over the long term.

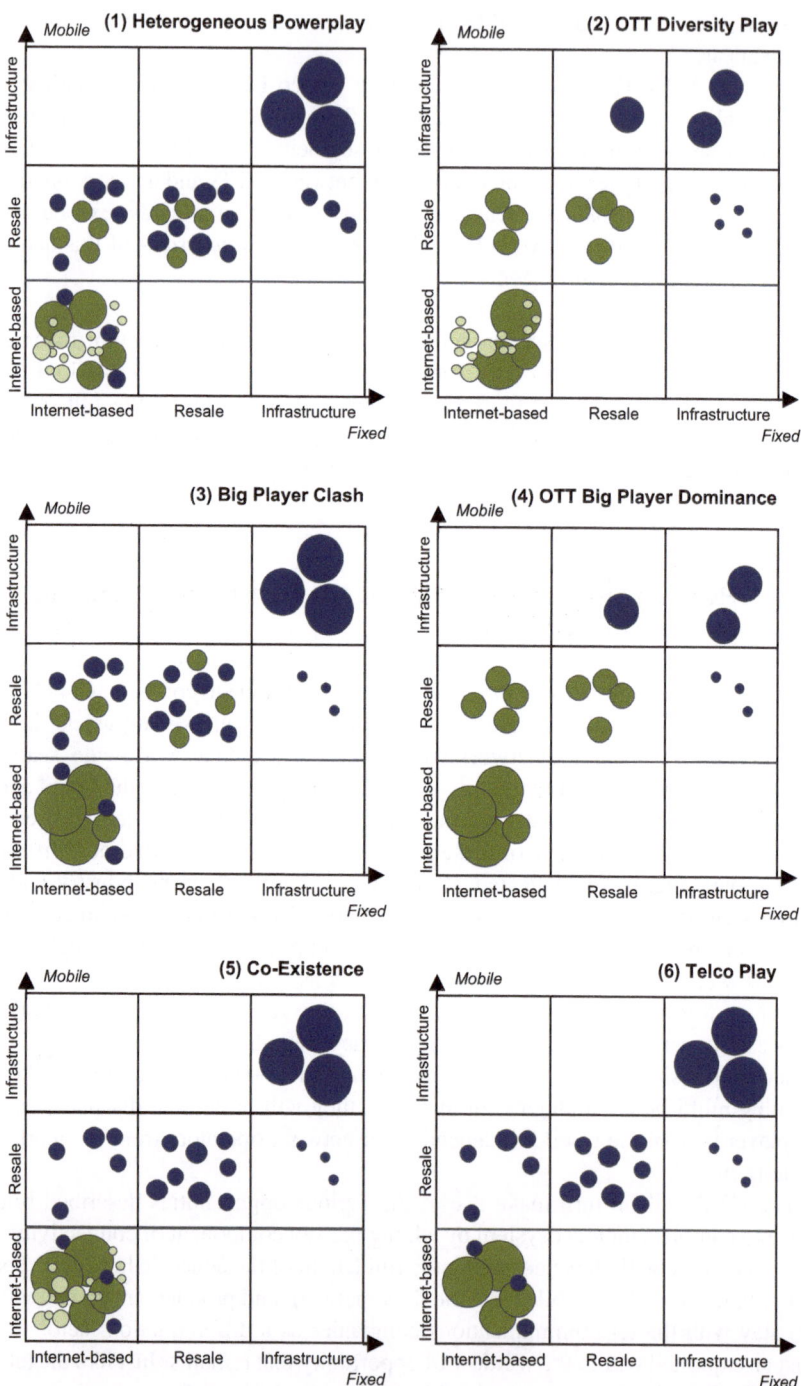

Fig. 4 Market structure matrices—market scenarios

But, only certain carriers (i.e., integrated heavy asset players) will be able to realize the above mentioned beneficial, relatively probable and at the same time ambitious role in the *Heterogeneous Power Play*. In the future, mobile-only carriers will presumably have a difficult time of maintaining production efficiency and economic sustainability and offering a convincing portfolio of broad scope for the consumer to survive on the market in their competition with integrated telecommunications companies. They could be tempted to change or extend their business model and to pave the way for the OTTs to mobile connectivity.

In Germany this possibility is especially conceivable for Telefónica. We see two options for this player:

- Option 1: Telefónica succeeds in supplementing its mobile network with a fixed network by means of wholesale agreements for the usage of fixed network assets or by a further consolidation on the German market (in-country consolidation)—whether as the enterprise leader or a junior partner is irrelevant here. The list of possible candidates, however, is not terribly long, and they surely have their own agendas.
- Option 2: Telefónica gains the position of a *cheapjack* on the consumer market with a focus on private customers and mobile services. It uses aggressive pricing, multiple brands, MVNO and OTT enabling to heighten competition, or sells its German activities to another international telco player such as Hutchison Whampoa that operates the business with perhaps similar ambitions (i.e., connectivity-only provider with low-priced conditions for consumers and whole-sale provider/OTT enablers) and that has recently been concentrating on the acquisition of mobile companies in Europe (cross-border consolidation).

Figure 5 depicts a speculative consolidation scenario of this nature. The consolidation groups shown in this scenario are based on currently existing shareholdings as well as on an assumed fit and similar interests of the players.

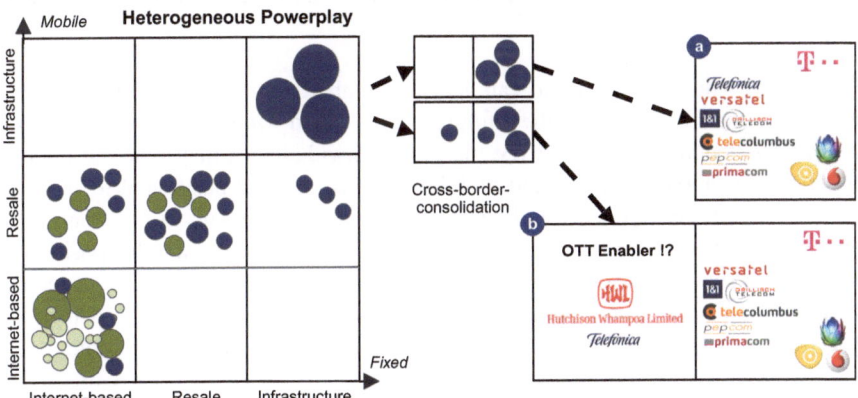

Fig. 5 Variants of the scenario heterogeneous power play for the German Telecommunications Market

In essence, either three or four telecommunications groupings will remain. In the first variant, these groupings would be Deutsche Telekom, Vodafone with KDG and Unity Media as well as a group of 1&1, the direct and indirect 1&1 holdings, and Telefónica. In the second variant, the groupings would be Telefónica together with an international player, for example Hutchison Whampoa, 1&1 with its current holdings as well as Vodafone with KDG and Unity Media and Deutsche Telekom.

5   Summary and Initial Recommendations for Network Operators

Based on the market scenario model, we derived six possible future market scenarios for the German telecommunications market. The scenario that is most advantageous from a network operators' point of view is the scenario *Heterogeneous Power Play*. The scenario and the positioning of the different player types in the market structure matrix can be briefly described as follows:

- Integrated network operators will have the chance to position as network based ecosystem providers. They will compete against large OTT EPs on all levels of the value chain ranging from connectivity, over platforms and services up to devices. They will offer QoS based premium quality services, based on network control as well as an open ecosystems of services and devices through intense partnering and cooperation. Only mobile operators will rather go for a bit pipe model or OTT enabler. Their business model is probably a dead end street for telcos that aim at shielding end customer relation, due to the fact that they will not be able to offer integrated services and to maintain production efficiency.
- Resellers will disappear from the market and be substituted by OTT EPs and network operators respectively network based ecosystem providers.
- OTT SPs will be affected by consolidation or cooperation efforts of large OTT EPs and network based ecosystem providers but innovation dynamics will create always new start-ups.

Network operators must and can act in a number of ways so that this scenario becomes the reality of the future. So called *integrated heavy asset telecommunications companies* (i.e., large nationwide telcos that own and operate a performant integrated fixed and mobile network) are best positioned to successfully realize the Heterogeneous Power Play for themselves and to position in the future as a *Network Based Ecosystem Provider*. If these players want this position in the Heterogeneous Power Play to become reality, they must undertake respective measures. These measures can be grouped into seven categories or levers which are enumerated below and further elaborated in the following parts of this book:

- Integrated Networks (Part I): Expansion of network capacity and coverage (outdoor and indoor), integrated planning, installation and operation of networks capable of handling the immense traffic volume

- Modern Network Concepts (Part II): Introduction of modern network concepts and principles such as software defined networks (SDN), virtualization of network functionalities (NFV) for the reduction of CAPEX and above all of OPEXs as well as edge computing for the provisioning of differentiated quality
- Innovation (Part III): Maintenance of push towards own innovations on the service level in fields where OTT providers would have difficulties to compete, e.g. security, data protection, quality of services, integrated network based offerings in order to shield end customer relations.
- Partnering (Part IV): Partnering and innovation are two sides of the same coin for telecommunications companies. Being attractive to consumers requires innovations that the companies have either developed themselves or that have been developed through more or less close cooperation with partners and monetarized to the benefit of both sides.
- Regulation, wholesale and wholebuy (Part V): The subjects of wholesale and partnering do not lie far apart from each other. Global as well as local reach and assured quality cannot be achieved without partnerships or business models in wholesale and retail trade and without a regulatory framework that creates fair conditions of competition.
- Customer Centricity (Part VI): individualization, emotionalization, security, mobilization, convenience, quality, digitalization of the customer interfaces are important changes in customer behavior or in customer expectations which telcos need to address.
- Internal Enabler (Part VII) The six points of action above are of fundamental, transformational significance for telecommunications companies. But they will lead to success only if the seventh lever is also applied: the creation of internal agility. This will be assured by radical change in traditional structures, cultures, processes, and the consequent adoption of agility in IT.

References

Bundesnetzagentur. (2017). Annual report 2016. Accessed Oktober 20, 2017, from https://www.bundesnetzagentur.de/SharedDocs/Pressemitteilungen/DE/2017/08052017_Jahresbericht2016.html

destatis. (2017). Price index for telecommunications services. Accessed October 20, 2017, from https://www.destatis.de/DE/ZahlenFakten/GesamtwirtschaftUmwelt/Preise/Verbraucherpreisindizes/Tabellen_/Telekommunikationspreise.html?cms_gtp=146542_slot%253D

Ericsson. (2016). 5G System. Accessed October 29, 2017, from https://www.ericsson.com/assets/local/news/2015/1/what-is-a-5g-system.pdf

Facebook. (2016a) Introducing Facebook's new terrestrial connectivity systems—Terragraph and Project ARIES. Accessed October 29, 2017, from https://code.facebook.com/posts/1072680049445290/introducing-facebook-s-new-terrestrial-connectivity-systems-terragraph-and-project-aries/

Facebook. (2016b) Introducing the Telecom Infra Project. Accessed February 22, 2016, from https://newsroom.fb.com/news/2016/02/introducing-the-telecom-infra-project/

Federal Register. (2015). Accessed November 20, 2017, from https://www.federalregister.gov/
    documents/2015/04/13/2015-07841/protecting-and-promoting-the-open-internet
Funkschau. (2017). Accessed October 29, 2017, from Mobiler Datenverkehr steigt weltweit um das
    Siebenfache. www.funkschau.de/mobile-solutions/artikel/139231/
Internet.org. (2016). Accessed October 29, 2017, from https://info.internet.org/en/
Opencompute. (2016). Accessed October 29, 2017, from http://www.opencompute.org/
Rewheel. (2016). Tight oligopoly mobile markets in EU28 in 2015. Accessed October 29, 2017,
    from http://research.rewheel.fi/insights/2016_jan_premium_tightoligopoly_eu28/
Welt. (2016). In 29 Jahren sind die Probleme der Menscheit gelöst. Accessed October 29, 2017, from
    https://www.welt.de/wirtschaft/webwelt/article152198869/In-29-Jahren-sind-die-Probleme-der-
    Menschheit-geloest.html

Assessment of the Telecommunications Sector from the Capital Market Perspective

Interview with Andreas Mark, Union Investment Privatfonds GmbH, and Wolfgang Specht, Bankhaus Lampe KG

Andreas Mark and Wolfgang Specht

The turbulence of recent years has not abated—the telecommunications sector is still confronted with elementary challenges. Almost all of the large companies are listed on stock exchanges and must regularly give an account of their actions to their shareholders. How does the capital market assess the current situation, and what are the prospects for the future? What strategies and actions are regarded to be effective in achieving the objectives? Andreas Mark, fund manager for Union Investment (Union Investment is the investment company of DZ Bank Group and a part of the cooperative FinanzGruppe. It is one of the largest assets managers in Europe), and Wolfgang Specht, analyst at Bankhaus Lampe (Bankhaus Lampe is one of the leading as well as one of the few independent private banks in Germany. Founded in 1852, the traditional institution is wholly owned by the Oetker family. Business activities concentrate on comprehensive customer support for well-to-do private customers, midsize companies, and institutional investors), answer these and other questions.

You have both been involved with the telecommunications industry from the capital market perspective for many years. What do investors and analysts think about this sector at the moment?

A. Mark: The telecommunications sector was characterized in the past by high debt, declining revenues, weak profitability, and reductions in dividend payments. The stabilization of profitability, coupled with improved prospects for dividends and (in some cases) a return to revenue growth, led to a more positive assessment of the sector. As a consequence of the regulatory rejection of further consolidation steps on

A. Mark
Union Investment Privatfonds GmbH, Frankfurt, Germany

W. Specht (✉)
Bankhaus Lampe KG, Frankfurt, Germany
e-mail: Wolfgang.Specht@bankhaus-lampe.de

© Springer International Publishing AG, part of Springer Nature 2019   25
P. Krüssel (ed.), *Future Telco*, Management for Professionals,
https://doi.org/10.1007/978-3-319-77724-5_2

the European mobile market, the interim higher assessment of the sector is again completely priced out. Moreover, inconsistent price trends and rising interest rates have had a sobering effect. Today, the sector is charged with a valuation discount in comparison with the market, but offers an attractive return from dividends of over 4.5%. For about the last 2 years, telecommunications stocks have been more or less treading water in terms of price development, which has been significantly worse than that of the European stock market as a whole. Overall, the sector is lacking in upward potential.

W. Specht: Until 2014, the sector suffered from weak revenues and declining margins. A mix of decreasing pressure in regulatory requirements, consolidation steps, and declining price competition brought about a stabilization in the number of customers and in revenues for many companies in the industry from the middle of 2014. The most important assessment criteria improved because of the heightened discipline in costs and investments. All of this has resulted in a "well-deserved" re-assessment of the industry in 2014/2015. Investors currently look mixed on the telecommunications industry; this is also a consequence of growing concerns regarding investments and shareholder returns.

What kind of influence do the capital market and banks have on the telecommunications industry when it comes to the challenges of the future?

A. Mark: Challenges mean two things. On the one hand, huge investments must be made in the expansion and performance capability of the networks. On the other hand, the business model must simultaneously be secured in an environment that is changing rapidly. The capital market plays an important role in the financing of these changes. Successful business models and mergers are valued more highly on the stock markets. In contrast, less convincing IPOs have had to be canceled because of the lack of interest on the part of investors. In the USA, records have been set in the placement of corporate bonds. Bond investors assess telecommunications companies as attractive because of their relatively stable business models. Corporate investments such as those required for the expansion of optic fiber networks are questioned critically by investors looking at future profitability. Investors focus on indebtedness and creditworthiness of a company. The credit rating is today an important element of corporate strategy and has led to stricter capital discipline.

W. Specht: Virtually all of the large players in the industry are listed on stock exchanges and make extensive use of the financing opportunities on the capital market. Investors in stocks and bonds today look more closely at the specific business model and strategy. The difference in the time horizons is often problematic. Most investors prefer fast realization of their returns. Moreover, they like glass-clear transparency in the assessment of the recoverability of their investments. This is often contrary to the long-term investment cycles related to investments in the network infrastructure and the uncertainties that remain because of regulatory decisions.

As things stand at the moment, might an unlisted company actually have an advantage on the market?

A. Mark: The transparency regulations applicable to unlisted companies are less stringent, and these companies have the advantage that, under certain circumstances, they can follow a longer-term planning horizon in their investments without having to be so conscious of short-term profitability. Listed companies are constantly under pressure to deliver good results for each quarter.

W. Specht: I agree with the argument that an "unlisted" status can be advantageous during certain development phases of a company. This is especially true when companies are in a "transition phase" such as times of complex integration measures or extensive restructuring processes.

Does the capital market display a high level of willingness to make investments in new network infrastructures? What would have to be done to raise this level, and what investments are considered to be meaningful?

W. Specht: Speaking broadly: There is plenty of money out there. If investors are convinced of the return of an investment, their willingness remains at a high level. Unfortunately, they have suffered a number of disappointments in this respect in the past. Technological trends have proved to be shorter-lived than originally expected, and regulatory decisions have on occasion devalued investments; one example here is the mandatory provision of wholesale products to competitors by incumbent operators.

A. Mark: The successful creation of infrastructure funds in recent years is evidence of the fundamental willingness of the capital market to accept such investments. Willingness to invest in network infrastructure exists only if there is a prospect of attractive returns on the capital. That requires planning security and an investment-friendly regulatory framework. The fall in prices for some of the companies after the announcement of additional optic fiber investments reveals the skeptical attitude about the profitability of these projects. Government policies and regulatory authorities can and must take actions for improvement if the goal of full-area coverage of a high-speed data network is to be achieved. Basically, new providers such as Deutsche Glasfaser also show interest in participating in the expansion of regional optic fiber networks. This will work only if the framework conditions for all involved parties are acceptable.

Do you see any risk that additional investments will not result in the desired yields because (for example) the threshold of customers' willingness to pay is too low or the regulatory authorities institute unfavorable rules?

W. Specht: Price sensitivity continues to be an issue for telecommunications services, just as in the past, but from the standpoint of the providers and in view of the price development of the primary services, the worst appears to be behind us. Service revenues of mobile service providers have recently stabilized after years of decline, and subsidies for the acquisition of new customers are being cut back. Regulation in Germany and Europe has also become significantly "more plannable". The so-called "ex ante regulation" that was applied even before the introduction of a service has been reduced in part and revised in favor of "ex post control".

A. Mark: Investments make good sense above all when they create added value for which customers are ultimately prepared to pay. A sudden change in demand behavior and customer needs can have a negative impact on the profitability of investments. In the telecommunications industry, changes in regulations continue to represent the greatest risk for investment decisions. So it is understandable when companies delay in making investments if the profitability of the investments is not assured. The risk of bad investments is lower today because companies are far more cautious in how they use their financial resources. New optic fiber cables are laid in many places only if a minimum number of lines has been assured in advance. Investments in uncertain projects are started initially with only small investment of capital or not realized at all.

Do you believe the subsidies the German government has decided to provide, especially for the expansion of broadband services in areas where coverage has been poor, are reasonable?

A. Mark: No one disputes the necessity to provide high-performance Internet access for everyone. In the meantime, broadband access in rural areas, which means for German midsize businesses as well, has become a very important location factor. This bottleneck factor impairs future growth. Private businesses cannot be expected to bear alone the investments for coverage in inadequately supplied areas when they will clearly suffer a loss. The state should step in with support in this area so that it can reach its goals according to the Digital Agenda. Driving the expansion of broadband must be given the very highest priority, significantly enhanced, and designed more efficiently. New initiatives such as the Digital Strategy 2025 from the Ministry of Economics are accordingly very welcome.

W. Specht: The current subsidies of 2.7 billion euros that will be provided by the national government by 2018 are an important building block and, along with the subsidies from states and municipalities, will stimulate expansion in the inadequately supplied areas. The release of the funds, however, has been relatively late, and meeting the political objectives by 2018 will consequently be touch and go. Moreover, this objective of full-area coverage with bandwidth of at least 50 Mbit/s cannot be more than an interim goal. To this extent, I assume that further subsidization funds will be created and politicians have already been vocal about it. I expect the announcement of a new broadband initiative to be one of the first measures of the new government. Beyond 2018, however, the subsidies are likely to be geared more in the direction of FttB/FttH while the current actions are technology-neutral.

What should a company on the market think about when deciding on price strategy?

A. Mark: Company size, network quality, market structures, market positioning, product variety, and target groups are important parameters for deciding on the pricing of products and services on the market. Integrated telecommunications companies can take advantage of the synergies from the operation of mobile and fixed networks to provide the network with the best quality available on the market. This position makes it possible for them to demand corresponding price premiums on the market. Moreover, the provision of attractive products and services from a single source creates additional differentiation and strengthens customer loyalty. Large integrated players attempt to raise the level of customer loyalty by offering attractive bundles of services encompassing mobile services, fixed network, and TV. Such offers can be sold only if they include a discount, however. Fixed network-only providers and mobile network-only companies will attempt to attract customers through quality, service, and attractive pricing. Specialized, smaller providers with free capacities can try to gain market share through aggressive pricing. No matter what the company, the focus is above all on the increase of revenue per customer. This can be accomplished by encouraging higher data use through attractive rate plans, supplementary products, or (as has been observed recently) through price increases in conjunction with supplementary services in so-called "more for more" offers. New price structures have been added, especially in mobile services, with the introduction of unlimited data rate plans that against payment of a surcharge allow users unlimited use of selected Internet applications or streaming products that are not charged to the agreed data volume. Free network capacities and intelligent network management are the prerequisites for the success of these products and open up opportunities to individual market players for differentiation in competition.

W. Specht: There is no such thing as a panacea—unfortunately. A number of factors, including individual positioning, targets, and behavior of the competition, are decisive. No major changes in the price strategy of the various players have been observed on the German market in the last 12 months. Price increases have been carried out here and there, but they have often been coupled with the granting of more service. The time periods in which the lower promotion prices apply have in part been curtailed, and there is less emphasis on subsidizing hardware.

We have seen a number of different consolidation activities in Germany in the past 2 years. Do you think they are sensible? Will there be any more?

A. Mark: We have seen various consolidations in the last few years, and from a company perspective, all of them—when viewed individually—have been sensible. The acquisition of the cable network operator Kabel Deutschland by the mobile services provider Vodafone was aimed above all at strengthening the latter's own fixed network activities and in closing the gaps in its own product portfolio. The merger of E-Plus and Telefónica Deutschland, two mobile providers, demonstrates how a lack of size in an intensely competitive environment must be compensated by mergers. In the cable network sector, Telecolumbus has acquired Pepcom and Primacom, further consolidating the network provider segment, with the objective of generating important synergies through size. The acquisition of Drillisch by United Internet should be the end of consolidation on the German mobile market. The increase of United Internet's holding in Telecolumbus, on the other hand, indicates that there is still potential in the fixed network. There is also a possibility that Vodafone and Liberty Global, the owner of Unitymedia, will come to an agreement on a further consolidation step for cable networks.

W. Specht: This appraisal of the four most important transactions is highly accurate. I expect to see further consolidation activities in this country, especially among the cable networks and smaller fiber operators. The history of this segment means that there are still a lot of small network operators whose business model will face difficulties in the middle term because of the need for investments and a critical size of the customer base. Various network operators at the local level, the so-called city and regional carriers, will join with larger units. We see Versatel in the role of an active consolidator here.

Both integrated carriers like Deutsche Telekom or Vodafone and specialized carriers like Telefónica Deutschland or Unitymedia are active on the German market. Do both of these business models have a future?

A. Mark: The popularity of bundled products is on the rise, and fully integrated carriers have quickly recognized that offering such products can reduce the willingness of customers to change providers. They can exploit the full breadth of their strength as complete providers in this category and offer these products from a single source and at attractive prices. The growth in market penetration reduces the customer potential of the specialist providers. While they are also in a position to offer these products, they are at a disadvantage because they must purchase the individual product components from competitors as advance services at higher cost and with less flexibility. Telenet, a Belgian cable network operator, has moved to a complete acquisition of a mobile network-only operator for these reasons. Specialized carriers, however, can offer convincing alternatives to fully integrated companies in terms of price, service, and quality within their core competence. Even specialized business models have a future in an increasingly convergent world. There will always be customer groups who want to purchase each of their products, i.e., mobile or fixed network, from different providers. Both business models are subject to change in a competitive environment that is in a state of flux and will have to be adapted according to their future orientation.

W. Specht: Integrated business models that have reached a critical mass within their market and that are well managed certainly have a future. I have a tendency to prefer the "fixed network-based" carriers among the specialists. My reasoning is that "mobile-only" network operators can become subject to pressures on the cost side of business. If traffic growth forces them to raise the density of their antenna structure significantly, they run a risk of not being able to install the additional antennas at a favorable cost because they cannot access antennas with their own fixed network infrastructures, but have to rent them from third parties.

What do you see as the most important drivers and hurdles related to consolidation among the telecommunications companies in Europe? Are we about to see a "consolidation wave"?

W. Specht: At the moment, "cross-border mergers" are hampered especially by a lack of synergy potential. This could change in the middle term because of a gradual standardization of transmission technologies and a uniform regulatory framework. I would also not exclude the possibility that in the future buyers whose core business is outside of Europe will appear as buyers in Germany. One reason for this could be the desire for a regional differentiation, consigning synergy expectations to a subordinate role.

A. Mark: So far, we have seen only in-market consolidation in Europe. Most of the mergers have been between mobile and fixed network operators among themselves or the combination of fixed network and mobile companies. Major drivers have been improved profitability achieved by scaling effects and an expanded product portfolio. Cross-border mergers have been hindered up to now by the lack of synergies and the differences in national regulations and legislation as well as by political opposition. Nor is further consolidation of the national mobile network markets from four to three providers desirable from the perspective of EU regulation. Such hurdles could be taken with the harmonization of regulations across all of Europe, especially in relation to the awarding and terms of mobile spectrum. Subsequent to the Brexit vote, the probability of pan-European mergers involving British companies has declined even further. As things stand at the moment, there is no reason to expect a wave of consolidations.

What dos and don'ts should players take to heart if they want to be attractive to the capital market in their consolidation scenarios?

A. Mark: As far as the capital market is concerned, the important point during consolidation is a strategy that is inherently consistent and logical. It must be able to generate value over a long term. Strengthening a company's own assets by acquiring complementary products or services can increase value just as well as in-market consolidations. Justifying expansion into other regions and business areas outside of the company's own footprint, on the other hand, is difficult. The acquisition of media content is viewed as equally critical, and its monetization and added value for the company is regarded as doubtful.

W. Specht: The "passive role" can also be interesting. If a company is able to position itself as an attractive target for acquisition, investors will often add an "acquisition bonus" to the fair value.

How important are cooperative ventures among telecommunications companies as a kind of substitute for a consolidation? In what areas do cooperative ventures among carriers make good sense?

A. Mark: Cooperative ventures help to compensate for a lack of size and for gaps in the product portfolio. They help to reduce costs through joint procurement, collaboration during network expansion and operation, development of products, or even joint acquisition of media content. Items that are lacking in the product line can be acquired from a provider on the basis of wholesale agreements. The make or buy decision is highly dependent on the terms and conditions that can be agreed so that it is possible to offer a product at a profitable price.

W. Specht: In the future, there will presumably be closer collaboration between mobile-only network providers and fixed network providers as a means of balancing out the advantage held in some areas by integrated providers. There is already a reciprocal relationship of this type between Telefónica Deutschland and Unitymedia, for instance and an extensive wholesale contract between Deutsche Telekom and Telefónica Deutschland.

How great a threat for various carrier business models do you see coming from OTTs?

W. Specht: The threat from OTT business models is in the meantime probably greater than the competition among the carriers. Among other factors, this has come about because the first "attacks" were extraordinarily successful—messaging platforms versus text messages or streaming platforms versus carrier libraries, for instance—and because the financing opportunities created by good core business have become even better—Alphabet or Facebook are such examples. The speed in the development and dissemination of new OTT models has recently accelerated, and this has made the response opportunities for telecommunications companies more complicated.

A. Mark: OTT players are both a blessing and a curse for the telecommunications industry. On the one hand, providers such as Netflix or Spotify, whose services previously did not exist on the market, are massively stimulating the use of data and improving the sale of higher-quality broadband connections and large-volume mobile data rate plans. But on the other hand, the introduction of disruptive services such as WhatsApp has eroded the business models of voice and text messages. Integrated telcos are better able to respond to this threat than pure plays, i.e., mobile-only or fixed network-only providers.

How can the relationship between carriers and OTT players be successfully structured from the carrier perspective? How important are vertical partnerships with OTT players?

A. Mark: Successful carriers have the chance to secure a competitive advantage for themselves by entering into partnerships with OTT players. Such a partnership should be aimed at bringing about a win-win situation. The OTT profits from the network quality and reach of the carrier that provide access to new customer segments. The carrier can attractively and innovatively design its product portfolio and benefit from the brand awareness of the OTT player. Mobile network services have recently been customized above all for high use of the most popular Internet services by the addition of new elements such as zero-rating and stream-on. The partnerships between OTT players and carriers have gained substantially in importance and enable differentiation on the market. Concentration on the top players has proved to be rewarding for the carriers.

W. Specht: Unfortunately, most carriers must be satisfied with focusing on "connectivity", although they do not have to give up completely their desire to offer some additional content. The carrier provides the "transport platform" and aggregates a number of different services simultaneously for the customers, but most of these services have been procured from partners. The result (for example) is a package of network access, data volume, security solution, and content. It is important to retain the customer or billing relationship.

In contrast to many other markets, a large share of the market in Germany is held by resellers and virtual network operators. How would you appraise the outlook for these market players?

A. Mark: There has already been some consolidation among the resellers in recent years. That is indicative of the competitive pressure that exists in this industry. It will become more and more difficult for players to maintain their independent positions in an increasingly convergent world. If the mobile product is not differentiated by its quality, it threatens to develop into a commodity, and price competition will move to the forefront. In that case, size and distribution will be the decisive elements on an extremely competitive market. freenet has already begun diversifying its product portfolio or investing in other telecommunications companies. United Internet's (1&1) acquisition of Drillisch is the largest merger among the resellers. It entails scaling effects, purchasing synergies, and an expansion of the product portfolio with an eye to convergence. The new company is in an excellent position to face the coming challenges.

W. Specht: Consolidation has indeed advanced a long way in this area. freenet and Drillisch are the only two players with relevant market shares who are still around. I would classify other "virtual network operators" such as "Aldi-Talk" or "Fonic" more as sub-brands of the network operators. I consider these business models to be sustainable, although they face the need of permanent re-invention. The recent merger between Drillisch and the B2C Telco Assets of United Internet is likely to form a strong "new" player in the market.

On the subject of regulation: Do you expect any major changes for the German telecommunications market? Who could profit, who is less likely to benefit?

A. Mark: We expect a continuation of the investment-friendly regulation in Germany as a way to encourage the expansion of broadband coverage that is needed. The consolidation on the German mobile market is over. The primary focus of attention will now be on driving broadband expansion and the creation of investment incentives. The large integrated players will be able to benefit from these activities because of their network coverage. The government subsidies and the related provision of wholesale products will also open up access for smaller companies and resellers. Regulation has a particular responsibility with respect to pricing. Now that the EU roaming regulations have gone into effect, telephone calls when abroad in countries outside of the EU could become a point on the agenda.

W. Specht: The recent decisions (about vectoring, for example) and the regulations for the awarding of subsidy funds tended to favor the large operators. In view of what we hear from politicians, the drafts for regulation directives, and statements at the European level, I also believe that future regulation will probably be "more incumbent-friendly".

Part I
Integrated Networks

At the Forefront: Network Expansion Oriented to Customers and Needs

Interview with Stefan Rinkel-Holgersson, Vice President Network Strategy at Telekom Deutschland GmbH

Stefan Rinkel-Holgersson

The dramatic growth of traffic in telecommunications networks entails a need for transmission speed and capacity at all network levels which is growing at an equally rapid pace. Telecommunications companies must find a way to accomplish the required network expansion, despite the immense financial resources it will devour, in the face of modest prospects for earnings. The challenge appears to be as daunting as the squaring of the circle. Stefan Rinkel-Holgersson, Vice President Network Strategy, is responsible for the network strategy at Telekom Deutschland GmbH. In the following interview, he answers questions about the approach taken by a large European incumbent to these challenges.

How important will the influence of technology strategy be for the overall success of Telekom Deutschland?

S. Rinkel-Holgersson: Our technical platforms are a key foundation on which Telekom Deutschland's services are built, and services are what the customer ultimately desires. Securing the high quality of our network means constantly investing in our network. While some of these investments have a very long-term character, the innovation cycles for technology are actually becoming shorter and shorter. Therefore, the right technology and network strategy are key success factors for performance and quality across networks and for sustainably investing in the future of the corporation.

What will be the greatest challenges for Telekom Deutschland over the next five years from a technological perspective and how are you addressing them?

S. Rinkel-Holgersson (✉)
Telekom Deutschland GmbH, Bonn, Germany
e-mail: Stefan.Rinkel@telekom.de

© Springer International Publishing AG, part of Springer Nature 2019
P. Krüssel (ed.), *Future Telco*, Management for Professionals,
https://doi.org/10.1007/978-3-319-77724-5_3

S. Rinkel-Holgersson: One major challenge concerns the optimized broadband rollout for both fixed and mobile networks. The basis of our rollout is the deployment of fiber optics towards the customer location, following the customer's needs and reflecting competition. Further fiber deployment is one of the biggest challenges given the high cost and limited resources in the market. At the same time we are continuously increasing the data rate on our existing copper-based network by using modern technologies such as Super Vectoring or G.fast so that high data rates can be made available quickly over a broad area. With regards to mobile networks we maintain our leadership by expanding our current mobile networks and preparing for the new technology generation 5G. There are a variety of technical and economical challenges that we address for 5G, e.g. fiber backhaul, small cell sites, creating new ecosystems, service monetization and others. On the basis of our fixed and mobile networks, we will further enhance the broadband experience of our customers with convergent products. Other elements in our broadband strategy include cooperations with partners and participation in public tenders for subsidized broadband rollout.

The transformation into more flexible and modern networks is a further challenge. This IP transformation is creating a simple and efficient network architecture that substantially increases our flexibility for product introduction and enables simpler processes when (for instance) customers make changes to their product. Today, we are in the midst of this transformation process and by the end of 2018, we will have largely changed over from legacy technology to the new IP technology. In addition, we are opening our networks further to wholesale customers, for example by offering various bitstream access services. Moreover, network modernization allows us to shut down older platforms in consideration of technological and market-related aspects. This legacy technology phase out contributes to efficient network operation.

The third challenge is the steady improvement of production efficiency for our services. For example we are working on the virtualization of certain network functions, increased automation and integrated planning across all network layers; the latter taking into account the optimal interaction of fixed and mobile network rollout and relevant technological developments. By interlocking product development, technology, marketing and sales much more tightly, we ensure that our network platforms capacity utilization ramps up more quickly and we amortize our investments faster.

What principles is Telekom Deutschland using as the basis to optimize further expansion of network capacities with a special view towards efficient use of investment budgets?

S. Rinkel-Holgersson: The expansion of our network is oriented to demand. In other words, we provide the network our customers need so that they can make optimal use of the services they require—at the right place, for the appropriate device, with the necessary data rate and quality.

This also means, that we do not necessarily provide the highest data rate for every device at every location but we offer at least the data rate the customers requires to run his services in excellent quality.

This has led us in a first step to focus on FTTC expansion (fiber to the curb) rather than FTTH (fiber to the home) because FTTC does not require as many resources while providing data rates up to 250 MBit/s to customers. This way significantly more customers were connected to our broadband network in a much shorter time than it would have been possible with FTTH. Also, a significant part of this newly created fiber infrastructure lays the foundation for the expansion phases that will now follow, when we will bring optic fiber even closer to customers and increase the data rates even further.

How are the technical requirements to network infrastructures changing?

S. Rinkel-Holgersson: The evolution of existing services as well as the development of new services and use cases are the reasons why the requirements to the network infrastructure are steadily rising. The data rate is important but not the only consideration; there are other important parameters such as spatial coverage, latency, mobility, seamless connectivity when switching between different networks, device density at the access point, energy efficiency, resilience and data security. Video-based services in particular drive the demand for higher data rates. On the other hand the requirements for connected driving, especially when highly automated, rather focus on low and above all predictable latency, high mobility of the network connections and very high resilience. And sensor networks (e.g. smart metering, "smart agriculture", or the monitoring of parking spaces) demand only the very lowest data rate and generally do not have any special requirements for latency. Instead they require energy-efficient devices, data security and wide-area network coverage. These examples demonstrate just how diversified requirements are.

How high do you expect data rate demand for typical private households and business customers will be in the next 3 to 5 years? What applications will be the primary drivers of traffic growth?

S. Rinkel-Holgersson: There is increasing talk in political discussions of a demand of 1 Gbit/s and a call for networks that can meet this demand. This level of demand can be reached for business customers, public facilities such as schools, administration offices or health care facilities, whenever many users are present and accessing cloud- and video-based applications in parallel. For these customer segments optic fiber networks in particular should be accelerated.

Primary drivers for higher data rates in private households are video-based services such as high-definition video streaming; again especially when multiple services are in use at the same time. Take the following scenario (although it is certainly not representative for most customers) as an example. The demand for a four-person household in which all four people are simultaneously using different video applications in 4K quality, two people are communicating via video telephony parallel to this, and another two are also using cloud applications can still be met with a connection providing 200 Mbit/s in downstream. Generating a demand of 1 Gbit/s in private households would require e.g. multiple parallel applications of 8K quality as a minimum (along the lines of virtual reality, for example) or new disruptive services that as of today are not clearly discernible on the mass market.

In the mobile sector the main concern is an increase in network capacity and not so much the need for peak speeds. While we will make it possible for peak transmission rates in this sector to exceed 1 GBit/s as well, it is more important that a large number of customers can rely on having the data rate that they need for their use cases.

What technologies play the most important role in fixed and mobile networks to fulfill the technical requirements of future services?

S. Rinkel-Holgersson: There is no single solution that economically delivers the diverse technical requirements described above. Rather the complementary use of various technologies is necessary. The technologies differ with respect to technical performance capability, availability in terms of time and area coverage, and mobility. By the way, I am of the opinion that our debate in Germany is driven too strongly by technology. FTTH is not always and everywhere the best solution. For example, by taking the FTTC approach and bringing the optic fiber to a distribution point that is significantly closer to customers, we can achieve a high degree of wide-area coverage much faster than would be possible with an FTTH-only expansion. By 2019, Telekom will have reached coverage of about 80% of German households with its access network and provide transmission speeds of at least 50 MBit/s and up to 250 MBit/s. This has given us a good position in an international comparison. German coverage with fast broadband—i.e., over 30 MBit/s—is actually higher than in EU countries that depend strictly on FTTH and FTTB such as Sweden or Estonia (EU Digital Progress Report 2017). From this basis we are now bringing optic fiber closer to the customer as needed. Given the high cost and time associated with the FTTH/B rollout we are also assessing the viability of Fixed Wireless solutions.

At the same time we are expanding our mobile network and modernizing it continuously so that we can satisfy the growing need for mobility. For instance, we are increasingly closing gaps in network coverage so that interruptions while traveling by train or car are avoided and the mobile communication within buildings is improving. The introduction of the next mobile generation 5G will bring modern antenna technology and new frequencies that enable us to increase capacity, enhance data rates and decrease latency. Furthermore 5G will improve deep indoor coverage or enable very high device density, thereby also paving the way for the Internet of Things (IoT).

The common element of these technologies is that the optic fiber share in the transmission route is steadily increased.

What do you think of the chances for hybrid or integrated network connections?

S. Rinkel-Holgersson: The combination or bundling of connection technologies opens the door for additional performance enhancement. Our hybrid product combines the fixed with the mobile network offering higher data rates which we will be increasing in the future. The core benefits, however, are the expansion of broadband coverage, faster provision and increased resilience. There are also further bundled variants that we are evaluating, for example the bundling of multiple WiFi systems, bundling WiFi with mobile networks or the bundling of multiple copper twisted pairs.

How important do you believe the following technological trends and technologies will be for Telekom Deutschland with respect to the not-so-distant future: SDN, virtualization, data analytics, robotics, small cells?

S. Rinkel-Holgersson: All of these subjects will continue to gain in importance—in part each for itself, but in part they will depend on or augment one another. That is why it is very important to integrate them into the technology portfolio early. With this in mind, we at Telekom are highly active at international conferences and in standardization work. Key questions that must be answered are for example when new technologies will be ready for the market and in what scope rollout appears to be optimal from an operational and an economic viewpoint. We will virtualize more and more network functions and utilize various technologies even more holistically so that we can satisfy the high demands of our customers.

What major factors limit the selection of the optimal technology, and just how great is their impact?

S. Rinkel-Holgersson: Besides the issues of financial and personnel resources, there are limitations with respect to the availability of services we procure on the market, ranging from civil engineering to IT support services. Another important factor is the willingness of customers to pay. Furthermore political/regulatory obligations can limit the selection of technology. From our point of view technological innovation and competition through innovation should be stimulated rather than inhibited by law makers. For us this also means that we choose the mix of technologies that economically fulfill technical requirements.

Another issue is when new technologies ("carrier grade") will be available, i.e., when they will be ready for the market and technologically mature and can be offered by multiple manufacturers.

What convergence topics will be the most important in the future for you: convergence of network and IT, convergence of fixed and mobile, convergence of network layers or other areas?

S. Rinkel-Holgersson: Deutsche Telekom is an integrated fixed/mobile network operator. We offer our services both in fixed and mobile networks, across all network layers. Therefore, convergence topics play a key role for differentiation from our competitors. As networks are converted to IP technology and virtualization continues to grow, the convergence of network and IT as well as the convergence of network layers already play an outstanding role. The convergence of fixed and mobile networks is already realized at the planning and roll-out levels and will continue to be expanded. We introduced the first true hybrid product on the German market, our hybrid router. As developments continue to move in the direction of 5G, convergence of fixed and mobile networks will become more and more significant.

What decisive changes in organization and processes will convergence bring about?

S. Rinkel-Holgersson: Telekom Deutschland combined its fixed and mobile network divisions into one organizational unit years ago. The joint alignment of technology and IT units is also assured by close interlocking of personnel. Moreover, project-driven team organization is frequently applied to new topics with an early integration of cross-functional competencies to address convergent aspects. Network operation is also becoming cross-national where this is economically sensible.

What importance do you attach to the realization of partner models with OTT providers and what is critical from a technological perspective for taking advantage of opportunities or controlling risks?

S. Rinkel-Holgersson: Attractive services are what make our networks so meaningful. Many services are developed and provided by Telekom directly. But in this highly dynamic environment we want to incorporate the creativity of third parties as well and become their preferred network partner. Our goal is to quickly integrate partners and their products into our network and our processes in the most flexible manner.

How important is the topic of cooperation with others for realizing coverage with the required data rates for Telekom Deutschland, and how is this being approached?

S. Rinkel-Holgersson: Cooperation is becoming increasingly important. The fast and broad area coverage with FTTH/B in particular requires a lot of resources in both capital and human labor and is not possible without partnerships and cooperation. Cooperation agreements extend over all levels of the value chain. They include, for instance, partnerships with suppliers such as system manufacturers or even civil engineering companies as well as reciprocal open access on a bitstream basis with other network operators. Different operating schemes or co-investments with other carriers are also under consideration.

Integrated Network Planning: A Key Success Factor for Network Operators

Lutz Fritzsche, Mathias Schweigel, and Rong Zhao

1   Vision of a Future Network Infrastructure

The future network services are not limited to traditional triple players for consumer domain, but extended to support industrial applications, e.g. massive IoT, M2M, critical communications. The life cycle of a new service or technology from its development to implementation is becoming shorter. New requirements for the network infrastructure appear at closer intervals. The use of cloud computing has undergone intensive tests in many countries and has been identified by marketing and end customers as an extremely interesting alternative to current data storage concepts. More and more private and business data are being stored on the Internet.

In 1984, Nielsen proclaimed "Nielsen's Law of Internet Bandwidth", which states that the Internet bandwidth will double every 2 years (Nielsen 1998). This prediction has held true with very few exceptions: bandwidth per customer has risen from about 120 bit/s in 1983 to as much as 240 Mbit/s in 2016. Internet traffic has increased in comparably dramatic fashion as the number of customers has ballooned.

The primary objective of the Digital Agenda of the European Commission is to secure sustained economic and social benefits from a digital single market based on high-speed Internet and interoperable applications. One of the pillars of the policy envisions activities in the telecommunications sector to secure the Gigabit broadband targets for 2025, i.e. "at least 100 Mbps (up to 1 Gbps) access for all EU households", "Gbps connectivity for main socio-economic drivers" and "5G connectivity for major transport paths and urban areas" (European Commission 2016).

The primary concern for end customers is that they are satisfied with the purchased services or the provided bandwidth regardless of the technology behind its realization through a fixed or mobile network connection.

L. Fritzsche (✉) · M. Schweigel · R. Zhao
Detecon International GmbH, Cologne, Germany
e-mail: Lutz.Fritzsche@detecon.com; Mathias.Schweigel@detecon.com; Rong.Zhao@detecon.com

© Springer International Publishing AG, part of Springer Nature 2019   43
P. Krüssel (ed.), *Future Telco*, Management for Professionals,
https://doi.org/10.1007/978-3-319-77724-5_4

The expansion of broadband service is of national significance because of its economic impact, prompting governments to initiate programs aimed at this objective. The Federal Network Agency has declared in its broadband strategy for Germany that the aforementioned goals will be achieved "... by a technology mix and in competition." (BMWi 2015) An important parallel requirement for the broadband strategy was the exploitation of synergies among various infrastructure owners.

Market observers note that the revenues of network operators continue to stagnate. This is in part a consequence of the level of market saturation. Moreover, the trend to flat rate models limits revenues. This situation can be resolved by the development of new customers. M2M communication, for instance, might well create new "customer groups". But since these customers do not need sleep or rest, there will be a permanent change in traffic flows and load relationships. It is more than doubtful that it will be possible to secure an increase in the price per bit as the measure of the utilized network resources.

There are several goals regarding the future network architectures which can be observed: convergence of the various network levels, convergence of fixed and mobile networks, convergence of IT and network technologies, and convergence of static and dynamic architectural paradigms, softwarization and virtualization based on SDN and NFV (Schnitter and Bornhauser 2014; Gonsa et al. 2014).

2   Challenges in Planning

The planning of telecommunications networks is a means of managing and expanding available network resources such as transmission capacities while simultaneously complying with cost limits, performance indicators, and threshold values for reliability. There will be little change in these fundamental objectives in future telecommunications networks. The challenges in the planning of future networks will arise to a far greater extent in the constraints related to realization.

As a rule, telecommunications networks consist of a number of partial networks such as signaling and transport networks, whereby the latter can function either circuit-or packet-switched. The distinction between fixed and mobile networks is another example. This differentiation serves in part to reduce complexity because the partial networks are planned and operated separately from one another. But this separation was and is a consequence of their history as well because certain technologies became available later than others and were simply added to an existing infrastructure. The systems for operation of the partial networks naturally differ from one another; they have been specifically adapted to the existing technology, and in general merging them into a single system requires substantial effort. However, this separation presents an obstacle to finding a cross-layer solution for the entire network at optimal cost. It is often the cause of multiple redundancies of security measures against network failure at various network levels or can even result in a lack of precisely such redundancies. Path diversity in a packet-switched transit network, for instance, can use the same underground cable. If there is a cable break, the service quality may suffer, a situation which could result in the payment of penalties or loss of income.

The merger of the various partial networks is a reasonable step and is being driven forward, as shown by the integration of the transit network technologies IP/MPLS and WDM. The convergence of the previously separate fixed and mobile networks and consequently their joint usage of transport network capacities and locations can be viewed in the same context. In the long run, this integration will simplify planning of the telecommunications networks.

As far as technologies are concerned, the technical constraints are just as significant as local influences. For instance, the characteristics of the ground—whether rock or sand—for laying underground cable or the possible installation of an above-ground cable connection and the related labor costs play an essential role in the comparative analysis of the various transmission technologies. Conclusions cannot be transferred directly from one geographic area and market to another. Regulatory aspects, competitors, customers' expectations, and special features of geography have a major influence on planning results. Planning is always a local decision.

One challenge for the planning, however, is and will remain the integration of the network data from active and passive network technology. When the appropriate protocols are used, it is possible to record automatically and process configuration and capacity utilization information from the active network technology, and this can be used to create highly accurate network models for planning purposes. Passive network technology cannot be monitored by this kind of automation, so the quality of these data and subsequently the planning results will continue to be highly dependent on the commitment of the responsible departments in the future.

The progress in the integration of inventory systems for network operation opens up new prospects for the planning of telecommunications networks. Consistent data storage of the current network is an important prerequisite for cross-layer network planning of various technologies which takes the specific properties into account. This information must be combined with certain command variables such as maximum load for network analysis; it also provides a valuable foundation for strategic planning for the initiation of technological conversions in combination with tactical considerations.

New technologies, such as C-RAN, will drive requirements for high-bandwidth and low-latency connections to a large number of access nodes. Network architectures and topologies need to reflects these demands.

But even integrated planning in future networks will be confronted with challenges comparable to those facing network planners in the past. The constraints will change, and new interdependencies will arise. However, improved data storage, more powerful computer performance, more precise network models, and improved algorithms will help to master the greater complexity. The general problem of recreating the subject of the examination "as precisely as necessary" inherent in any modeling will still be around.

As types and number of services increase rapidly, their demands on networks entail various combinations of high data transmission rates, low latency, massive devices, etc. (Aumann et al. 2017) In pursuit of the vision of transformation into "Network as a Service", 5G will be designed with special emphasis on these features:

- Flexibility in the allocation of resources, functions, even topology (fixed and mobile access agnostic)
- Compatibility to legacy telecommunications systems (fixed and mobile)
- Scalability for the support of enormous interconnectivity of machines
- Integrated systems of "secure connectivity between applications over secure networks" for privacy, authentication, etc.
- Interoperability for various operators—(third-party) vendors.

An important concept in 5G is Network Slicing to support different use cases in terms of specific business requirements, where SDN and NFV are the key enablers.

3   Approach for Successful Integrated Planning

3.1   Constraints

The challenge in creating uniform planning for an entire network results from the large number of different constraints which must be processed while giving due consideration to the possibilities of the existing network or the planned technology. These constraints can be classified in the dimensions of time, space, and degree of detail.

Each of these dimensions is subject to its own constraints and possible solutions which steer the planning process. It must be noted that these dimensions can in part be used independently of one another for the input parameters and the planning results.

The time level determines how quickly the planning results can be realized. Obviously the scope of action for short-term immediate measures differs from that of long-term projects. As far as the input parameters of the planning are concerned, time plays a role in the assessment of the current situation and the history of the network (measurement data).

The spatial level defines the distribution of the resources over an area in the results. In terms of the input parameters, this can refer to the availability of free resources (movable, immovable), user distribution, or traffic volume in the area. The definition of the planning area—parts of the network or the complete network—also falls in this category.

The degree of detail determines whether the planning is carried out at the user or at a higher abstraction level, e.g. service level. The degree of detail in the input data often varies, so they must be adjusted and matched on the basis of suitable assumptions. The degree of detail in the results will naturally be determined by the input value with the lowest degree of detail. It is desirable for the planning process to have information which is as detailed as possible. The abstraction and compilation of this detailed information in the extent required for the planning must be carried out by the planning tool which is used.

The following are examples of various combinations of the dimensions mentioned above:

- Best possible exploitation of free resources which are spatially restricted. This includes free transmission capacities such as wavelengths and fibers as well as free interfaces in installed devices.
- Optimal distribution of available resources which can be distributed spatially without restrictions. Examples are interface cards in the warehouse or contingents for leasable transmission capacities.
- Expansion of the network in alignment with demand at locations with the highest forecasted market potential.
- Remedy of a spatially limited overload on a device or a transmission line.

3.2 Realization of the Integrated Planning

The realization of integrated planning in the future will be even more heavily dependent on appropriate software solutions than in the past. The ability of these programs to process large quantities of detailed network data with configuration information, measurement data, and load values will serve as the basis for calculating the quantity of resources in the future. Big data analytics methods can help to improve the planning input quality by searching for pattern in mass data. Additionally those software solutions are necessary for the upcoming automation of detailed short-term network planning tasks. However, parameters which can be quantified and measured only with great difficulty will continue to play a role in the future, especially in the middle- and long-term network planning. It will not be possible to automate planning as a creative process completely, especially for strategic considerations, in the immediate future.

Rising to the challenges of standard network planning will require a concept which both integrates the network planning smoothly into the network operator's processes and itself exerts an integrating effect, e.g. on the convergence of the former mobile and fixed network sectors.

Moreover, this integrated planning must encompass to the greatest possible extent all of the network layers from the services to the infrastructure. Only this approach ensures for example an exact planning of fiber optic cables as a shared medium for various applications. This is especially important for upcoming FTTH and 5G network deployments to achieve the cost efficiency. Integrated planning should consider market and technology data and unify various planning horizons such as those existing for strategic planning or project planning.

3.3 Planning of Various Time Horizons

One approach is a three-stage planning process comprising strategic planning, detailed concept, and implementation. Strategic network planning entails looking ahead over a period of 5–10 years, and its objective is the development of future-proof network concepts and structures with a special focus on commercial aspects. The input data and influencing factors for strategic planning are manifold, including

Table 1 Degree of detail for network nodes, lines, and demand in the various phases of integrated network planning

	Strategic planning	Detailed concept	Implementation
Network nodes	Aggregated nodes	Single network nodes as devices	Devices with equipment information (Components)
Lines	Aggregated line bundles between aggregated nodes	Single lines connected to devices	Single lines connected to ports
Demand	Aggregated demand between aggregated nodes routed over line bundles	Specific demand between single nodes routed over single lines	Specific demand between single nodes routed over through single lines and ending in tributary ports

items such as marketing forecasts, technology trends, service concepts, and the existing networks. Although this requires sophisticated strategic planning, it offers in return high potential for optimization—when drafting concepts for new technologies, for example.

The detailed network concept underpins the results of the strategic network planning with technical details and continues to develop it over a period of 1–3 years. While strategic planning works with aggregated "high level" network objects in defined regions of a network, the network concept covers the realization in spatially concrete network objects such as devices and lines (see Table 1). The implementation steers and ensures the realization of the planning results.

Generally, the degree of detail in the planning results rises parallel to the progress of the depicted planning phases. The planning results will become more and more detailed and precise the closer the implementation phase comes.

3.4 Cross-Layer Planning

A cross-layer perspective is important during all of the phases. In the past, established network operators set up many widely diverse technical platforms which were often planned independently of one another in different departments. Occasionally the utilization of other platforms was hidden, which led to subsequent costs which had not been budgeted. For instance, the existing optical fiber infrastructure must be taken into account right from the strategic planning phase of a new transit network platform so that any investment costs which may be required for its expansion can be budgeted. Cross-layer planning is not possible without comprehensive and correct documentation of the operated network platforms and their interdependencies as well as of the offered services and extending to the utilized infrastructure across all layers.

The nowadays discussed concept of Network Slicing assumes a cross-layer view since a common physical infrastructure is the basis for different business purposes (slices) for consumers and industries.

4   Prerequisites for Integrated Planning

The prerequisites for integrated network planning include above all comprehensive network documentation of high quality, the preparation of forecasts of the greatest possible precision, and the support of the planning process by the processes established in the operator's business.

4.1   Comprehensive, Contemporaneous, and Cross-IT System Network Documentation Is the Basis for Every Planning Project

An outstanding example is a standard ID concept for locations which represents the basis for every kind of network documentation. A unique ID for locations, whatever form it may take, must be applicable to all of the network operator's departments, to every IT system (regardless of whether OSS or a planning or order system), or to every type of location (regardless of whether the technology at the location is active or passive). Building on the standardized identification of the locations, a unique ID must also be assigned to the network objects such as active and passive devices, transit lines, cable, or routes. The creation of an ID which includes an indication of the geographic location is helpful. Cross-layer documentation is often a problem, but it is indispensable during the planning phase. It should display the connections of point objects with one another such as the allocation of devices to locations or of components to devices as well as information about lead and configuration information for linear objects such as the configuration of the fiber pairs with DWDM systems or the laying of cables along routes. This cross-layer documentation is absolutely mandatory so that statements about the load on specific network objects, network areas, or the complete network can be made or so that non-disjoint routings can be identified.

Comprehensive network documentation must also include the customers on the mass and individual market along with their input data relevant for planning such as the ordered services.

The number of documentation systems in use should be limited. A standard inventory extending in coverage from access to aggregation to core networks and encompassing all of the technologies is optimal.

4.2   The Most Precise Forecasts Possible Throughout All of the Diverse Phases of Planning

Beyond a doubt, the preparation of forecasts belongs to the essential input data for every stage of the planning process. A distinction must be made here between the forecasts for internal and external customers. Whereas internal forecasts designate, for example, the need for internal capacities, external forecasts are characterized by predictions for customers on the mass and individual market and in wholesale business. The forecasts must be provided in correlation to the time period for every planning phase. Naturally, forecasts for 10 years in the future will be less precise than those for the coming year, so regular review is necessary. Forecasts data such as population, structural development, average income, or penetration by competitors should be prepared for the early stage of the strategic planning.

4.3   The Living Processes in the Network Operator's Business Must Encourage, not Hinder Integrated Network Planning

Integrated network planning will prove to be a success only if the established processes in the network operators' business have a supportive effect. Redundant planning must be avoided to enable a smooth transfer of the planning results between the various phases. In addition, responsibilities in the planning process must be clearly defined and the results must also be realized in network operation!

5   Examples for Application of Integrated Planning

5.1   Optimal Network Rollout and Expansion Strategy

Finding the optimal network expansion strategy is a complex task. The available technologies give a network operator the chance to realize services using a broad range of diverse technologies. The local influences, sometimes on street level, must always be taken into account. They include market data like the customers' income position and the competition as well as technical realization. While investors are often concerned with the costs for the construction of a completely new network, the most precise mapping possible of the existing infrastructure, along with configuration and load information as the basis for a precise estimate of the network expansion and extension costs, is important for established network operators.

Big data algorithms enable the development of market models on urban district level. Considering market and topology data on this scale in a nation wide planning considering access and backbone network results in large data sets to be processed.

Integrated, cross-technology, end-to-end network planning allows to draw up the optimal network expansion strategy at any granularity. In preparing this plan many constraints need to be considered. Mobile and fixed network technologies must be combined, the existing infrastructure must be given due consideration, and projected market data over a period of several years must be processed.

This type of planning results in recommendations for detailed actions describing what technologies should be used in what areas so that the network expansion is driven forward while maintaining cost efficiency. In the same way, various rollout scenarios can be examined and compared with one another. Synergy effects from the combined usage of various technologies can be examined in detail.

The results can easily be tested for plausibility when a geo-based network planning tool is used. The possibility of breaking the calculated costs all the way down to specific devices in combination with visualization of spatial spread of network elements and configurations guarantees transparency and ensures the applicability of the planning results.

5.2   Strategic Planning

If existing infrastructure is to be developed in long term perspective for an uncertain future, strategic planning is required. Carriers faced with the task of establishing strategic network planning for a period of up to 10 years with the aid of a suitable software program should follow the procedure below (see Fig. 1).

First, they should begin by analyzing the as-is network. The next step is the identification of the existing logical and technical platforms and the corresponding data sources. A suitable planning tool is to be used to import and aggregate the as-is network consisting of individual devices and lines. Aggregation is based on each connection area with one aggregated location as the compilation of all of the existing locations and devices and as the terminus of the line bundle. This procedure drastically reduces the volume of processed data. The aggregate network created in this fashion provides the foundation for strategic planning which can be carried out in the form of scenario calculations. The major challenge of this aggregation step is reducing the amount of data while maintaining the required details.

Both platform decommissioning and the setup of new platforms can be examined this way. In addition, the methodology for the scenario calculation can be implemented by the development of modified building blocks for automated calculations.

The building blocks should treat the aspects customers and services (e.g., with respect to the distribution of nationwide forecasts to the connection areas), platforms (such as the automatic generation of network topologies), dimensions and cost estimates.

It then becomes possible for the network operator to determine the development of the network topology over a period of 10 years (including the required investment costs), to describe the development of the load on the network, and to identify required investments, for example.

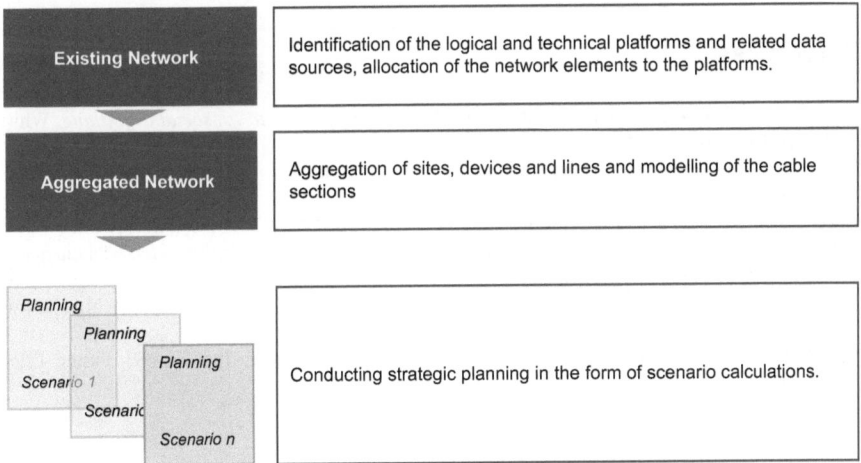

Fig. 1 Scenario planning

6 Benefits: Network Transparency and Efficiency

Effective integrated planning is capable of creating transparency across all of the different network areas, maintaining control of network costs, and maximizing the commercial benefits. The most important advantages for network operators can be summarized as follows:

Comprehensive view of fixed and mobile networks Integrated planning makes it possible to obtain a complete picture of fixed and mobile network systems from the technical as well as marketing perspective. Integrated planning can include a broad range of highly diverse technologies for the realization of needs or customers' products.

Crossover network planning and analysis In contrast to "segmented" strategic planning, integrated planning creates transparency among the various network levels and layers. This includes above all a consistent view from services to transit layers to cable network and route infrastructure.

Sound selection of future network developments Integrated planning makes it possible to calculate scenarios especially for middle- and long-term planning periods, ensures their comparability—particularly with respect to expected costs—and thus simplifies the selection of the best variation. The calculations incorporate market as well as technical data.

Close cooperation with the IT infrastructure Integrated planning requires close meshing with the IT infrastructure in the company. It paves the way for the simultaneous realization of planning tasks and the necessary adaptations in IT so that fast product development can be achieved.

References

Aumann, C., Zhao, R., & Zhao, L. (2017). *5G: From smart phones to . . . social everything*, White Paper, Detecon.

European Commission. (2016, September). *Digital single market – 5G for Europe action plan.* Brussels.

German Federal Ministry of Economics and Technology (BMWi). (2015). www.bmwi.de

Gonsa, O., Chrestin, A., & Reith, L. (2014) *Virtualization is transforming the telecommunications industry* (pp. 56–71). Future Telco (Book).

Nielsen, J. (1998). Nielsen's Law of Internet Bandwidth. Retrieved April 5, 1998, from http://www.nngroup.com/articles/law-of-bandwidth, Nielsen Norman Group.

Schnitter, S., & Bornhauser, U. (2014). *Future network architectures* (pp. 34–49). Future Telco (Book).

5G and Satellites: A Viable Ecosystem?

Hans-Peter Petry and Saher Salem

1   The Coverage Dilemma

Imagine you are on a trip to a fairly remote mountainous region. Of course, you have planned your route carefully and you are equipped with a modern Smartphone and a powerful App to track your route on the embedded high resolution map. As an experienced user and with certain Apps supporting this feature, you would even be in the position to investigate how many satellites your device is receiving in order to determine your position. As a rule of thumb, this number will in most cases vary between 5 and 10, the more the better in terms of accuracy. So far so good. But what would happen in case of an emergency? Sure, you would initiate a phone call or send a message. But now, you need a bidirectional connection and you have to use the terrestrial network and if you are lucky, you may connect to a single radio base station. In mountainous and remote regions, however, the probability is high that your device shows "no signal". Professionals in such cases carry an additional satellite telephone communicating with a "base station in the sky"—a geostationary or non-geostationary satellite. But the devices are bulky, the service is expensive and consequently no widespread acceptance can be expected.

Similar problems may come up in less exceptional situations, e.g. driving in a car. The observed phenomena are exactly the same: where the navigation system is still working well (showing the usual 5–10 positioning satellites), the signal from the terrestrial network shows strong variations along the driving path. This can be directly observed from the signal indicator in the car radio or more professionally by investigating global measurements of terrestrial network coverage performance provided by respective institutions (e.g.: www.opensignal.com). Of course, the effect is especially present in remote areas but also happens quite often in urban and dense urban areas.

H.-P. Petry (✉) · S. Salem
Detecon International GmbH, Cologne, Germany
e-mail: hans-peter.petry@detecon.com; saher.salem@detecon.com

© Springer International Publishing AG, part of Springer Nature 2019
P. Krüssel (ed.), *Future Telco*, Management for Professionals,
https://doi.org/10.1007/978-3-319-77724-5_5

Now, what are the reasons for these observations and how could the situation be improved? As a matter of fact, there are two main aspects leading to a coverage dilemma as we experience it daily: physical reasons and economical reasons. The physical reasons have their origin in the propagation behavior of electromagnetic waves: the higher the carrier frequency, the more a line of sight connection is necessary, only carrier frequencies below a certain limit (around 3–5 GHz and less) allow reliable connections without line of sight. These frequencies are rare, hence the licenses expensive and the available bandwidth is limited. This dilemma could be overcome by building a considerably higher number of mobile base stations but such a coverage improvement strategy is limited in terms of cost for the equipment, the sites and the backhaul. It is therefore only viable in dense areas improving both coverage and capacity but the problem outside such an area still exists.

So, finally, nobody can really be blamed, not the network operator who has to optimize his business case and will only invest in areas where he has a fast return on invest, nor the government subsidizing telecommunication services in underserved areas. Government support certainly can improve the situation but will not be able to solve the problem completely due to costs that grow asymptotically with growing coverage. Hence, relying on terrestrial networks only, we will continue to live with a limited coverage forever. Satellite networks, however, offer a much better situation in terms of visibility due to the high elevation of the propagation path widely de-correlated from the horizontal propagation path of terrestrial systems. Consequently, a cooperation has a remarkable potential for synergy.

2   5G: The Magic Box?

Mobile communication technology standards have continuously evolved over the past decades. We are today in the marketing phase of the 5th generation (5G) with 2G being the first digital technology and 3G/4G representing the current built. Initially designed for voice communications, mobile systems have incorporated messaging and data starting with relatively low data rates ramping up to up to 100 Mbit/s for 4G systems in good signal conditions today.

Now, why then would we need even a higher performance 5G solution? For sure, the bandwidth hunger will still grow exponentially and new markets for the technology are coming up such as the "Internet of Things", where billions of devices are going to talk to each other. Standardization bodies therefore are looking for further enhancements of air interface technologies and system architectures to cope with the ever increasing demand. In fact, latest versions of 4G such as e.g. LTE Advanced Pro are performing quite well, so the headroom for further improvement is quite small.

Looking at some details of typical characteristics of mobile communications (see Fig. 1), a future 5G solution is a gradual move from 4G adding improvements in mainly four areas. First is the adaptation of the air interface, system architecture and the protocols for IoT applications in order to make the system compatible with the overflowing number of new users. This is mandatory but would not alone justify to

Main Characteristics	2G		3G			4G			5G
	GSM	GSM EDGE	UMTS	UMTS HDPA	UMTS HSUPA	LTE	LTE	LTE	5G
Applications:									
• Voice/Messaging	☑	☑	(☑)	☑	☑	(☑)	☑	☑	☑
• Data		☑	☑	☑	☑	☑	☑	☑	☑
• Broadband			(☑)	☑	☑	☑	☑	☑	☑
• IoT									☑
Air interface:									
• Single carrier	☑	☑	☑	☑	☑				
• Multicarrier						☑	☑	☑	☑
• High Frequencies									☑
• Adaptive (HO) modulation				(☑)	(☑)	☑	☑	☑	☑
• SU - MIMO							(☑)	☑	☑
• MU – MIMO									☑
• Extremely low latency									☑
• Peak user data rate [Mbit/s]		0.2	2.0	10	10	100	150	300	1000 +
• Average user data rate [Mbit/s]		0.2	0.2	2	2	40	60	80	?

Fig. 1 Mobile communication generations

talk about a new generation. Second is the utilization of higher frequency bands, preferably in the Ka-Band (20–30 GHz) in order to get access to larger channel bandwidths for capacity improvement. In fact, available channel bandwidths there are considerably larger leading to a proportional increase in channel capacity and improve available data rates even more. Theoretically possible peak values are therefore entering the Gbit/s area but experience form the past shows that real world values perceived by the user are normally considerably lower. This is mainly due to the adaptive behavior of the air interface that can adapt data rates to signal quality. This leads us to the coverage problem again which will not at all improve with 5G, just the contrary: frequencies in the Ka-Band require line of sight connections to a large extent, consequently, these higher data rates will only be achieved in such a situation. The 3rd category of improvement measures (SU-MIMO and MU-MIMO) has better chances for success as it is using enhanced spatial filtering technologies together with sophisticated antennas on both sides of the link, a technology that is already proven but at the expense of a higher hardware cost and system complexity. Last but not least, a low system latency is certainly beneficial for a lot of modern applications where high performance real time operation is required.

Overall, the improvement potential of 5G is certainly present but should not be overestimated. In particular, the coverage dilemma will be the same in frequency bands where 3G/4G are used today and get even worse if higher bands are used. So, these higher bands will be limited to areas where line of sight is possible e.g. in very dense urban (pedestrian) areas or situations where the mobile base station is mounted indoors. Special attention has to be paid to the IoT application. As long as the mobile base station is located outdoors, the coverage problem becomes even more critical. IoT devices often require what is called "deep indoor coverage". Here, extremely low frequencies are the optimum solution, but these are traditionally occupied by classical systems that may not be switched off immediately. There is some hope that lower bands can be made available but this will be a long term regulatory process with many hurdles. As a consequence, IoT applications will additionally require intelligent system solutions delivering a sufficient signal quality form the outdoor network to the indoor devices.

3   Autonomous Driving: A Killer Application?

Imagine you are on a trip again, this time you are driving with your convertible through a beautiful landscape on a bright and sunny day. Would you think about autonomous driving in this moment? Definitely not and as long you are in a remote area, the necessary communication infrastructure is most likely not present yet—we remember: the coverage problem.

But without any doubt, autonomous driving is one of the big hypes today with a lot of opportunities to improve our daily life and with a lot of future market and business chances for the industry. Following numerous scientific analyses, autonomous driving—we really mean autonomous and not automated—can improve and optimize the ever increasing traffic in terms of efficiency and safety. Together with alternative and sustainable vehicle concepts with low or zero emission, our world could be improved substantially. So far theory, but are such future perspectives realistic and feasible technically and ethically?

The answer of the industry is: yes. A lot of major players—not only from the automotive industry—are investing substantial amounts of money in startups and own development. Activities actually center around the necessary car infrastructure and the IT system environment, first demonstrations show astonishing results, early versions of automated driving add-ons even in mid-class cars are already available today, still with some limitations but more and more impressive.

This situation has triggered a lot of discussions mostly in the ethical domain: what happens, if we handover the driving responsibility to a machine and an in-build algorithm? How will the algorithm decide in an emergency case, what are the ethical standards in particular if the decision is not only technically based and when will the command be handed back to the driver based on what parameters? The list of difficult questions certainly is much longer. Consequently, in Germany as one of the first countries, the government has started an initiative to create a corresponding ethical framework and associated guidelines (www.bmvi.de). The key findings can be summarized as follows:

- Autonomous driving in a network environment is ethically desirable if the associated systems are producing less accidents than human drivers
- In an emergency situation, human life protection has always the higher priority
- In case of inevitable accidents, any individual qualification of human persons (age, gender, constitution) is inadmissible
- For any driving situation, clear rules and indications need to be established, who is responsible for the driving task. The actual situation needs to be documented and recorded
- Associated data are in the ownership of the driver

These initial considerations certainly need more discussions and refinement but one aspect for general acceptance is evident from the very beginning: safety. We all know that technical systems can't be 100% safe but a total safety level has to be approximated as far as possible otherwise, a possible technical progress will not be accepted. Looking at similar situations that already exist (e.g. aviation) we can derive conclusions how to make autonomous driving technically as safe as possible. Flying in an airplane is one of the most secure ways of travelling today. This high level of security is achieved by two basic principles: redundancy and global networking. In aviation, redundancy is mostly applied in the aircraft where the decisive technical systems are doubled or even tripled. Global networking and supervision is based on a global (terrestrial) aviation network, admittedly still with some large "coverage" holes far outside the land masses—we know the problem already. But lots of efforts are underway to improve the situation with the help of—satellites.

As far as we understand it today, a global autonomous driving "system" will most probably be a lot more complex as aviation, in particular looking at the traffic density which is order of magnitudes higher and the degree of deregulation: nobody will accept such strict rules in our daily traffic as we know that from aviation. As a consequence, an even more enhanced overall system concept will be necessary to make all these dreams come true.

Solutions today are centering around a powerful local (car based) infrastructure incorporating numerous sensors. They are mainly acoustical (ultrasonic), optical (lasers, infrared, cameras) or electromagnetic (radar). Together with a sophisticated onboard intelligence ("sensor fusion"), remarkable technical solutions are already available today and if the road infrastructure is upgraded accordingly, impressive results can be achieved in favorable situations. But here, the problem area starts: can we assume that any road infrastructure can be easily adapted and what happens in case of ever present changes? And even more: system solutions based on local sensors—even if the information is completely correct—will always be short range and reactive, i.e. an action is initiated after the problem has come up. This automatically leads us to the problem of ultra low latency which limits the headroom for decisions and imposes strict requirements on hardware and software performance. Consequently, a more proactive system behavior is highly desirable. With a more global approach, a high percentage of potential conflict situations can be detected much earlier and critical situations can be avoided before they really occur. Low latency is no longer an absolute must.

One of the solutions widely discussed today is based on a regional infrastructure which is present anyway: broadband mobile communication systems in particular a future 5G solution. Such a solution would even have the potential to act globally based on handover and roaming algorithms that we know very well from our broadband communication experience. But we also know that the coverage problem remains. The key question therefore is: can an ever present limited coverage be accepted in terms of overall system security and reliability? The authors think: no.

For sure, 5G is offering much lower latencies compared to previous generations and is compatible with the local sensor approach where this is decisive. But the move to higher frequencies that we already have discussed, is counterproductive. Applying

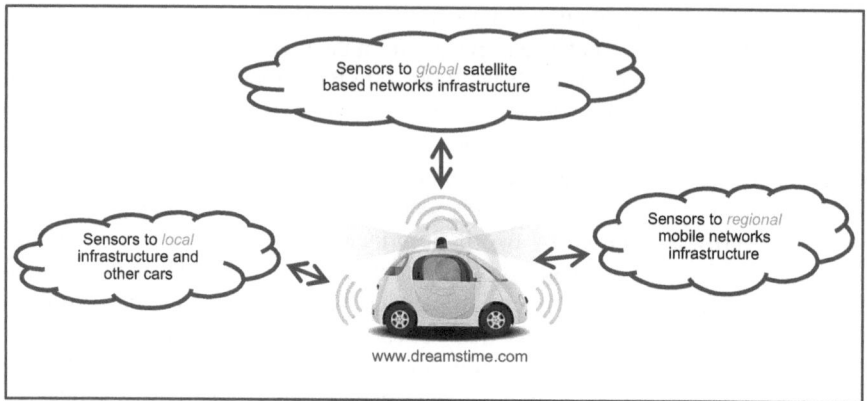

Fig. 2 Potential high safety scenario for autonomous driving

5G also at lower frequencies, therefore, would make a lot of sense, not only for automotive applications but also for IoT. First initiatives are being discussed but will only marginally improve the situation.

Therefore, an additional system component based on LEO satellites can represent the missing link to a really global system approach with the highest safety level possible. As mentioned, the coverage characteristics of satellite systems and terrestrial systems are widely de-correlated: where horizontally oriented transmission of electromagnetic waves suffers from shadowing effects, satellites can step in, where satellite communication is difficult (e.g. dense urban areas), the density of terrestrial cell sites is high, just ensuring a good coverage by the numbers.

Figure 2 shows a system approach where three different communication systems are combined will be able to offer the highest level of overall system security and functionality mainly based on an optimized coverage. Furthermore, a very high degree of system redundancy is automatically ensured, a temporary failure of a single system is not resulting in an overall malfunction.

4   LEO's: Friend or Foe?

For a long time, satellites are contributing substantially to our modern world, starting with the well known TV distribution, globally available positioning systems, earth exploration, specialized systems for bidirectional communication and much more. Most satellite communication systems today still are working with specific user terminals, no application has found its way into a modern portable broadband device except global positioning. Broadband bidirectional communication and in particular mobile broadband based on satellites therefore is still a small niche market with a huge potential for future evolution. Imagine what would happen if Silicon Valley companies succeed in building such a broadband satellite infrastructure enabling a full or partial bypass to terrestrial infrastructures. Not only would this contribute

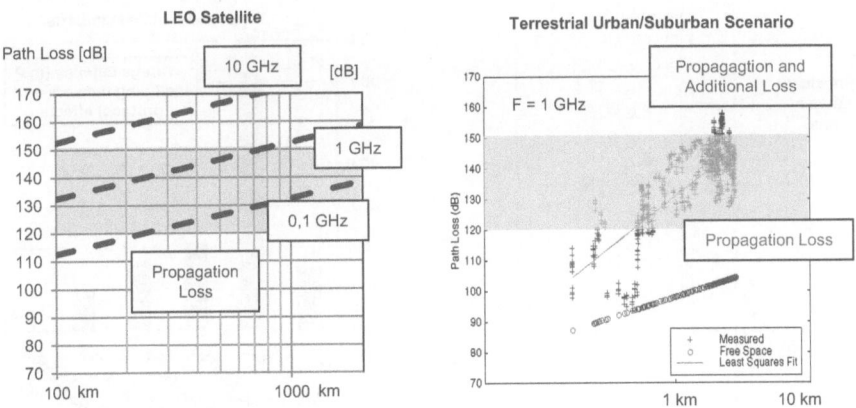

Fig. 3 Link budget comparison LEO—terrestrial

substantially to a more secure autonomous driving experience—as pointed out—but also create a tremendous threat to the monopoly of terrestrial network operators who already have identified the autonomous car and its passengers as potential new broadband customers.

Until recently, nobody really had to worry about that. Communication satellites exclusively were GEO's (GEO = Geostationary Earth Orbiting Satellite) with link budgets and latency parameters incompatible with compact portables and modern protocol requirements. Although GEO's have substantially evolved over time in terms of broadband performance and flexible coverage, they are still extremely expensive, linked to specialized satellite communication providers and associated manufacturing and qualification processes are complex and time consuming and in the hand of a few sophisticated players. Not astonishing, several attempts to solve the problem this way have miserably failed in the past.

This is going to change. First players using Low Earth Orbit (LEO) satellites in large numbers (so—called "constellations") are entering the market, still with specialized terminals in most cases but the step to compact mobile terminals is in reach: for certain frequencies and orbit heights the link budget parameters become comparable (see Fig. 3). For the LEO satellite, it is mainly determined by the distance and the frequency. For the terrestrial network, the free space attenuation is of course much lower but there is additional attenuation due to NLOS shadowing, especially at the cell edge, often this is even the determining part and also dependent on frequency. Admittedly, this is a rather simplifying consideration but it shows that decisive parameters of such a LEO link (mainly user terminal transmit power and antenna performance) start to be compatible with the values actual devices have. There is no chance that these parameters can be improved in the future. The everlasting constraints are mainly given by limited transmission power due to electromagnetic compatibility regulations and battery lifetime as well as a limited antenna performance based on the device form factor and the coupling with the human body.

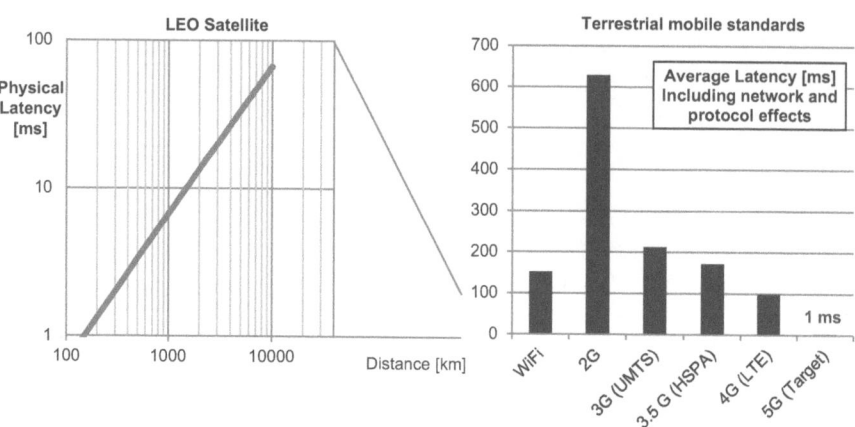

Fig. 4 Latency comparison LEO—terrestrial

On the LEO satellite side the headroom for establishing compatible system parameters is a little bit higher but also quite limited. Comparing the decisive parameters of a modern terrestrial radio base station with a future LEO satellite in a constellation with many others we have to focus again on antenna performance and transmission power i.e. the effective isotropic radiated power (EIRP). Due to the limited size of a LEO, the maximum available power certainly will be limited compared to medium or even large satellites, consequently, the antenna performance will become the decisive part. Physically, the antenna performance (mainly gain and beam width) is given by the geometry (size). Together with intelligent active antenna concepts all antenna parameters become dynamic and can be adapted to various link configurations. Nowadays, this is well understood in terrestrial communications, especially in 5G, such antenna solutions will become an important building block. For the usual satellite frequency bands (including Ka-Band) the sizes of such antennas will be comparable with those in terrestrial base stations and as such be compatible with LEO sizes. Not astonishing, a whole bunch of innovative antenna startups have shown up, investigating not only exciting new antenna concepts but also new antenna materials not only for the space segment but also for future terrestrial terminals, in particular cars.

The next main hurdle to be considered is latency. Figure 4 shows the end-to-end latency which is given by the physical latency of the transmission path and the network latency which is determined by network protocols. For a GEO, the physical latency alone is prohibitive and much higher than all mobile standards including network, air interface and protocol effects, for a LEO, the physical latency is substantially lower and compatible with existing and future mobile standards. Whether the pending requirement for 5G (latency < 1 ms) is realistic, needs to be discussed in particular looking at the promising features of future proactive satellite based solutions for autonomous driving. In any case, a pending 3GPP 5G standardization should seriously consider satellite based system scenarios in order not to close the door for attractive applications. Fortunately, such activities are underway.

While the shorter distance of a LEO compared to a GEO is beneficial to link budget and latency, the relative speed of transmitter and receiver has to be considered depending on the orbit geometry. The resulting Doppler shift is a function of the carrier frequency, the satellite angular velocity and the elevation angle from the ground based equipment to the satellite. This effect is especially critical for modern air interface standards that are being used for 4G and 5G terrestrial mobile communication systems (e.g. LTE). For high elevation angles (zenith) the Doppler frequency shift is small or compatible with the values anticipated in the 3GPP standards for terrestrial systems, for lower angles the effect increases substantially. The deteriorations therefore can be mitigated by a limitation to high elevation angles or a pre-computed compensation based on geometrical data and a sufficiently precise knowledge of the LEO satellite position. Again, corresponding boundary conditions have to be implemented into future 5G standardization activities in order to fully exploit the opportunities LEO constellations are offering.

So, technical evolution is pointing into the right direction. The question remains whether such approaches will be commercially viable or even more competitive as legacy satellite based systems. The answer is given by what is called "New Space Economy". In a nutshell, the world for space applications in general is in motion, driven by protagonists like Elon Musk and Steve Bezos and supported by considerably large investments. According to their plans, the future space economy will no longer be dominated by national or international agencies (NASA, ESA) but by private enterprises. Space equipment will no longer be exclusively manufactured by specialist companies based on time consuming development, manufacturing and qualification processes and protected like a Fort Knox treasure but based on a larger degree of standardization and off the shelf components. By the way: this is a philosophy that has made GSM (the first worldwide mobile telephone standard) so successful.

As a consequence, associated system solutions are anticipated to become much lower in cost and faster in time. Although some of the initial ideas seemed to be quite aggressive and optimistic, a lot of recent successes—admittedly after some learning curves—show the viability of the approach. Starting with launchers and manned spacecraft the approach will over time also cover other areas of the satellite business such as earth exploration and for sure satellite communication, in particular LEO's and LEO constellations. With a LEO concept, the entry barrier to build a satellite becomes much lower. Consequently, we are seeing a whole bunch of new players entering the market such as small specialist startup companies and even universities where students are developing, building and launching innovative LEO's. The innovation window here is wide open and all major investment companies have realized that. Not astonishing, large OTT players have also detected these opportunities and are working towards own access infrastructures in order to bypass the terrestrial monopoly.

Last but not least, a serious aspect of future LEO constellations has to be considered: the threat of further low earth orbit pollution by space debris. At a first glance, this seems to be even more critical than all technical aspects mentioned before. Large constellations are aiming at hundreds to thousands of satellites per constellation which is absolutely necessary for excellent coverage and large capacity. Short satellite lifetimes and a permanent exchange of individual satellites will create lots of small debris type of objects. The proper handling of obsolete LEO's, therefore, has to be closely monitored. Corresponding global activities for such a "space debris mitigation" are already underway and need to be an integral part of such a future system approach.

5   A Necessary Ecosystem

Undoubtedly, broadband and ubiquitous mobile communication infrastructures are a major prerequisite for our future digitized world. While initial attention was focusing on broadband, a global coverage is nowadays emerging as the even bigger requirement and challenge. Relying on terrestrial solutions only is not only a dead end street for remote and non-accessible areas but also lacks commercial viability even on a longer run. Together with new safety critical global applications, an affordable global infrastructure even will be mission critical.

This is a tremendous chance for future satellite communication to leave the past and present market niche and become a respected and required member of a future 5G ecosystem, not only a competitor to the terrestrial network. But we are not there yet, both the satellite community and the terrestrial players including regulation authorities have to undergo some more paradigm changes, still lots of technical challenges have to be resolved. However, this is not far from being realistic. Terrestrial-Satellite-Convergence has gained a lot of momentum lately when the German Telecom Company (Deutsche Telekom) joined forces with Inmarsat to offer broadband internet connectivity for European flights. On the other side of the globe, Middle East, Etisalat launched a new bundle together with Thuraya Communications to offer telephony and SMS services via satellites where Etisalat customers face a mobile white spot. The outlook is bright: satellite systems—in particular LEO constellations—and terrestrial radio based networks (fixed and mobile) have strongly de-correlated properties. Their individual advantages can and have to be combined and integrated in order to make a ubiquitous and affordable broadband communication infrastructure come true. So, "Welcome to the 5G Ecosystem" is a positive and highly innovative slogan for all parties.

5G: New Opportunities?

Falk Schröder

1 Introduction

Starting from the 1990s when the mobile network industry has started to develop, people used to follow the natural way of cellular network evolution. The key role in the change, was always played by the network operators who introduced 2G, 3G, 4G. All systems are still in use (even 2G has not been retired yet). However the market has changed significantly during last almost 20 years (mainly due to OTT big players) and what is now going to happen is far beyond standard evolution, it will be the 5G revolution.

When does the 5G story begin? The new technology discourses are ongoing in the technical community and in case of 5G the first organization in Europe officially involved in it is 5GPP (5g-ppp.eu), which was set up in 2013 but so far the most significant input in 5G milestones evaluation is considered to be published by Next Generation Mobile Networks community (NGMN, www.ngmn.org). This whitepaper was issued in the beginning of 2015 and it is a kind of 5G manifesto where all major issues are addressed including various requirements, directions, use cases and complementary technologies.

The NGMN vision is still valid:

> 5G is an end-to-end ecosystem to enable a fully mobile and connected society. It empowers value creation towards customers and partners, through existing and emerging use cases, delivered with consistent experience, and enabled by sustainable business models.

The standardization of 5G network started in 2015 with official 3GPP timeline announcement in March, just before the freeze of Release 13. The first official standard that included 5G 'data' was defined in the Release 14, where main focus

F. Schröder (✉)
Cluster Networked Infrastructures, Detecon International GmbH, Cologne, Germany
e-mail: Falk.Schroeder@detecon.com

© Springer International Publishing AG, part of Springer Nature 2019
P. Krüssel (ed.), *Future Telco*, Management for Professionals,
https://doi.org/10.1007/978-3-319-77724-5_6

was put on the basic requirements for the future cellular networks. Detailed description of the 5G can only be found in Release 15. Today the standardization of Radio Access Network is much more advanced comparing to Core Network, however the basic Core Network architecture models are:

- Non-Standalone (NSA) 5G NR designed to utilize the existing LTE radio and core network as an anchor for mobility management and coverage while adding a new 5G carrier. This is the configuration that will be the target of early 2019 deployments (in 3GPP terminology, this is NSA 5G NR deployment scenario Option 3).
- Standalone (SA) 5G NR which implies full user and control plane capability for 5G NR, utilizing the new 5G core network architecture also applied in 3GPP.

In the Standalone (SA) solution we can identify two basic approaches to the CN—Reference Point Architecture and Service Base Architecture. We will see a 'revolution' by introducing the so called cloudification and softwarization of 5G (SA) CN, which will significantly change the way of handling security, mobility management, session management and other aspects. As this is not yet standardized only 5G NSA will be available for the first deployments.

A latest good example of 5G NSA mode HUAWEI SON provided a solution which was heavily awarded at one of the most important events at the NFV SDN World Congress in October 2017. In fact, the architecture itself is already claimed as 5G Service Based Architecture, however applied Core Network is still Virtualized EPC—it means pure 5G NSA.

In order to follow current standardization of the 5G Core Network, the Architecture study 23,501 will be the base, which was also underlined directly by Georg Mayer 3GPP CT Chairman, who said: "...If You want to follow our work, look at this document (23,501), if You read it Yesterday it is really outdated..."

Since 5G RAN is more advanced in development comparing to CN, we can expect many implementations of Non-Standalone (NSA) 5G NR in early 2019. Therefore and according to 3GPP Enhanced Mobile Broadband (eMBB) and Ultra Low Latency Communications (URLCC) use cases will be covered and seen in the first place.

In the future the role of the standardization bodies will be different, especially within cloudified Core Network implementations. As seen above the current discussions around 5G are focused on Broadband and also partly on massive usage of m2m services. We did not invest enough time in understanding the machine critical application and its requirements for operators and its interfaces. Furthermore much more automatization is expected to come for e.g. services, testing and operation.

Nowadays many operators all over the world run 5G trials and started to prepare schedules for the commercial implementation of NSA 5G system. In Asia operators like NTT Docomo and SK Telekom started trials already in 2017 and are planning commercial launch in 2020. China Mobile is even earlier, starting non-commercial start in 2019 to be ready for 2020.

In India 5G Network Development is supported by the government, the newest decision from September 2017, regarding setting up "Special committee for 5G rollout by 2020", may help the country with the 5G revolution. The Indian Minister of State for Communications Manoj Sinha said: "We missed the opportunity to participate when the standards were being set for 3G and 4G, and don't want to miss the 5G opportunity. Now, when the standards are being set for 5G across the world, India will also participate in the process."

In North America we see plans for commercial launch in 2019 by T-Mobile and Verizon and in 2020 by Telus/Canada. In Europe we saw already some trials, latest one in Berlin in September 2017. However commercial launch is also planned for 2020 (DT and Orange).

The RAN trials are either ongoing or are already done, it is foreseen that in 2019 the first E2E 5G NSA solutions will be commercially available for customers.

This leaves another only 2 years for understanding and connecting to a new so called 5G era.

According to statista we will see in 2021 already 100 Mio. Subscribers on 5G worldwide and about 1.1 Billion subscribers in 2025.

2   The Essence of 5G

The requirements on 5G systems are diverse paying tribute to the fact that various use cases have to be considered. This illustrates the challenge ahead of the standardization bodies in defining a global telecommunication system that does not only employ and incorporate new technological concepts aiming at improved performance to be experienced by customers and realizing economical superiority in terms of network operation.

Historically, every new generation of mobile technology was designed to deliver better performance beyond what the evolution of its predecessor could bring. 5G should improve the performance envelope for mobile communication by a magnitude that might not be feasible with LTE and its evolution. The most common performance requirements as identified by leading industry experts are summarized below. Partially, the requirements have been defined relative to the (initial) 4G technology capabilities:

- 1–10 Gigabit per second (Gbit/s) per connection
- 1 ms latency
- 1000× better capacity
- 10–100× better connection density (number of connected devices per geographical area)
- (Perception of) 99.999% availability
- (Perception of) 100% coverage
- 1000× more capacity at half of the energy consumption
- Up to 10 years battery life for low power, machine-type devices

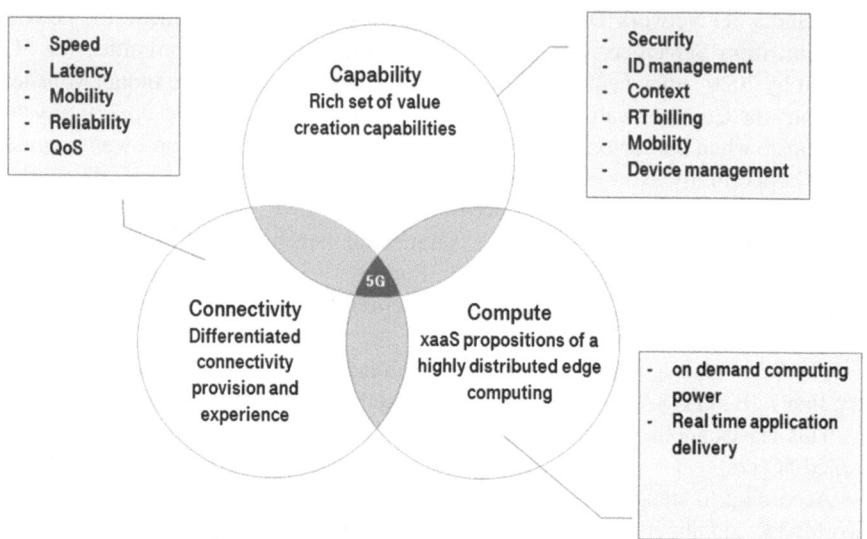

Fig. 1 5G key assets

The abovementioned performance requirements may need a redesign of the network. For example, adding more capacity to the system may require getting access to spectrum beyond 6 GHz (centimeter and millimeter waves). This requires more research in terms of channel characteristics as well as new waveforms and protocols. Realizing 1 ms latency requires further optimization of the radio stack as well as leveraging EDGE computing (see below). Furthermore, densification of the network will require a fundamental and radical new approaches to deploy small cells at an economically viable way.

Obviously connectivity will always be at the core of what 5G has to offer. However, connectivity has to be enhanced and differentiated including on-demand capabilities like latency, speed, reliability and the likes. This so-called Enhanced Connectivity needs to be enriched by capabilities like security, privacy and a more consistent experience. Capabilities like QoS, Charging and context will enrich many services and applications at the benefit of 5G users.

5G will have to encompass a comprehensive and E2E view on the network, respectively on the service. Just offering better connectivity will not satisfy the overall set of requirements. 5G should therefore be built for a *beyond connectivity* paradigm. 5G should covers the 3C's; Connectivity, Computing (at the edge) and a Capability set that all together are delivered on demand and in a very flexible way to match the requirements of a specific use case within a certain context in highly digital society. This is reflected in Fig. 1:

The 3C's that constitute the 5G key assets can be assembled into a holistic and virtual network optimised for a certain purpose. This is referred to as *Network Slicing* and is explained in the next section.

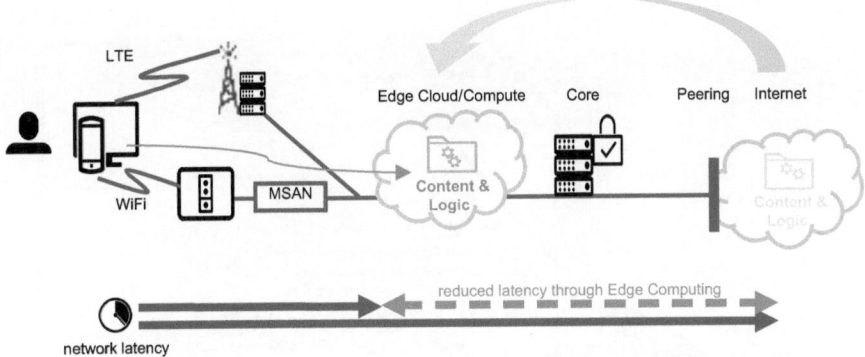

Fig. 2 Edge computing

EDGE Computing provides the option to have a data centre close to the user, therefore close to the antenna, shown in the Fig. 2. By connecting a defined area to one EDGE compute system services can be provided with low latency. According to feedback from the industry it is required to have a kind of managed latency, which in essence is a maximum delay. An operator is in a premium situation to provide this guarantee. However the operator has to learn how to manage and operate an EDGE Compute system. Services on the devices are able to foster using a service placed on the EDGE Compute system close to the user allowing short times for reaction and therefore low latency.

On top of an individual EDGE system it is essential to get from a service prospective a global reach. This can only be achieved by having a centralized function hosted outside any operator, which works as a peering point for new low latency based functions. Adding here functionalities like AI, ML and SDK options helps to address the services to large number of developers.

3   5G Network Slicing

NGMN's architecture, as reflected in Fig. 3, consists of a three-layered structure including infrastructure resources layer, business enablement layer and a business application layer to enable various use cases in parallel.

The infrastructure resources layer composes and abstracts the underneath physical resources and provides them to the upper layers in a virtualized manner. The upper business enablement layer is in charge of the management of the repository of the function building blocks and the corresponding access Application Programming Interfaces (API). The business application layer provides the specific applications

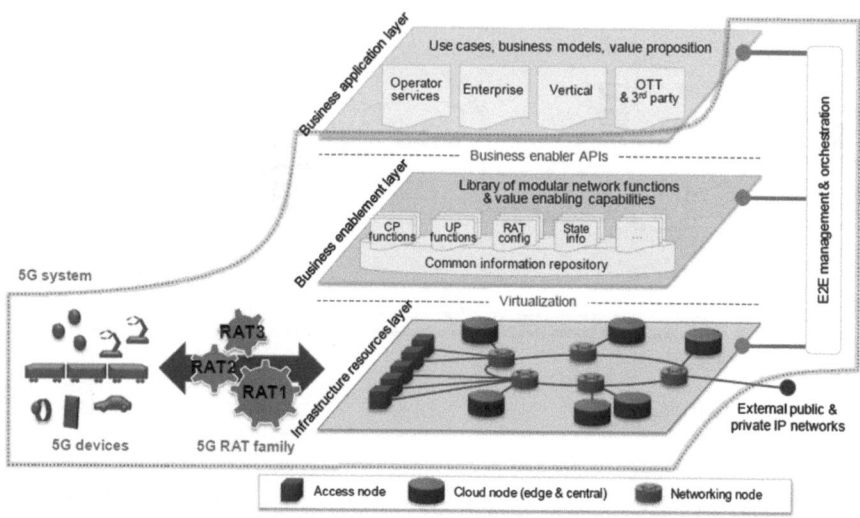

Fig. 3 NGMN 5G architecture

and services to the customers (enterprise, verticals etc.). There is an end-to-end management and orchestration entity in the proposed architecture dedicated to overall control and providing the interface dealing with translation between business requirements and network functions/slices.

Network slicing is one of the most prominent concepts being discussed in the context of 5G. Network slicing promises a *just-in-time* and *what-you-need-is-what-you-get* delivery of network capabilities to the actual recipient of the resources—may this be in terms of a segmentation that addresses the internal needs of a mobile operator or of its (business) customers. Network Slicing has been presented as an architectural approach to flexibly create a number of virtual networks on a single network platform in a generic way. This allows to create networks that accommodate a diversified 5G business context. The traditional *black box approach* consisting of purpose-built hardware and software (as it has been applied in the telecoms industry for decades) will transform into a software-driven approach based on commercial off-the-shelf hardware and open-source software. This transformation is enabled by technologies like SDN and NFV. Both technologies SDN and NFV will be one of the key enablers for the proposed architecture.

Network slicing will only be available from about 2022 as the standardization has not been finished yet. The number of existing slices for one operator may vary heavily, however, it is mainly driven by industry needs and by the operator's capability to fully automate the usage and operation of network slices. Without automation it will not work on a broad level.

The high-level requirements for a viable 5G architecture have been defined earlier.

4   5G Use Cases

Today's mobile networks provide capabilities that allow for a consumer experience across various applications in a Smartphone-centric eco-system. However there are specific use cases that need explicit improvements in terms of network capabilities. These use cases cover the consumer space, B2B as well as the emerging digitalization of vertical industries.

Today's digital natives will be tomorrow's 5G customers. These customers are always connected. They do not only consume, they also generate a big amount of content and are highly interactive and constantly online. The amount of devices that consumers will have around them will dramatically increase. Those devices will differ in terms of network capabilities, functions as well as the type of data they collect or generate. The development of consumer electronics will further open up completely new use cases and experiences like Virtual Reality and Augmented Reality. All these developments does and further will dramatically change the way we communicate, the way we educate ourselves and in general the way we experience many aspects of our lives.

Enterprises will continue to drive for more agility, increased productivity and higher production efficiencies. At the same time the relevance of security and business continuity will remain very crucial for the future enterprise. Likewise, industry verticals are further mobilizing their existing applications and digitizing all aspects of their business and customer processes. Additionally, new services and offerings will be provided to industry customers to position themselves more flexible and digitized.

5G is positioned as a key enabler in delivering these new digital experiences in a highly digital society.

4.1   VR/AR Applications

Virtual and Augmented Reality (VR, AR) will enable new experiences and allow for new business cases in various industries. Already today we can see some tangible implementations with impact to day to day business. Once AR glasses are in use, information can be displayed on top of what the user can see. Bringing the surroundings in a new context and providing additional information it can provide benefits in various directions.

AR remote maintenance is an area which is already provided. At DTAG glasses can already been used for network engineers to detect cables and connect them. In this case, the glasses are connected to a system, which identifies the objects and brings it in the context of a task. For instance reconnecting a customer to more bandwidths. The technician gets a visual information by a red pointer or by an arrow, showing which cable to get off and how to reconnect. This works already end to end. Assuming that other industries can do the same, why not fixing large engines by using AR based glasses instead of moving experts from place to place? The information can be provided automatically by a system or by a specialist who does not anymore needs to be onsite. This use case need some data rates, which can probably already be provided by LTE. However by having accurate display information Low Latency becomes important. Additionally some compute power might be required for calculations. This needs to be moved to the EDGE Compute system. DTAG and Zeiss have announced to work on smart glasses with AR functionalities addressing for instance the remote maintenance case for various industries on one platform.

Finally AR based teaching systems are under evaluation. Most of them will be home based and therefore connected to the fixed line. However, on the long run users will use it as well on the move, in trains etc. These interactions also require Low Latency capabilities.

Another example of an early implementation are developments for drones, where on the move (or on the fly) related information based on a film by an onboard camera need to be evaluated and an action by the drone has to follow. For instance in wine yards it is essential to identify insects which harm the fruits as such and eliminate them by spray or similar. This compute power and the relevant latency requires technologies been provided by 5G.

All cases above are under implementation or short to get implemented. They all need low latency and some of them even require computing capabilities in the EDGE. Maybe on top some cases may require in the future dedicated Quality of Service (QoS) capabilities like guaranteed data rates on the move. These QoS solutions are under evaluation today and not yet in place. However it will become crucial to get common solutions on QoS reflecting market needs.

4.2   Automotive: Evolution of Cars and Its Digitization

Talking about cars, there is one use case making the impact of the digital transformation really clear: connected cars with its most extreme case of digitization of the car, autonomous driving.

Since many years cars have different embedded systems to assist the driver and make driving easier and more safe, e.g. the anti-lock braking system to let you control the car even in case of a full braking, firstly introduced in the automotive market with the Jensen FF in 1966.

But in the last years, connectivity came more and more into focus for these embedded car systems, as the information of these embedded systems is very valuable for the car producer, e.g. to ensure the quality of the car and its systems. This communication type was used for the internal usage by the car producers. It needed only low data bandwidth and had no real time requirements as it was mainly used to check and verify the embedded systems. This state changed only a few years ago, when the first car producers developed more communication services to the customer, including latest traffic information or internet access for the passengers, controlled by the car. This evolution was impacted mainly by the decrease of prices for mobile internet connections as well as the computerization of cars, while

IT-systems getting more and more powerful together with the general development that more and more things are getting connected in the internet of things.

Today the car is changing more and more into a smart device on wheels, consisting of many IT-systems and at least one SIM card to provide mobile internet access for the connected services. While the traditional car producers like Volkswagen, General Motors and BMW are facing the challenge of the integration of such services in their traditional way by equipping their traditional product with more and more IT-systems, there are new players entering the market like Tesla or the Future Mobility Corporation FMC, starting to build up car as a device and grouping the ingredients of a car, like wheels around this IT-system. Tesla, for example, equips all its cars with the same hardware and enables new features and functions on demand via the firmware of the car.

The vision of all car producers of the future car is, despite of the different starting points, very similar: an economic car (mostly using electricity to drive the engine) that is connected and integrated to and with the IT-systems in its environment, e.g. the smart city, having the ability to drive autonomous and provide additional services.

For all connected services, where information is exchanged, the car needs connectivity, even if it needs to drive safely autonomous without connectivity for a specified time in areas with low or no cellular coverage.

But what type of connectivity is needed for which connected services? Many connected services in the car we are using today are a kind of infotainment services and do not have real time constraints, e.g. streaming music to the sound system in the car. These services are only limited in their connectivity by the bandwidth they need. Another example are on demand downloads of digital maps by entering e.g. a big town including latest adoptions. Software updates for the car and all its various IT-based subsystems could be another issue, as these updates are expected to be done in a few seconds by the customer but meanwhile the firmware of a car can exceed volumes of terabytes of data. Of course this needs to be finished within the expected time.

This changes dramatically if we think a bit ahead, using real time information to make driving more safe, e.g. sharing the information about an obstacle on the street with other cars in the immediate vicinity, e.g. a tree or animals on the street or the tail end of a traffic jam. In these cases, every millisecond counts: even with today's 4G network with its round trip latency under 80 ms it could not be fast enough. Also the bandwidth of 4G cells could be congested easily by many cars sending big volumes of data to the central cloud.

5G is a lot more promising for connected car services with real time constraints: aiming for a round trip latency of 5 ms it will be fast enough even for autonomous driving controlled by a cloud application which implies the sending and analytics of big data in real time and a software update of the car's firmware could be done while waiting at the traffic lights. However, we will first see autonomous driving cars based on onboard sensors. But by introducing connectivity, services will be enriched.

Recognizing the importance of 5G for the evolution of the car and its future connected services, many car manufacturers have founded an alliance, the "5G Automotive Alliance", 5GAA together with leading chipset vendors, network equipment vendors, car OEMs and network operators. This alliance has the mission to use the capabilities of the upcoming 5G standard as best as possible to connect the cars and make them drive autonomously, integrate them into the ecosystem of the emerging smart cities and intelligent transportation and drive in general the penetration of 5G in the global market.

As the most important market players participating in the alliance, 5G has the best chance to get the connectivity standard for the future car, taking care of fast and reliable communication which provides also interoperability between different ecosystems in our future connected world. Last but not least we should not forget that coverage is still an issue as the best technology only works by having access. Anyhow, some features can still been used if the car 'knows' the coverage map and is adopting its service accordingly.

4.3   Digitization

The manufacturing industry is by far the industry with the highest IoT investment as it accounts for about one-fourth of the IoT investments according to IDC Worldwide Semiannual Internet of Things Spending Guide. In here manufacturing operations IoT spend is $102.5 billion resp. 57.5% of total manufacturing IoT spend. These are high numbers which are reflected in various reports in the market promising high investments on connecting the manufacturing industry with services in various regards. Some are pointing to provide robots controlling systems from an EDGE server. Other reports are more pointing to providing various information for further computing and evaluation like smart metering, information about physical assets like emptiness or similar.

However reality shows that there is still a lot to learn. Operators need to understand the service needed and getting a glim of where to contribute, industries need to see the benefit of connecting machines and vendors need to provide related solutions. The market is there, but solutions still have to be found. To push this activity some m2m or IoT manufactories are testing the capabilities of e.g. EDGE and low latency for potential services. However clear use cases still have to be provided.

4.4   Drones

Today we see an exploding market of drones for very different use cases from support in the agricultural environment to taxi services like in Dubai moving humans from one place to another automatically. In the industry drones are needed, which can fly automatically with no line of sight. These drones could automatically fulfil certain tasks like observing special areas or buildings. In these cases compute power need to be moved to the EDGE and on top services like precise positioning but also low latency become an issue. However some regulatory aspects need to be solved as well.

5   Partnership

As described above, 5G is more a kind of introducing a new ECO system. This system needs Partners and a new way of cooperation.

Taking the introduction of AR/VR as an example, we need new products, which are working on a global scale offering for example remote maintenance solutions. Here we need Hardware which is programmable easily by a large group of developers to get a quick take up. The Hardware will be produced by a typical HW producer, the product might be enlarged by an eSIM to get connected to one or various users on the fly. Finally the real visual capabilities and functionality to use e.g. a remote maintenance case need to be developed by developer communities. On top capabilities like Low Latency have to be provided and used. Operators need these communities to get the full empowerment of the developer knowledge and abilities in place. On the other side operators have to commit providing easy to use environments.

Thinking even further for the AR/VR case, the service need to work globally, which implies the cooperation within all operators to support and not to fence solutions. Here certain standards like the OEC—Open EDGE Community is fostering for low latency based services need to be pushed and further developed.

As seen in the use case for automotive a close alignment between operators and car producers are also required. Having autonomous driving based also on mobile communication in place includes certain QoS reliability, Low Latency provisioning etc. requires cooperation as well. Regardless who the customer 'owns', the service need to be king and common solutions where car producers and operators cooperate will win. In a scenario where an operator sees car producers as a customer only may fail.

To get the full empowerment of the services the cloudification or softwarization of the technology needs to get in place. Here operators need to come along with solutions by datacenter providers like face book or others. TIP (Telecom Infra Project) is already a good direction to address these needs.

6   Way Forward for an Operator

A look back in terms of key characteristics of the previous generations in mobile communications will help to classify the industry debates concerning the role that 5G is assumed to play in the future. Despite all the attention for 5G, it is envisaged that LTE will be the main mobile technology for many years to come and will also evolve to address low latencies and better performance.

Supporting the ongoing digitization wave, it requires operators to think beyond connectivity and embrace capabilities such as big data, computing, security, etc. 5G will address new use cases with a fundamentally new design around low latency (below 1 ms), enabled by computing at the edge of the network. Many operators will move to network clouds in the coming years to leverage SDN and NFV. Edge computing will be a natural step to bring cloud to the edge and with that leverage a highly distributed computing capability.

5G is a move towards a new era of communication, which could re-open the discussion of adding new types of services and business models. Other needs triggered by IoT or industry 4.0 are not yet well enough understood to provide the right solution portfolio design. Critical to the design of 5G is an early engagement of operators with the relevant industry verticals to understand the role of 5G and the required technology architecture. However, the potential will differ from country to country and from operator to operator depending on the maturity level, size of the country and local market conditions.

For operators it will be decisive to review their overall strategy in terms of being prepared for the shift of paradigms connected with the upcoming introduction of a 5G standard. This encompasses the business as well as the technology strategy. The latter one should provide unambiguous answers to the questions whether and how the already existing network technology assets can be embedded into a consistent architecture framework. 5G needs to be justified and therefore part of the overall business strategy of an operator. New concepts like SDN/NFV will most likely happen before 5G gets implemented. Operators will most probably position 5G implementation as the next logical step beyond the transformation triggered by SDN/NFV and address new business opportunities. At the same time, 5G should leverage existing assets as much as possible and provide a comprehensive framework for doing that.

Last but not least, it remains very instrumental at this stage of 5G development to drive for global standardisation. 5G should become a technology that is based on one global standard. Operators have a duty to drive all industry players to deliver a 5G eco-system that is free of fragmentation and open for innovation.

Even today some 5G capabilities can be introduced based on limited functionality, like EDGE computing. It is a perfect possibility to experience the power and get partners on board for developing e2e views and services. Here low latency use cases can be tested in an early and financially feasible frame. New ideas for the IoT market need to be identified and solution challenged for usability and benefit for the market. This benefit can only been understood by an early engagement with partners from the industry.

By working out new concepts, these have to be embedded were required into the standardization—this might be important for inter operator cases.

References

GSMA Intelligence. (2014, December).
IDC Worldwide Semiannual Internet of Things Spending Guide.
NGMN Alliance. (2015, March). 5G White Paper.
Statista. https://www.statista.com/statistics/691623/5g-mobile-connections-worldwide/

Small Cells: A Critical Assessment

Peter Krah and Tillmann Eckstein

1   Introduction

The rising numbers of smartphones and tablets being sold worldwide as the digitalization of society proceeds fast represent a challenge to mobile network operators as they scramble to still the customer needs for broadband services.

It is well known that Small Cells are an optimum measure to increase the network capacity in an area in terms of $bit/s/km^2$. Nevertheless, we cannot observe that the advanced network operators deploy Small Cells on larger scales as predicted since recent years.

The "small boxes" are much more complex to handle than one would expect. This complexity might be one of the reasons why Small Cells cannot be implemented to reasonable costs so far.

2   Are Small Cells Really Demanded?

Small Cells are not easy to deploy in big numbers due to many limiting or impacting factors. In order to access the overall demand on Small Cells in given network environments the following study was performed.

The intention of the study is:

- To define a set of key relevant use cases for a capacity-driven Small Cell roll out
- To develop a set of method modules for estimation of capacity-related Small Cell demand in existing mobile networks
- To verify capacity demand as the biggest driver for Small Cells in nearby future

P. Krah (✉) · T. Eckstein
Detecon International GmbH, Cologne, Germany
e-mail: peter.krah@detecon.com; tillmann.eckstein@detecon.com

© Springer International Publishing AG, part of Springer Nature 2019
P. Krüssel (ed.), *Future Telco*, Management for Professionals,
https://doi.org/10.1007/978-3-319-77724-5_7

- To quantify the Small Cell demand in an existing mobile network as a function of traffic growth

Small Cells as such are a suitable measure to improve coverage spot-wise at the cell edge. Small Cells would locally improve the cell performance from the user perspective, respectively. However, the focus of the current study is on Small Cells as a use case for a **capacity** driven roll-out.

The study was performed following a step-wise approach:

Step 1: Specify and set-up

Step 2: Prepare and execute

Step 3: Assess and analyze results

Step 1: Specify and set-up

In the first step, scenarios and use cases were defined and very advanced and mature network operators in European market environments were addressed. Based on the situation today different traffic growth developments were considered like

- Conservative traffic growth: up to 50% of today's traffic,
- Aggressive traffic growth: more than 100 times of today's traffic.

On top of that traffic generated by Internet of Things (IoT) devices was incorporated as well. As part of the study that is based on realistic traffic data, an appropriate network environment was selected which represents a comprehensive sample of a mobile network in terms of size, network density and traffic load. In the given exemplary case, a European metropolitan area was selected which is covered by networks of very advanced and mature network operators. In detail the selected network area is identified by the following characteristics:

- Network covering GSM, UMTS and LTE services
- Area of coverage: approx. 900 km^2 covering approx. 3.5 m inhabitants
- Market share: 1/3
- Network characteristics: around 900 macro sectors with an average inter-site-distance of 1 km

Traffic volumes were collected for each cell in the network. Furthermore, the allocated frequency ranges and bandwidths were identified per cell.

Step 2: Prepare and execute

Based on the previously collected data, a spatial distribution that led to a traffic map was prepared. On top of that a comprehensive model was developed to address

- The overall traffic growth and traffic mix based on the existing network and the spatial distribution of population and customers

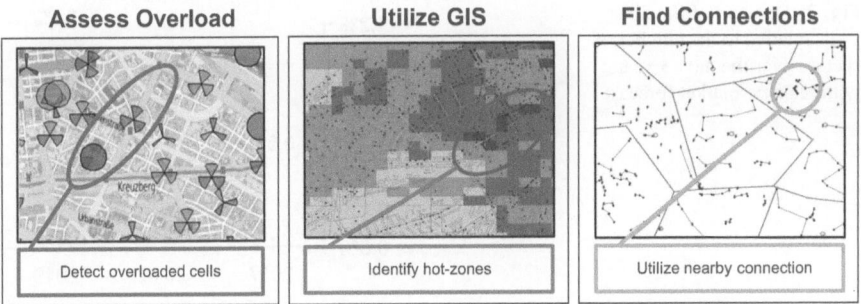

Fig. 1 Snapshots on executed planning on how to identify potential Small Cell spots

- The dimensioning of the radio and backhaul network based on existing network grids and already existing connections or access points. Fiber optics and microwave connections were considered.
- Further capacity enhancement options to increase capacity where needed.
- Cost figures for each deployed technical measure were considered.

Figure 1 shows several snap-shots on how the planning tasks were executed.

Potential technical measures to improve spectrum efficiency and to increase radio network capacity were applied if and where suitable. The following options were taken into account:

- Adding MIMO in order to improve performance on up- and downlink
- Activate carrier aggregation in order to increase peak data rates
- Add spectrum in order to increase overall cell performance
- Implement high order sectorization in order to increase offered capacity
- Densify macro network up to certain limits
- Add Small Cells in order to add capacity

Step 3: Asses and analyze results
Various scenarios were calculated and sensitivity analyses were conducted in order to find out how many Small Cells would be required due to applied traffic growth rates in order to satisfy the capacity demand. Key sensitivity aspects in terms of

- Various traffic growth rates
- Available spectrum
- Site density

were considered.

Fig. 2 Demand of Small Cells relative to the number of sectors over the next 3 years with realistic market forecast

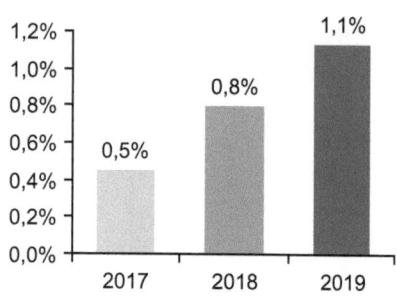

The following key results were achieved:

Sensitivity aspect—Traffic Growth

Figure 2 shows the amount of Small Cells needed for a realistic traffic growth in order to cope with the capacity demand. That capacity demand which needs to be offered by the existing network under the selected boundary conditions for an operator with a comfortable amount of spectrum up to 80 MHz. In this scenario, it is assumed that future traffic growth will happen with similar growth rates as experienced in the recent years.

The demand of Small Cells in 2019 out of the total number of cells in the network is 1.1%: Every 100 cells require one Small Cell to fulfill the capacity demand.

Only if the traffic demand did increase extraordinary, there would be a reasonable demand on Small Cells. In the given example, there would be a demand on Small Cells of 33.4%, if the traffic growth was 107 times higher than today's traffic demand. That 33.4% is relative to the total number of cells, which means that for every macro site in the network "one" single Small Cell shall be added to provide the needed capacity in the network. Figure 3 shows the demand of Small Cells relative to the number of sectors versus the traffic growth.

Consequently, it has to be concluded that the **capacity-driven** Small Cell demand is very small under the existing mature network conditions and comfortable spectrum availability of up to 80 MHz.

Sensitivity—Spectrum

The demand on Small Cells will change and will become much more probable if the available spectrum is limited. In order to exceed 10% of Small Cell demand

- With plausible mid-term traffic growth of 5-times of today's traffic, the available spectrum has to be significantly less than 40 MHz or
- With extreme high traffic growth of 20-times within the assumed available spectrum of 80 MHz.

Fig. 3 Demand of Small
Cells relative to the number of
sectors—Sensitivity towards
extreme traffic growth

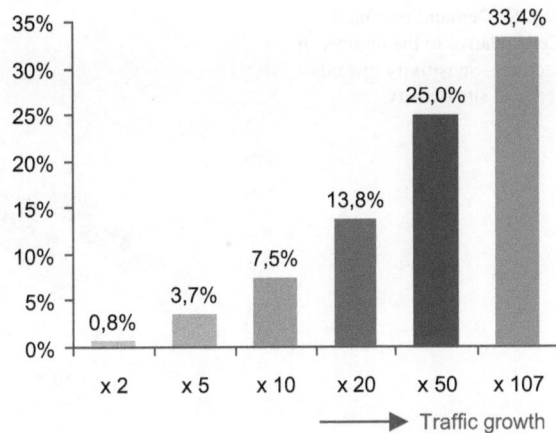

Fig. 4 Demand of Small
Cells relative to the number of
sectors—Sensitivity towards
limited spectrum

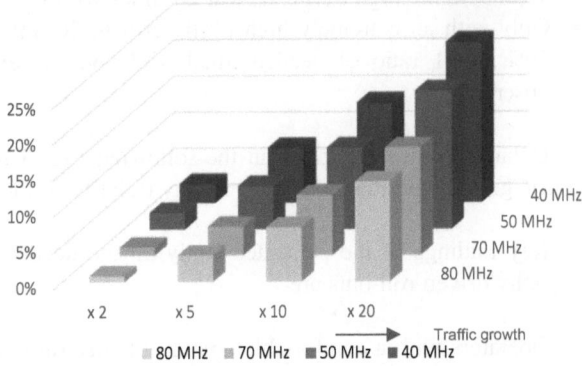

- Only with extreme high traffic growth of 20-times and available spectrum less
  than 40 MHz a 1:1 ratio of needed Small Cells versus existing macro sites can be
  observed.

Detailed results are given in Fig. 4 for various scenarios of the traffic growth and
the amount of available spectrum.

It can be concluded that there are two factors which lead to a high Small Cell
demand: an extremely high traffic growth like in emerging market or only limited
available spectrum.

Sensitivity—Site density

A capacity-driven Small Cell demand becomes more likely with a lower macro site
density (less mature network):

Fig. 5 Demand on Small Cells relative to the number of sectors—Sensitivity towards reduced site density

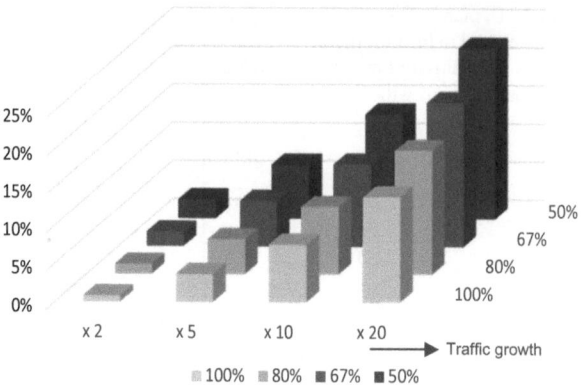

- In order to exceed 10% of Small Cell demand with plausible mid-term traffic growth of 5-times, the site density has to be significantly less than 50% or
- With extreme high traffic growth 20-times within existing site density.
- Only with an extremely high traffic growth 20-times and site density less than 50% a 1:1 ratio of needed Small Cells versus existing macro sites can be observed.

Detailed results are given in the following Fig. 5 for various scenarios of the traffic growth and the macro site density (relative to the initial macro site density).

Key findings of the presented study on the needed amount of Small Cells on capacity driven roll-outs are

- Operators with a comfortable amount of spectrum have a very limited need for Small Cells to provide capacity.
- The less spectrum available the higher the need for Small Cells.
- Only an extreme traffic growth as e.g. by a factor of 100 rectifies a reasonable **capacity driven** demand for Small Cells.

Even if today's demands on Small Cells due to capacity upgrades are not reasonably high, the long term traffic growth and the evolution towards next generation mobile access may expect massive amounts of Small Cells. Additionally, Small Cells are a best fit to improve coverage spot-wise and performance at cell edges or fill small coverage gaps. Therefore, further aspects of Small Cells need to be considered as well.

Furthermore, it should be seen that the continuous upgrade of existing sites with additional services, sectors, carriers and frequency layers will bring these sites to physical limits. Especially in urban environments where operators compete for the best sites and often share most suitable sites a typical roof location carry more than 9 mobile services. On the one hand, that amount will often bring sites close to constructional limits. On the other hand, the massive overall radiated total power per site would increase safety distances so much that sites are not suitable any more. Moreover, unwanted physical effects as e.g. passive intermodulation might occur which are very difficult to control.

To avoid that level of complexity and unwanted effects, Small Cells would be a good measure to accommodate.

3   Are Small Cells Simple to Deploy?

In principle, Small Cells are considered as small boxes of network equipment which are simple to deploy. In reality, this expectation cannot be met due to several reasons.

The following key deployment aspects are identified:

- Planning and Processes
- Network architecture and vendor selection
- Site location and issues on electromagnetic radiation
- Connection to the network

Planning and Processes

- The Small Cells must be integrated in the already existing network environment and process landscape for the whole lifecycle plan-build-run.
- The existing processes and procedures for planning of macro sites could also be applied for Small Cells. However, they are probably too expensive taking into account the limited traffic served by a Small Cell.

"Fire and Forget" would be a low cost solution for the deployment: The cheap Small Cell would be deployed at a site, which provides power and network connectivity as well. However, the drawback of this approach is that it may happen that the overall costs of the Small Cell are not spent in the optimum way. Considerable contributions to the overall costs are the rental costs for the site and the cost for the connection to the network. Consequently, the mobile network operator should look for the optimum location for a Small Cell:

- Optimum with respect to the traffic distribution in the network: The Small Cell should be able to offload the macro cell in a hotspot area.
- Optimum with respect to the cost: availability of the site, site rental cost as well as the connection to the network.

Lamp poles, bus shelters and street furniture are often proposed as sites for Small Cells. However, if a mobile network operator owns sites in the hotspot area, these sites might be the first choice. An example are telephone booths.

Therefore, the challenge is to develop planning processes which can on the one hand ensure that a Small Cell is deployed at an optimum location and which are, on the other hand, less complex than for a macro cell.

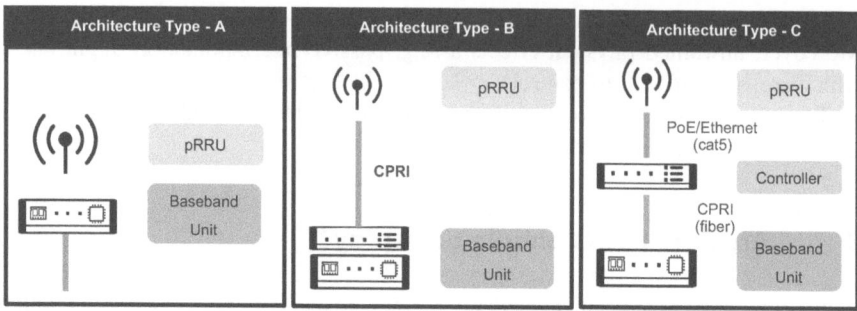

Fig. 6 Typical Small Cells architectures

Network architecture and vendor selection

Another point to be considered is the large vendor landscape for Small Cells; there are many vendors that do not have a portfolio of macro cells. This makes the optimum vendor selection difficult—possibly more difficult than for the macro cells. The mobile network operator should consider commercial aspects as well as the integration of the Small Cells in the existing network architecture—starting from the support of important features as e.g. dual connectivity up to the network management. Moreover, the various vendors offer different architectures for the Small Cells:

(a) The Small Cell is a small macro cell with the complete baseband functionality.
(b) The Small Cell consists only of a small Radio Remote Unit (pico RRU) at the radio site. The baseband functionality is in the baseband unit (BBU) of the macro cell in the area of which the Small Cell is located. This is very similar to centralized RAN (radio access network) which requires a fronthaul connection to the radio site.
(c) Some vendors propose a split between the baseband functionality between the Small Cell at the radio site and a—vendor specific—concentration node.

Figure 6 provides an architecture overview.

One third of the available vendors offer option (a) and the majority of the available vendors offer option (c). For in-building implementations, option (c) is the preferred solution.

So the challenge here is to select the best suited architecture and the right vendor after an in-depth vendor assessment. The architecture of the Small Cells and their support of important radio features is key to the integration into the existing network.

Site Location and Issues on Electromagnetic Radiation

Small Cells are typically deployed at lamp poles, bus shelters and the street furniture. That means that the antennas are much closer to pedestrians on streets than for macro sites resulting in a more stringent limit on electromagnetic radiation.

- That limit restricts the antenna gain, the transmit power and the number of frequency layers installed in the Small Cell and hence has a negative impact on the traffic which can be carried by the Small Cell.
- There is no harmonized regulatory framework for the electromagnetic field strength (EMF). E.g. in Germany, a permission of the national regulator for the site is required if the total radiated power of the site exceeds 10 W. If the gain of the antenna is considered, this is a quite stringent requirement. The alternative is to perform the same site approval process as valid for macro sites.

Connection to the network
The connection to the network is one of the crucial success factors. The "typical" Small Cell site offers a power connection but not necessarily a fiber connection to the network.

In principle, the following options to connect the Small Cell are available:

- A fiber connection is the preferred solution. In case of fronthaul it is the only suitable solution.
- Microwave or mm wave (in the 60 or 80 GHz frequency bands) might be an alternative. However, in frequency bands above 5 GHz, there is a need for a direct line of sight to a concentrator node. It should be kept in mind that a radio connection contributes to the overall electromagnetic radiation of the site as well.

Our conclusion is that planning is required to find the optimum sites for Small Cells taking into account the spatial traffic distribution in the network and the cost for the sites and the respective connection to the network. Just "Fire and Forget" could end up in a situation where the Small Cells do not considerably offload the macro cells in a hotspot area.

4 Can Small Cells Be Deployed at Reasonable Costs?

Small Cells will be a success only if they can be deployed at reasonable costs and if the Total Cost of Ownership (TCO) is justified by the traffic carried by the Small Cells. In the previous paragraph, we already mentioned some of the cost drivers. The key cost drivers are

- Planning and Processes
- Network architecture, type of product and operation
- Sites
- Connection to the network

Planning and Processes
In the previous paragraphs, we showed that planning is needed in order to ensure that the Small Cell is able to offload the macro cell in a hotspot area and to avoid high

costs for the site and the connection to the network. As a Small Cell can only serve less traffic than a macro cell, the planning of a Small Cell must consume less resources and time than the planning of a macro cell. Consequently, smart planning approaches have to be developed like already applied during the execution of our previously described case study.

Network Architecture, Type of Product and Operation

There are many vendors of Small Cells, that do not have a portfolio of macro cells, which makes the vendor selection difficult. It may happen that a Small Cell vendor offers a commercially very attractive solution, but the mobile network operator should consider the overall costs:

- Commercial introduction of an additional vendor in purchasing and service
- Technical introduction of a further vendor: interoperability tests, especially the interworking between the macro layer and the Small Cell layer
- Network operation: the operating teams must be trained in order to become familiar with the new equipment and the interworking with the already existing network

Sites

If a mobile network operator owns sites as e.g. telephone booths, these sites are the preferred ones as there are no recurrent cost for site rental (opex). Lamp poles, bus shelters and street furniture are often proposed as sites for Small Cells. It may happen that the owner of such a type of sites tries to rent all his sites in a city area in a single frame contract. A network operator would have to sign a frame contract for many sites in the city area although he would only need a few sites in a hotspot area.

Connection to the network

As already mentioned above, a fiber connection to the site is the preferred solution. On the one hand the civil works for laying out fiber in a city area are expensive and time consuming, on the other hand we observe that more and more fiber is laid out for high bit rate fixed network connections Fiber to the Home (FTTH) or Fiber to the Curb (FTTC). The latter can help to reduce the investment needed for the fiber connection.

For a microwave or mm wave connection to the network, the mobile network operator has to consider not only the cost of the equipment but the whole TCO with the following contributions:

- Need for a concentration site as second endpoint of the connection incl. power connection and connection to the network
- Installation of the equipment
- Frequency license in case that the system is operated in a licensed band

Our conclusion is that a mobile network operator should optimize the existing planning procedures in order to allow a less time and resource consuming planning of Small Cells. Moreover, a mobile network operator should carefully select the vendor of Small Cells taking into account the overall cost, especially for network operation.

5   Risks and Opportunities

Beside certain risks that need to be mitigated on the long-run Small Cells offer various opportunities to add capacity.

Risks

- Many countries require regulatory approvals on radio sites. Safety distances are defined due to the level of electromagnetic radiation. As of today, only Small Cells below certain radiated power levels do not need regulatory approvals.
- In case network planning cannot be optimized and planning efforts cannot be reduced significantly, Small Cell rollouts on large scale will not happen in the near future.

Opportunities

- Sites: many options for sites are available. However, a mobile operator should carefully investigate to determine the best suited sites.
- Higher frequency bands in next generation radio access allow for limited coverage areas only. Therefore, they are enabler for Small Cells.
- Massive fiber rollouts towards digitization of societies provide the opportunity to decrease the overall cost of a site connection.
- Assumed a Small Cell can be provided at much lower costs than today, Small Cells can boost significantly the spatial capacity of mobile networks.

6   Outlook

Even if today's demands on Small Cells are not reasonably high, long-term traffic growth and the evolution towards next generation mobile access will require massive amounts of Small Cells. Small Cells are a precondition for the utilization of higher frequency bands as the 3.5 GHz or even bands in the mm wave range.

One key success factor will be a smart planning approach in order to increase operational efficiency and to reduce operational cost items. Big data analytics might help to reach that goal.

In order to come up to the TCO targets for Small Cells, mobile network operators should carefully select the Small Cell product as well as revise and adapt their deployment processes and implementation strategies.

Operators need to be prepared in order to catch up immediately if the demand on Small Cells will rise. Therefore, we recommended that they assess their demand closely in order to identify rapid traffic growth at early stages. We recommend further that the network operator should prepare in parallel all necessary steps in the domains as highlighted above.

That shall ensure that Small Cells can be deployed within reasonable budgets.

Part II

Modern Network Concepts

Managed Services 4.0: On the Journey from Cost Efficiencies to Business Transformation

Patrick Hung-Fai Ma and Manfred Schmitz

1 Development of Telecom Network Managed Services

1.1 Managed Services Providers: From Selling Cost Efficiency to Adding Values

Managed Services for telecom networks have gone through different phases as shown in Fig. 1. From the initial objectives of lowering operations costs and, particularly among start-up telco, filling the gap of technical staff, MS have developed towards more focus on improving network quality and customer experience.

The service offering of Managed Services Providers ("MSP") has evolved accordingly to meet such objectives. MSPs have repositioned their value proposition from simply assuring OPEX cost efficiency to adding more values in a closer partnership with the network operators. They have broadened their value propositions to cover expectations like:

- Improving customer experience;
- Improving E2E network and service quality;
- Shortening the time to market from product ideas to technical solutions;
- Getting access to expertise to cope with the increasing technology complexity;
- Optimizing network planning for better network quality; and
- Optimizing CAPEX efficiency for better Return of Investment.

P. Ma (✉) · M. Schmitz
Detecon International GmbH, Cologne, Germany
e-mail: HungFai.Ma@detecon.com; Manfred.Schmitz@detecon.com

© Springer International Publishing AG, part of Springer Nature 2019
P. Krüssel (ed.), *Future Telco*, Management for Professionals,
https://doi.org/10.1007/978-3-319-77724-5_8

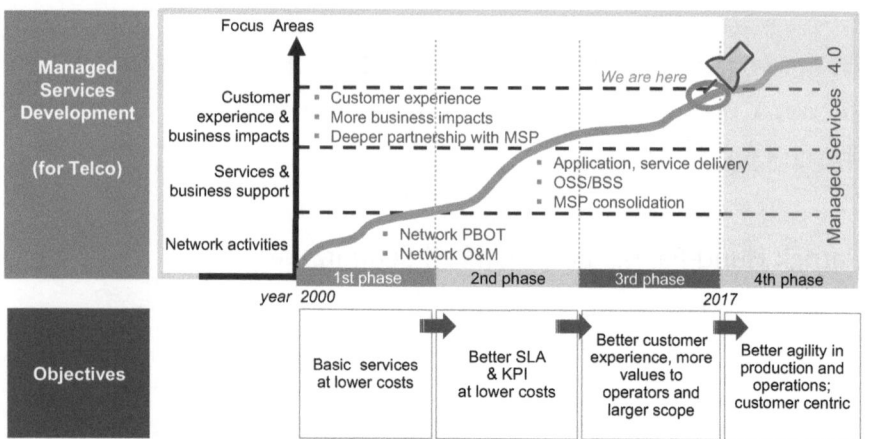

Fig. 1 Phases of telecom network Managed Services

In the past few years, new Managed Services offerings have been actively promoted. Some of the examples are:

- MS on planning, design and engineering, covering activities such as full network design, proof of concept, system verification, site infrastructure design, site acquisition, third party engineering management, end-to-end capacity management and network asset lifecycle management;
- Service Operations Centres, which monitor the service performance in real time with network probes and other tools; and
- Experience Management Centres, which continuously track the end-customer experience on various services offered by the operators.

1.2   Operators: Cost and Quality Are Key

Throughout different phases of Managed Services development, the key motivations for network operators to use Managed Services remain to be cost related: a lower and more predictable OPEX. Such objectives have been common among the operators in our previous consultancy projects. Our 2012 and 2015 surveys also confirmed this, as detailed in Fig. 2.

When deciding a partner, the proposed service quality, capabilities and project references of the MSP become more important, as revealed in our Detecon MS Surveys in 2015 and 2017 (see Fig. 3). Such considerations from the demand side, together with the dominance of tier 1 telco equipment vendors on the supply side, explain the 85% Managed Services market share by Ericsson, Huawei and Nokia (c.f. ABI Research 2014).

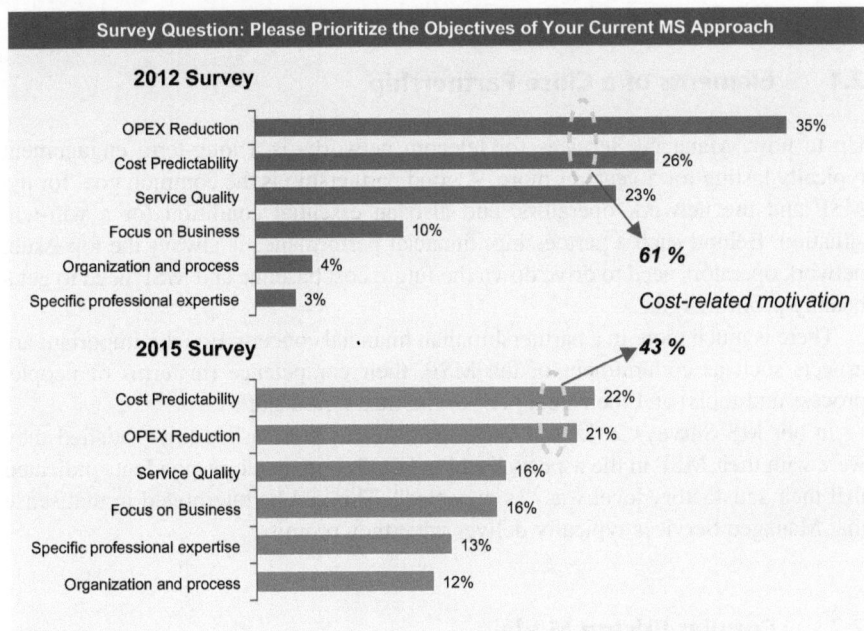

Fig. 2 Key drivers of using Managed Services—Detecon MS Surveys 2012 and 2015

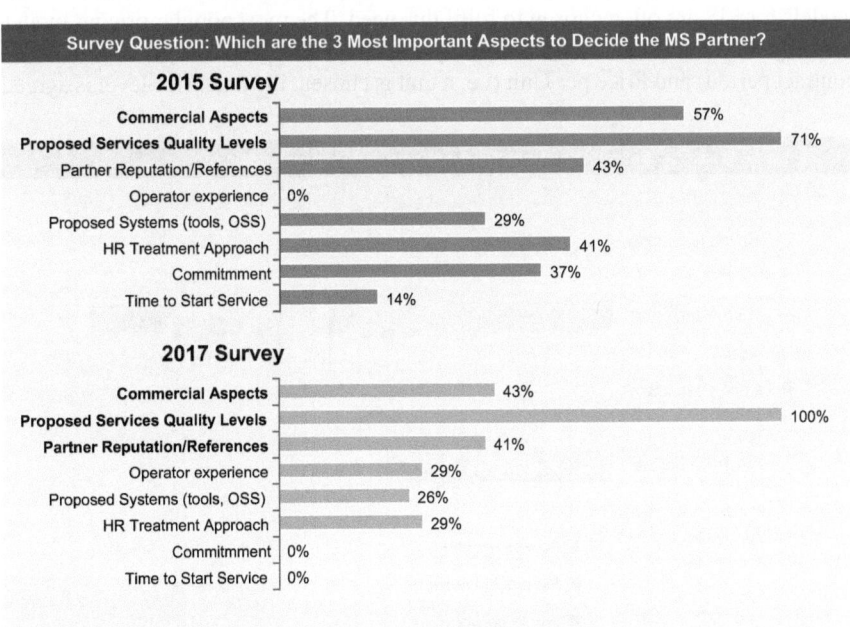

Remark: 100% means all respondents chose it as one of the top 3 important aspects

Fig. 3 Key decision factors on MSP selection—Detecon MS Surveys 2015 and 2017

2 MS Partnership

2.1 Elements of a Close Partnership

Up to now, Managed Services for telecom networks is a long-term engagement typically lasting for 5 years or more. A good partnership is the common goal for the MSP and the network operators, and also an essential condition for a win-win situation. Behind such a partnership, financial performance is always the top issue: network operators need to drive down the future cost baseline and MSP need to get a healthy profit margin.

There is much more in a partnership than financial concern. Equally important are aspects such as commitment of the MSP, their competence (in terms of people, process and tools) and their treatment of the transferred staff.

In our MS Surveys 2015 and 2017, participants were asked how satisfied they were with their MSP in the aspects listed in Fig. 3. In general, respondents indicated that their satisfactory level was "as expected". This can be interpreted in that sense that Managed Services typically deliver what they promise.

2.2 Popular Pricing Models

OPEX predictability is a major objective among network operators, and pricing models for MS are often chosen to fulfil this need. The most popular pricing models (see Fig. 4) are Fixed Price (i.e. a lump-sum price is agreed for each year of the contract period) and Price per Unit (i.e. a unit is chosen, the unit price level is agreed,

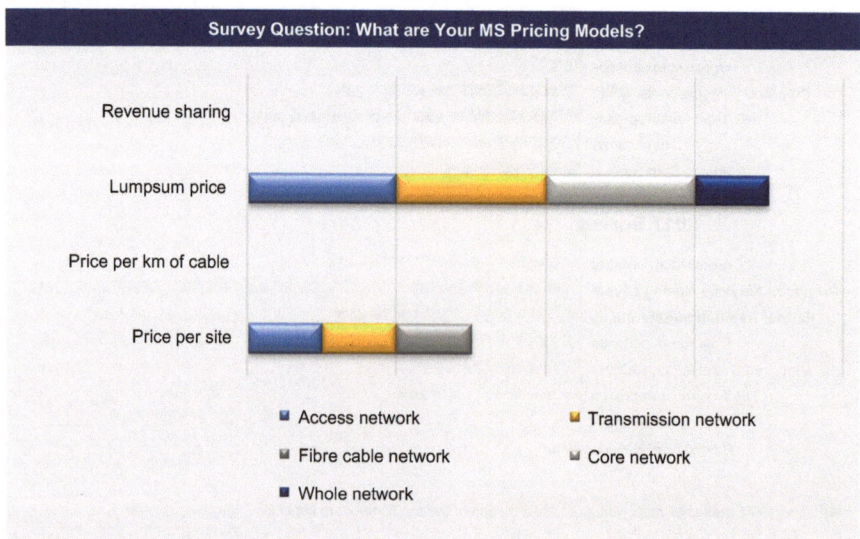

Fig. 4 Popular pricing models—Detecon MS Survey 2017

and the payment amount will depend on the actual number of units). For the latter pricing model, the unit chosen is usually equipment site such as radio access site and transmission site. In some MS deals, smaller units instead of sites were considered, such as network elements. However, agreeing on the unit price per network element during contract negotiation, and resolving disputes on the actual number of units in each billing cycle, can be very tedious and difficult. Use of small units such as network elements will likely lead to more discrepancies between the actual figures and the forecasts, and therefore larger deviation from the initial OPEX projection.

Revenue Sharing models are often discussed as they embed the desire that the MSP's reward shall stronger be linked with the success of network operator based on the usage of Managed Services. But there are two major pitfalls:

1. It is very difficult to define the value and impact of the Managed Services related to the envisaged revenues. This is in particular true if the Managed Services scope only incorporates specific technologies such as transport and dedicated processes like operations.
2. Despite the name revenue sharing, it implies risks for both parties as the predictability is more difficult than for other pricing models. The risk for the operator is then that they might pay far more for the service than they planned, whereas the MSP is concerned that they have much lower revenues while still having big OPEX at the same time.

2.3   Win-Win or Win-Lose?

Despite the simplicity of the favourite pricing models, some network operators consider that OPEX predictability is not always easy. As shown in Fig. 5, some respondents to our MS Survey believe that their MSP are charging too much, and that some MSP would rather pay penalty than putting more efforts to achieve certain SLA/KPI targets.

On the other hand, we are also aware of MS deals in which the MSP's actual profit margin fell below the expected healthy level, especially where pricing models other than fixed price or price per unit were used.

Regular discussion among the partners is important. Contract review focussed on SLA/KPI target and price level is common. However, focussing on these two aspects alone, which both parties want to defend for their own interests, may sometimes cause the partnership to drift even further apart. For a more comprehensive review of a partnership and its maturity, we should examine the MS arrangement from six different dimensions on both the MSP and the network operator. Such an assessment is explained in Sect. 3.

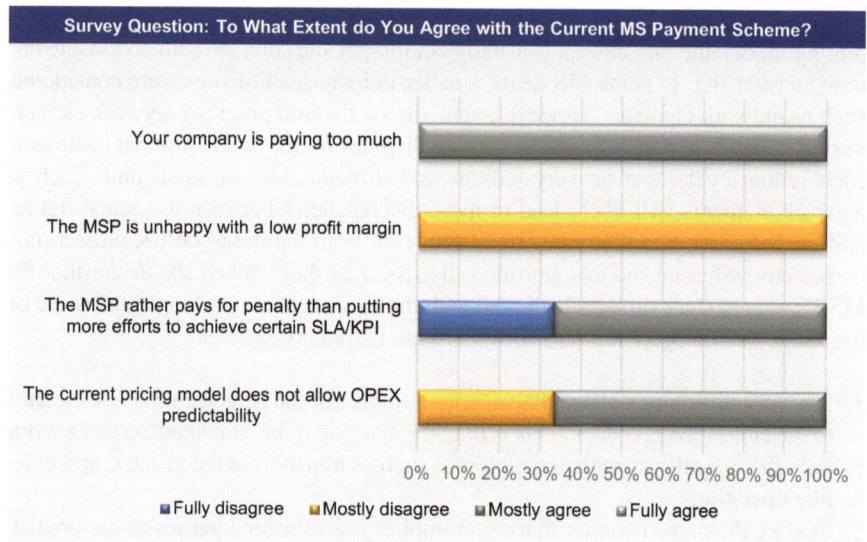

Fig. 5 Concerns on pricing models—Detecon MS Survey 2017

2.4   How About a Deeper Partnership?

After several years of stable Managed Services, network operators who are satisfied with their MSP would consider extending the scope to cover further technology domains, such as fibre backbone, FTTH and IT. Some would put all technology domains under the same MSP for more price benefits and end-to-end network quality.

However, when it comes to extending MS along the network activity lifecycle (i.e. to cover Plan and Build apart from Operations), the considerations are a bit different. Plan and Build activities are closer to network operators' internal business process and therefore more strategic. Loss of such expertise resulting from staff transfer also demands more cautious decision making.

Some operators have shown their open-mindedness to extending MS to cover Plan and Build activities. As revealed in our previous MS Surveys in 2012 and 2015, the MS scope was expected to extend to such activities in the next couple of years. On the supply side, it has also come to our notice that some MSP have been actively promoting MS on Plan and Build in the past few years, trying to leverage their existing MS partnership in Operations.

However, the reality is that most of the MS contracts today are still about Operations and Maintenance activities, as illustrated in Fig. 6. According to our estimation, the number of real MS contracts[1] covering Plan and Build is around

---

[1]By "real MS contracts", we refer to long-term arrangement usually accompanied by staff transfer, with MSP taking up all day-to-day and management responsibilities. One-off professional service contracts for turnkey projects are not counted, even if they last for several years.

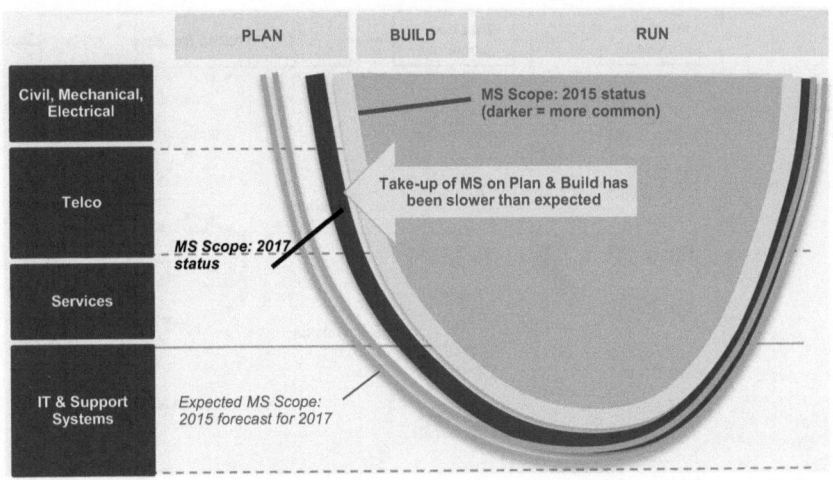

Fig. 6 MS scope extension trend—Detecon MS Surveys

30 (or even fewer),[2] equivalent to less than 3% of the total number of on-going telecom network MS contracts.[3]

The reason for such is likely strategic: operators may not want the MSP (equipment vendors in most cases) to be so close to the planning decision particularly on the choice of future equipment. The risk of vendor lock-in will be higher if a MSP/equipment vendor has more opportunities to influence network planning and equipment purchase decision.

Besides strategic concerns, pricing for MS on Plan and Build is also an issue. We have observed that in some offers, pricing was based on the number of Plan and Build staff transferred, or on the forecast quantity of work orders in the coming years. Since network Plan and Build is not a day-to-day activity like Operations, and the network investment depends much on the demand forecast of the operators, their financial position and the economic environment, such pricing models can be very risky for both MSP and operators. There are also ideas of pricing based on agreed unit prices for a long list of Plan and Build activities. However, such activity-based costing approach is complicated and the price negotiation is often very difficult.

---

[2]The estimation of 30 MS contracts on Plan and Build is based on Detecon's knowledge and research of such contracts won by tier 1 vendors.

[3]Estimation of the number of on-going contracts is based on different sources:

Ericsson: 300+, same number stated in their Annual Report 2016;

Huawei: 200+, based on 450+ cumulative by end 2016 and assuming half are on-going (% of on-going to cumulative contracts ≈ 45% in year 2013);

Nokia: 200+, as stated on their website.

The three MSP had 85% market share in 2014 (see ABI Research 2014). Assuming similar share today, the total number of on-going telecom MS contracts is estimated to be over 900.

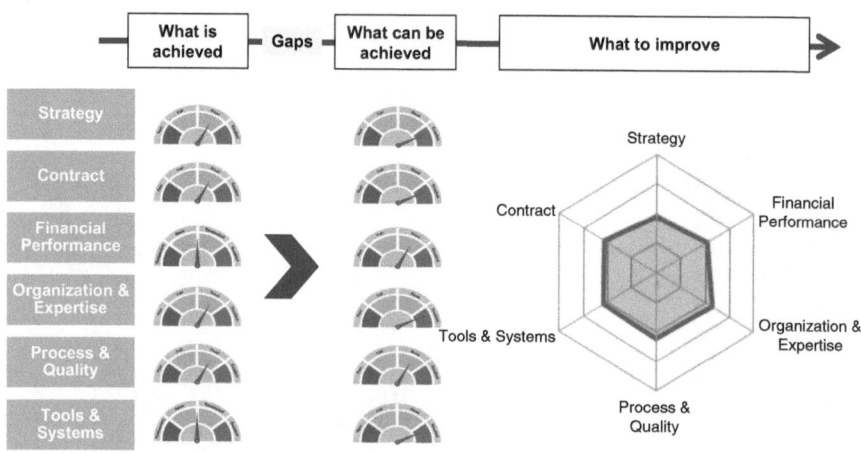

Fig. 7 Detecon Managed Services maturity assessment

3   Maturity of Managed Services Partnership

The purpose of Managed Service Maturity Assessment is to help telecom operators understand their MS achievement and to explore further potential for more business improvement. In such an assessment, different dimensions of MS partnership are examined: strategy, contract, financial performance, organization and expertise, process and quality, tools and systems. This is illustrated in Fig. 7. It is more comprehensive than simply looking at Managed Services pricing and SLAs, and gives valuable insights on what might have gone wrong in a partnership and on the improvement opportunities.

Desktop evaluation on the service scope, contract and performance achievement, as well as workplace discussion with stakeholders of both the MSP and the operator, are performed on the six dimensions stated above (see also Fig. 8 for detail). The level of maturity will be assessed through expert judgment and benchmarking. The way forward towards better MS delivery and partnership will be recommended.

4   Managed Services 4.0

Digitalization is affecting all business sectors and creating challenging demands on telecommunications services. Network technology is shifting towards a new paradigm with more virtualization realised by Software Defined Networking ("SDN") and Network Function Virtualization ("NFV"). Apart from new technologies, a new way of working is required, with seamless collaboration between development and operations teams, continuous integration and continuous deployment, more involvement of procurement to handle a variety of service contracts with long and short duration, partnership management of a large ecosystem (see Fig. 9) and so on.

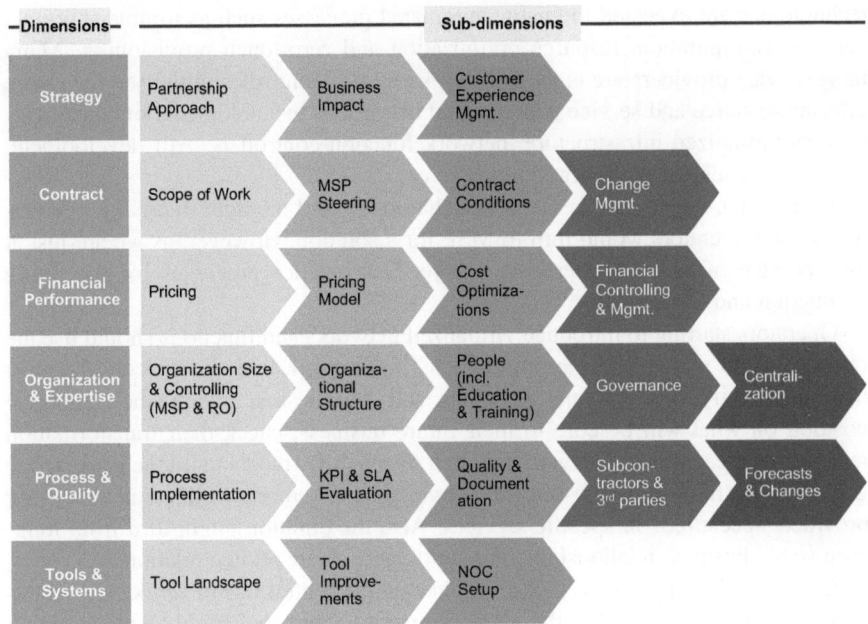

Fig. 8 Different dimensions of Managed Services maturity

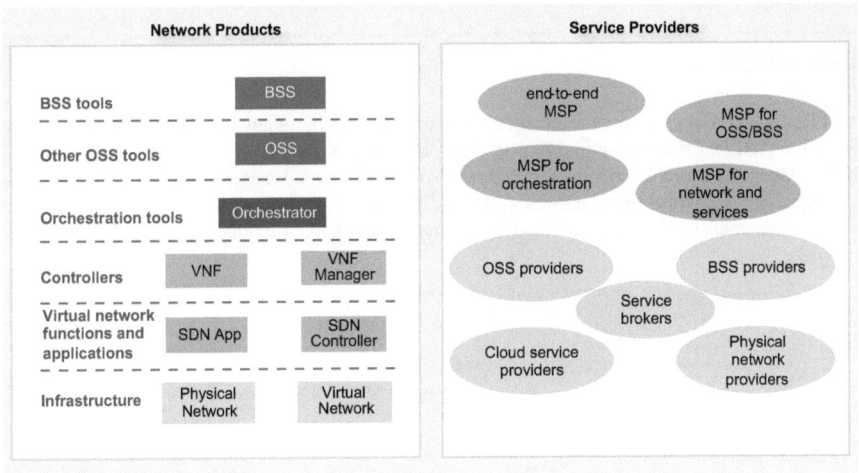

Fig. 9 Example of service providers and network layers under MS 4.0

Against this backdrop, Managed Services are stepping into a new phase, MS 4.0, aimed at supporting the telco operators to meet such challenges with the ultimate goal of delivering better service agility and customer experience. The heavy Managed Services focus on Operations and Maintenance today will change, as the technologies are expected to enable automated processes such as trouble shooting, network configuration, resource optimization and zero-touch provisioning. Many more service providers are emerging to take up different roles in the service chain, offering resource and service management on various technological domains/layers such as virtualized infrastructure, network function controllers, API development, network and service orchestration, and so on.

Lacking the know-how in the new technologies and the right internal processes, many telco operators would turn to MSP for a solution. However, as we are just at the doorstep of MS 4.0, there are not yet best-practice processes for the future production and operations.

Operators starting to introduce virtualized networks and functions should use the opportunity to understand the requirements for organization and process transformation, identify their internal strengths and weaknesses, review their strategic position on what will be core to their future business, check their transformation readiness and capability, and work out their own MS 4.0 model to fill the gaps. These gaps could be filled by micro-services provided by the current MSPs or new, smaller providers specialized in specific services. Also the duration might turn from long-term stable business relationships into short, agile, more ad-hoc relationships.

One of the key questions is who will take on the central role of service orchestration. It could be insourced to the network operators, but also could be the evolving role of the MSPs. At the end there is no right or wrong, and most likely multiple setups will be established.

References

ABI Research. (2014). Network managed services competitive review, December 2014
Ericsson. (2017). Ericsson Annual Report 2016, February 2017.
Nokia, Press and Stock Exchange. Releases on Nokia's website. Accessed October 02, 2017., from https://www.nokia.com/en_int/news/releases/2017/03/07/nokia-and-vimpelcom-sign-major-five-year-managed-services-contract-for-management-of-fixed-mobile-and-transport-networks

DevOps Telco: An Oxymoron?

Krzysztof Korzunowicz

1 "DevOps Telco"?

The title might sound incorrect: two nouns do not create an oxymoron, as it usually takes a noun and an adjective or a verb and an adverb. This is where looking at the word "DevOps" closer brings interesting results. "DevOps", as many a buzzword, is nowadays used as just about any speech function: nouns, verb, adjective and adverb. In this paper the usage will be restricted to either an adjective or adverb. As an adjective it will describe people—those that are introducing and nurturing the principles in a company. As an adverb it will describe how the main production processes of a company are set up.

2 Simplified History of Change in IT

Throughout a large part of IT history, the process of creating and maintaining value was based on two concepts: "stability" and "change". Stability is a state of low stress levels of each person maintaining a system (Ops) and is somehow related to two capabilities that they possess: whether they can see and understand what is happening with and around their system, preferably for an extended period, and what they can do to respond if anything out of the ordinary happens. The Ops were separated from the teams building those systems (Dev) and their goal definition was set to make sure the systems works, irrespective of anything else. The measurement of their efficiency and source of bonuses was how smoothly the system was running. As a logical consequence their interests were not focused on what value the system produced but rather what infrastructure it used, how it behaved in different

K. Korzunowicz (✉)
Detecon International GmbH, Cologne, Germany
e-mail: Krzysztof.korzunowicz@detecon.com

© Springer International Publishing AG, part of Springer Nature 2019
P. Krüssel (ed.), *Future Telco*, Management for Professionals,
https://doi.org/10.1007/978-3-319-77724-5_9

circumstances, what redundancy mechanisms were available and how many ways of measuring the state were possible.

The opposite of stability, obviously, is change. Ops had zero incentive to like any kind of change, as in the time of monolith systems—that is not modular and very interdependent internally, often including dependencies on hardware capabilities—it used to increase the probability of failure by a few orders of magnitude and dramatically increased their stress levels. Thus, Ops have acquired an additional skill: they learned how to block changes. That was the most effective card in the game to get their bonuses.

This game of cat and mouse of people, that had totally opposite goals and incentives was possible to be played for a few dozen years just for one reason: because the changes were slow. They were slow due to usage of waterfall methods that typically define a to-be state, that was substantially different than the as-is state. Two, the technology capability to build distributed systems was weak. Especially the tight coupling with hardware was difficult to overcome and caused a lot of headaches in areas of:

(a) Scalability, as it required a lot of reengineering when adding additional hardware.
(b) Cost control, as hardware typically had big jumps of capacity, causing heavy underutilization of resources for long periods of time.
(c) Service reliability, as redundancy based on a few simplified models of state exchange between hardware elements allowed illusion of control. They often had incorrect assumptions on behaviors during stress, due to never properly being tested, as such testing "caused risk in production". Thus, the whole thing was not trustworthy, but often there was someone else—the vendor—to blame.

Since the changes often failed, habits related to risk avoidance kicked in: change was done as seldom as possible. On the bright side, the bigger the program, the more important its manager, so people had a place to play their politics. Finally, changes affecting any of the above took a long time due to decision processes on investment or technology architecture. Often both.

Side note: The point of problems on service reliability is hidden in plain sight in telco industry: every network had enough outages to make the sacred 99.999% of availability an unachievable dream for a long time. On top, the calculation of SLAs is often absurd in how many things it omits, yet the industry speaks about "telco grade", as if that actually had a real meaning. The architectures were pretending to solve the problem and the habits related to viewing the problem from vendor-blaming and hardware perspective are one of critical areas of needed change.

Back to the point: People did their huge waterfalls with very low success rates on project completion, with the changes taking years before coming to production. Since there were huge blunders from time to time, it caused the vendors to adapt by introducing risk premiums in their prices. A world aligned with Ops incentives (most of the year) and very far from Dev incentives. 1:0 Ops.

Then about 15 years ago Dev started to strike back. They came up with the concepts of agile methods, increasingly focused on automated testing including—very importantly—customer experience oriented testing. Those changed the rules of the Dev versus Ops game dramatically, as Dev could push out changes quickly and cleanly, while clearly showing what value it was producing and how bad it was that Ops was blocking it. 1:1. Since the Dev had the methods to push out software quickly without worrying about quality, it had the possibility to create increasingly complex customer oriented functions. The complexity caused orders of magnitudes increase of usage of all hardware resources as well. That means that software had to scale properly with infrastructure and virtualization technologies enabling easy creation of environments for applications found immediate usage. But they also meant that the measurement of how well the system was behaving got substantially harder and even the definition of what the system was has changed: many application instances with "some" hardware capacity, "somewhere". On top, the hardware based redundancy mechanisms that Ops held so dear, were no longer applicable. The applications showed up and were gone at very short intervals of time, completely independent of hardware and application Ops wishes. Finally, Ops possibilities to respond to abnormal situations got limited, as they lost control of hardware. It became owned by dedicated infrastructure teams. That left application Ops struggling. They did what the only thing that was left in their situation: they blocked changes even more.

It quickly became obvious to people that knew both sides of the coin, that the setup like that just does not allow to produce anything valuable quickly. The Ops capabilities had to scale as well and it was not enough to buy some magic box. It required a redefinition of the whole way Dev and Ops cooperate. Starting at the incentives, by removing the clash of interests. Making Ops aware and involved in what really happens on the application level at the least related to: visibility of state, testability, deployability, security and redundancy. To make them able to impact what the Dev produce, not as a separate entity, but as a key member involved in the production process and value flow. Thus, a single place to weigh the importance of Ops related capabilities against new business opportunities was created, and it is there that the definition of priority of the next things to be done was set. Companies that balanced these two properly, became an order of magnitude faster in providing new value to customers.

3 Value Chains?

Today, DevOps is a go-to model of working for startups and increasingly to bigger companies. Many startups, which often provide value based purely on IT capability, are completely built like that. And that is possible, because they have a single and easy to understand value chain, especially if Ops part is provided by a public cloud provider. For larger companies, the answer to the questions about where DevOps principles would be generating most value is not an obvious one. The reason is because they look at their value chains holistically rather seldom, simply because it often is a very resource intensive task. That has become an issue. Technology lifecycles are getting shorter and thus capabilities that were crucial— done by the most experienced team a year ago—might be part of the public cloud or some software package today. Thus, the picture of the organization should change in details quickly, but keeping up with updates of the responsibilities inside the organization usually does not have an owner, as this requires gradual constant change, while organizations are used to huge shake-ups every year or two, mostly related to people taking over positions of power. What ensues is a long period of power play that helps nothing..

The key learning is that companies need to stop seeing themselves as a stable value chain that one needs to fit the organization to perfectly, but also that they need to be capable of doing meta-analyses of their value streams continuously and elastically adapt the scope of responsibility of teams in accordance to the environment. That means that the old metrics measuring how good they are at executing today's value chain should be extended with one related to how quickly it takes to change those chains effectively. That means there should be no drama associated with such changes.

This is when an Enterprise Architecture Management (EAM) proponent will say that its exactly what EAM was invented to deal with. And it does, in theory. The implementation of EAM as often seen in companies now is a constant chase to keep the as-is state up-to-date. With that approach the architects are just an audience to the changes with almost zero influence of the direction of the changes. That is bad even in a hierarchical organization, but gets almost unmanageable in an agility oriented one. There, each team prides itself with choosing their preferred tools for almost everything, often creating sets of internally competing parts of the Ops part of value chain. It could have a positive side, because it allows for learning and would be ground for Darwin theory implementation: "best solution wins". Unfortunately, human habits related to "never being wrong" and the love of power play come into the game. If they are not managed at a general level, the possible outcome is that instead of a free flow of ideas and people between teams, one gets a new version of silos. Just this time not even manageable by the hierarchy, as hierarchy is limited.

To look on the other side of the coin—the Dev related part of the value chain— here a possible method to figure out what the chain looks like is Customer Journey. It analyses the areas of interface between a company and its customers and allows creation of "as-is" and "to-be" pictures and prioritization of things to be improved. Unfortunately, the "as-is" vs "to-be" creates a beautiful waterfall friendly environment that many are trying to move away from. What should be considered is making Customer Journey analysis not a once-in-a-blue-moon and internal exercise only, but rather a continuous one with at least some orientation on competition and partners. That would be a very powerful capability allowing companies to identify areas where they can differentiate, and they immediately would know where applying DevOps principles makes sense. This continuity and external direction makes it beyond the scope of any particular DevOps team and could be considered the job of the new and improved Enterprise Architecture teams, which should include business people specialized in customer interactions.

4   Cloud: The "Ultimate" Generalized Ops Capability Set

The mystic "cloud" is the next piece of the puzzle of a DevOps environment. The cloud today is a lot of software with business level capability, put on a very powerful set of managed Ops capabilities, in turn put on shared hardware. Each DevOps wannabe company has a choice: build a cloud themselves, use public ones or go for a mixture of the both. This is still uncharted territory for many companies and with either option they need new skills. A decision to go either way should not be taken lightly, because it has immediate implications on current, as well as expected value chains.

A private cloud should be considered as a separate value chain, as a target, as it is supposed to provide value to multiple users, with as little particularity as possible. It just makes no sense to build the cloud otherwise: if each cloud is being used by only one group, then silos are back, but this time on virtualized hardware. Then cloud capability will not scale at all: Ops people will be fighting with the same problems multiple times. Still it is important to keep in mind, that even if they do get their value stream right and solve the problem once, the private cloud will not have as many generalized capabilities as public clouds do. The biggest providers simply dramatically out-scale the private cloud companies in the efforts of building them. Thus, the reason of usage of private clouds should not be based on "innovation", but other reasons: e.g. no lock-in, higher/different security standards or geographical unavailability of public clouds.

When building a private cloud, it is the easiest to have the infra teams work with product teams (their users) at the beginning, until at least after a few successful product iterations are launched, when both teams—building it and using it—learn enough. Then the value streams should be separated, under the assumption that the cloud is stable enough and new users barely bring any requirements into the cloud.

For public clouds the path is easier to start on. It takes having an infra person in each team in the main value chain. The path of upgrading the skills needed by the teams is much clearer, and it is easier to find people on the market with those skills. Then the decision is whether it is one cloud provider or multiple: whether the price and "freedom" is more important than provider specific capabilities. They do differ a lot, but are not a problem of this article.

A strategic decision should be made across the company to which cloud type to put the effort first, as they require people to learn a lot and the learnings are not directly transferrable between. In any case people should be left alone with this topic for a while to learn and given too many milestones, they will fail. Building cloud capability "somehow", without people dedicated to it, also does not work.

5   Assumptions Needed to Get DevOps Rolling

Let us assume that companies can:

(a) Identify their differentiators to choose at least initially the area of most value,
(b) Make the most sensible cloud scenario decision,
(c) Put the right amount of effort into setting up skills internally and leave the people alone for long enough, that they are successful in learning.

This is where hard work starts. There are two fields that make or break the concept:

(a) Technology stack
(b) Organization culture and setup

It would probably be the easiest for the reader to understand, if there was just a list of how the technology stack needs to change and how the culture and organization setup need to change. Then it would be obvious to whom in current organization setups to address those topics to. The issue with that is, that it really is impossible to explain them separately, as they are strongly dependent. Additionally, if it was possible to just address it to people that already have their roles defined, then the message about the need of organization change would sound somewhat artificial.

The following paragraphs will explain the expected changes in technology and provide a direction in which the organization, as well as its culture need to follow to successfully become a company efficient in both executing its current value chains as well as changing them.

The technology part is about fundamentals of how software that differentiates the company from its competitors is architected and built: the architecture change means no more monoliths are allowed and in the building part it is about taking full responsibility of releasing software in-house.

5.1   Responsibility for Releasing

It is scary to be responsible for releases and many have made a habit out of transferring such responsibility entirely, for a price. That habit will be difficult to change, as it will require a lot of learning on how to be able to release themselves, and/or how to set up relationships with vendors differently. Many people will not want to do this learning, especially in organizations that have a very clear separation of ownership of revenues and cost, and related budgets. This is because the effort is required by what is considered the "cost side". It includes a hefty investment in the beginning, on technology capabilities (people, indirectly) and creation of dedicated teams (people, directly), while most probable positive implications will be seen on the revenue side. Unfortunately, they are not easily guaranteed by a simple business case, since it is a kind of a leap-of-faith to a world in which the people working at lower levels are supposed to figure out better ways of producing value and making money. Thus, it will be very difficult for the cost part to convince the revenue part of its importance, especially in organizations that have low trust between those parts. Additionally, for many managers it will be considered a leap into a chasm head-first, because for decades they have been taught that management is there to make decisions and people are there to execute. As a starting point the company trust levels, management style and the revenue/cost separation should be reviewed. Agility frameworks provide grounds on how to address all the topics above and could be a good basis for changes. The proposal of the author is to consider LeSS (Scaled Scrum) for situations when interdependency of the elements of the value chain is low at start and Scaled Agile Framework (SAFe) when interdependency is high.

5.2   No Monoliths Allowed

To get rid of monoliths, it takes a change of the architecture, very often causing rewriting the software from scratch. The goal, and today's state-of-the-art is to have independent microservices with well-defined and relatively stable interfaces between them. To build microservices quickly the production model must allow on-demand building, testing, and releasing of software. Continuous Integration/Deployment/ Testing (CI/CD/CT) chains enable that on-demand capability. They need to be set-up and must be used. Ops make it a condition of acceptance of the pace of change. If there is cheating, Ops will find out the hard way: things will break in production.

Then the question is: how to start with this "no monoliths" rule?

The choice is to either make a CI/CD/CT work on cleanly separated part of the value chain, reshape the architecture and build first microservices, or to try to put it on the most valuable part of the value chain... and then immediately hit a cross-domain problem, as the most valuable part of the value chain will very likely be highly interdependent with others. The first DevOps team will be releasing too quickly for everyone else around, or rather everyone else will releasing too slowly for on-demand testing and releasing of the whole value chain, since others are not conforming to the ideas. Thus, the added value of speed on a small piece of the chain will be very limited. Setting up chains for all of them immediately is not be possible... It ends up a being a chicken and egg problem. The only way out is gradual improvement. But where?

6   Stop and Count

The first step is then to find that separated value chain and experiment, so that people can play around, get first learnings and show-off success with improvement of some typical change metrics. Then the second is to consider the topic a substantial change in the way people think and work, because their goal will be automation of everything that can be automated—that is what the CI/CD/CT chains are all about and that is the only way a substantial increase of speed can be achieved.

The next thing to do is to stop and count. Not only money, but rather the effort and time of as many things that the company does, as possible. Proper understanding of EA would help implementing measurements on the production model tremendously. The measurement is needed to asses where to start the automation, as implementing CI/CD/CT should be about automating the human tasks that:

(a) Introduce the most queues into the system by their nature (e.g. sharing of testing environments for multiple projects with different configurations needed by each of them and manual work to change them),
(b) Take the most time and are most error-prone,
(c) Are never/seldom done and by not doing them the risk of releasing unsuccessfully increases substantially, but simply there is no time to do it in the current model

This may sound easy, but unearthing the numbers on those types of tasks will automatically put the people involved in them under tremendous pressure. There is a risk that at least one of their managers, feeling "responsible", will consider any of these numbers inadequate and require immediate effort to improve them, preferably without sacrificing anything else. Blame game and deceit in measurement ensues, with the outcome completely useless as substance to make decisions on where to prioritize improvement. Thus, it must be done under proper supervision, with a lot of communication effort at all levels, to make sure the intent is understood. Such measurement, analysis and proposals of prioritization of improvements are other things that only a team with a very broad knowledge of the company and its capabilities can execute properly. Enterprise Architecture Management teams could start doing what the name implies...

EAM would become a trusted partner in the decision-making during planning. Not as a deciding party, but a party that knows the whole picture of the company and can provide information on what is available as general capability and what is not, but maybe should. It would help business make the prioritization for the teams to execute. The analysis should be done not per product, but per portfolio or even across portfolios, if possible.

To summarize the possible Enterprise Architecture Management Support team tasks in larger organizations after a DevOps transformation:

(a) Continuous analysis of internal value chains with focus on:
    (1) Partner/vendor capabilities that overlap with internal (help teams decide whether there is a point in continuing the work or not)
    (2) Capabilities of competition that should be copied
    (3) Ideas for improvements or new capabilities that would help differentiate based on e.g. Customer Journey and Design Thinking

(b) Measurement of the effectiveness of current value chains and proposal of improvements based on critical identified metrics: e.g. a mixture of velocity of change of the value stream, length of a full iteration of implementation and quality stability.

(c) Analysis of general (cloud) versus specialized (team scope) Ops capabilities as part of every product lifecycle—if a private cloud is used.

7 DevOps Telco Implications and Limitations

People in telco companies like to consider themselves "different". They like to use words like: "telco grade", "standard", "telco specific" and similar. This is an interesting outcome of the general value chain setup the telcos have used for years. The fundamental pieces were based on a priori agreement between numerous groups of market participants referred to later as standards. The risk on implementing those standards was pushed to specialized vendors. This market was closed, because the entry barriers were very high—implementing old standards took time and money and telcos wanted to talk only to big companies with market share. SDN and NFV allowed for an easier entry by companies from the adjacent IT market, but their methods and products made telcos painfully aware that their ways of evaluating/ comparing and then purchasing such products just does not fit. No "standards" guidance is available for RFPs and the expected methods of cooperation do not allow for a full transfer of risk to the vendor. An immediate reaction was a call to "standardize more" and make things more comparable so that they can execute their favorite RFPs. The next call was to make "telco devops" and this paper is an answer to such call. Two areas are considered for differences between Telco and general DevOps:

(a) Responsibility for production of software. For 99% of cases it will be outside of telco, as there is little programming capability in most telcos. Prioritization of requirements would be left within. Key area of change required in methods and processes is purchasing (enabling agile contracts) and ownership of the product: restrictive licenses will not allow the telco to avoid lock-in, so license to modify at will and ability to build things with different partner on top are a minimum. The best possible model is open sourcing, as many topics will be very complex and costly: e.g. CI/CD/CT with mixed physical/virtualized elements of the network. Telcos should avoid doing it each by itself, as the total effort will be absurd and they are in for a fight with very efficient IT companies. Otherwise the telco would need an increasing number of people with programming skills. That may bring them closer to their hardest competition, but is a big decision and one that is difficult to execute: the cost of learning the skills or acquiring them outside will be big. It can be expected that with time Telcos will be increasingly software oriented.

(b) Responsibility for releasing the service (created out of many systems): An alternative to the own releasing described above is that the responsibility is not taken by the telco, but rather by a third party, for a price. This could be considered as an even further step in outsourcing than (a). This model is very risky, because the telco will cede completely its future to a partner. It would be difficult to create a set of incentives that would make sure that the partner does not abuse such a relationship due to diverging goals: the partner's goal is to sell more, but there is no way of measuring whether the things being sold and services that are built are helping create a better future for the telco. This might work if there is an ownership relationship between the telco and the partner, otherwise it should be avoided.

7.1   DevOps Engineers

Even with partners building most of the software outside, a DevOps telco will need DevOps engineers to establish and run the CI/CD/CT based processes with partners. Thus, such DevOps engineers need a habit of thinking end-to-end: ignoring whose turf they are violating during this thought process, and how harshly some manager will defend it. The biggest issue is, that there are not many DevOps engineers out there and everyone wants them. Since there are very few of them compared to demand, good DevOps engineers will be picky when joining companies. They will check if the trust and freedom is present so that they can take ownership. If not, they will not join or leave soon.

That means that presence of trust and freedom are critical cultural habits of the organization for executing things DevOps successfully. The difficulty is that trust and freedom do not automatically scale. This is hard work of every group of people involved and especially management practices need to be updated to the "leading servant" approach, as that is the only one that fosters ownership and trust. This must be advertised so that prospective employees can see it! Otherwise it will still be difficult to hire those people. Unfortunately, this will have many enemies, because with an organization of ownership and trust, there is quite a big group of people that lose their personal power and a safe place of work.

7.2   Mid-level Management

Companies that have a complex value chain that has not changed for a long time, end up with a very strong separation of responsibility with many mid-level managers. Telcos are separated based on "OSI-standards-oriented" borders. The number of middle level managers and their teams makes any change to the value chain a substantial human resources problem. Many of the managers nor team members have not the skills to participate in a DevOps Telco as engineers nor enterprise architects, because for many years they were mainly learning to execute their piece of technology stack and fighting for its control... Unfortunately, there is not enough turf to give away in a DevOps organization to just take those managers and transfer to new managing roles. This will cause friction and the most crucial here are the possibilities for people at all levels to adapt—there should be general trainings programs for people across the organization for technology oriented roles updating the skills to be "cloud enabled", as well as product management oriented roles with mid management considered for the second role, assuming they learn to cooperate rather than fight for turf.

8   DevOps Telco: Completely or Partially?

A very interesting question remains: is DevOps as a model applicable to the whole company if it is not a start-up? In a Telco everything related to large scale infrastructure build-up should keep on using waterfall as agilization will not bring any immediate improvements. Thus the question applies only to places where softwarization can bring substantial differentiation. Still this question is very valid, as DevOps is "expensive", because it disallows the main trick organizations use to hide the fact that the number of effective specialists is not enough compared to the demand of the parallel projects running: multi-teaming.

Multi-teaming never allows to focus on critical projects since teams always need eachother to get complex project moving. With many projects running in parallel the outcome is that people mostly wait queued, as prioritization is not company-wide, but rather turf-wide. Their only option of trying to reduce the queues is to discuss with other teams what is missing to finish each of the countless projects that are being done, which makes the culture of the company "meeting oriented". Removing multi-teaming will immediately surface the skill problems the Telco have: they will find out that some people are doing 500% of others, plus there will be many people that are literally the only ones able to do certain complex software oriented tasks. If such people are put in a single team, every other team will need someone else with their capabilities. This will be a really big wake-up to Telcos and keeping this a secret might be an agenda point for many executives used to playing the low risk game.

Additionally, due to cultural aspects not everyone will want to take ownership for their area of expertise as part of a DevOps team. They might be comfortable with the way it was done so far, or not feeling confident enough with the knowledge they possess. This is another reason for effective training programs and very open communications at all levels.

In the end, using Agility/DevOps as widely as possible can be considered worthwhile, but it must be executed keeping in mind the cultural and skill aspects, because effective change in those areas requires that this is a long-running program with many stages, allowing people of different character traits to adapt. And if Telcos consider doing DevOps seriously they should make it a priority one topic that goes across the whole company, but with a clear starting area and scale-out plan, with a lot of communication and swift deadlock removal, otherwise it will be yet another "change management" topic that failed to deliver.

Reference

Hickey, K. *Role of an enterprise architect in a lean enterprise*. https://www.martinfowler.com/articles/ea-in-lean-enterprise.html.

Paradigm Shift! The Path to Brutal Automation

Insights from SDN NFV World Congress 2017

Lothar Reith

1   Brutal Automation

The term "brutal automation" has been coined at the recent SDN NFV World Congress in The Hague by Arash Ashouriha, SVP Group Technology Architecture and Innovation and deputy CTO at Deutsche Telekom. In his video recorded congress opening keynote (Ashouriha 2017) he pointed out the fact that the 2017 congress marks the fifth anniversary of the NFV journey and therefore an opportunity to look back what has been achieved so far. From day 1, the NFV journey had the aspiration to learn from the hyperscale companies, but the reality is that the telco industry is still far away from the original rather cloud native vision. He reported that the challenge associated with setting up an operator infrastructure for running VNFs in a telco grade way is just 10% of the overall challenge, with 90% of the challenge being about fundamental changes in organization involving people, processes, and procurement. Some of these fundamental changes will be discussed in the chapter "paradigm shift". He made the statement that this has brought "a lot of great new ways of producing flexibility and agility, but brutal automation is the only way to succeed". The term "brutal automation" turned out to be sticky with many delegates as it appears to reflect well the situation that telcos aiming to become digital service providers are currently in. Arash did not explain in detail what he meant with brutal automation, but it appears that traditional approaches to automation are destined to fail, if they lack brutality. My personal interpretation is, that the needed type of automation is brutal in the people dimension because many practitioners in the telco industry need to go through an individual paradigm change, having to un-learn old paradigms and adopt new paradigms. It will be explained in the subsequent chapters what a paradigm shift is and various examples of paradigm shifts relevant for "brutal automation" will be given. The SDN and NFV driven transformation of the telco

L. Reith (✉)
Detecon International GmbH, Cologne, Germany
e-mail: Lothar.Reith@detecon.com

© Springer International Publishing AG, part of Springer Nature 2019   111
P. Krüssel (ed.), *Future Telco*, Management for Professionals,
https://doi.org/10.1007/978-3-319-77724-5_10

industry is accelerating—5 years after the seminal NFV White paper, and 3 years after the author has published the article "Virtualization is transforming the telco industry" as co-author together with Osvaldo Gonsa and Arne Chrestin in the first band of Future Telco (Gonsa et al. 2014).

An example of the kind of brutality that may be required to be successful could be the Jeff Bezos approach to IT-governance. He kicked off the transformation of Amazon to a service oriented architecture by mandating—without exception—that all data shall be exposed via APIs. He closed his email mandating the use of APIs with the "brutal" words:

Anyone who doesn't do this will be fired.

The source of this story is a widely shared "rant" by Steve Yegge—meant for internal distribution inside Google, who accidently published it on his public Google Plus account. Rip Rowan provided an insightful comment in a post on Google Plus (Rowan and Yegge 2011).

2   What Is a Paradigm, and What Is a Paradigm Shift?

Paradigm According to the Oxford English Dictionary the basic meaning of the term paradigm is "a typical example or pattern of something; a pattern or model". In his landmark book from 1962 "The Structure of Scientific Revolutions" Thomas S. Kuhn (1922–1996), defines a scientific paradigm as: "universally recognized scientific achievements that, for a time, provide model problems and solutions for a community of practitioners". Applying this to the context of automation in the telco industry, this community of practitioners are the experts and practitioners in information and communications technology (ICT). Thomas S. Kuhn gave it its contemporary meaning when he adopted the word to refer to the set of concepts and practices that define a scientific discipline at a particular period of time.

Paradigm Shift Thomas S. Kuhn's definition of paradigm includes "at a particular period of time" and he referred to a "paradigm shift" as "a fundamental change in the basic concepts and experimental practices of a scientific discipline". Therefore a paradigm shift marks the transition between two periods of time with different universally recognized scientific achievements. This transition time is often a time of great confusion, in particular when multiple paradigm shifts occur in parallel or when two consecutive paradigm shifts occur, where some practitioners still adore an old "legacy" paradigm, while some advanced practitioners already adore a modernized paradigm which I like to refer to as Level-1 modernized paradigm. However, while this first paradigm shift from legacy to level-1 is still ongoing, another paradigm shift from level1 to level2 emerges and gets quickly adopted by advanced practitioners due to its superiority, which is however visible only to those enlightened by having lived through the paradigm shift from legacy to level-1. In the telco industry we are currently witnessing multiple paradigm changes happening in parallel, some of them are even of Level-2 nature, and perhaps even Level-3.

Unfortunately, in most cases paradigm shifts emerge without an explicit notification given to the relevant community of practitioners that are affected by the paradigm shift. It is like someone changing the ground-rules of the game, without telling this to the players.

Individual paradigm Shift As a result it is up to each player that is a member of the relevant community of practitioners to find out, and to perform his individual paradigm shift. This can be a painful experience, as it involves having to un-learn habits, knowledge, personal networks and in some cases a whole worldview acquired in years of professional experience. In fact there is a biological aspect of such a deep paradigm shift, as the paradigm has literally instantiated itself over time in the brain of the individual member of the community of practitioners. This occurs through synapse formation creating individual paradigm specific synapse networks which embody these Kuhnian "model problems and solutions" including these habits and the associated worldview. The term worldview can be interpreted in a neurophysiologic sense, as the visual perception of a human is heavily influenced by how the visual system of the practitioner has been trained in his professional career. In the moment when Nick McKeown stated in his keynote speech at the SDN NFV World Conference (Nick McKeown 2017) that "we had something like blinkers on", I realized that Nick referred to a level-2 paradigm shift "from vendor datasheet defined switch behavior to truly programmable switch behavior based on P4 Runtime". This is a level-2 paradigm shift because it follows on the shoulders of a level-1 paradigm shift that was triggered by academia around 2007 with the creation of OpenFlow, where the associated paradigm shift has been characterized as "from integrated control and forwarding to separation of control and forwarding". This is really what OpenFlow did—starting in 2007 at Stanford University. And the established telco-vendors had to learn, that the genie was let out of the bottle by OpenFlow—impossible to get it back in. This kicked off the SDN journey. I related the "blinkers" to an example that had originally made me understand the concept of an individual paradigm shift during a course dedicated to paradigm shift recognition. It was a report of a Swiss teacher that could not find an explanation why an otherwise mathematically versed primary school pupil could not solve a very simply mathematical problem, which was: "it is now a quarter past one, how late is in half an hour?". She found out that the reason for the young boy's inability to solve this simple problem was, that he had learned the concept of time only with a wrist-watch with a digital display. So his concept of time was not associated with something that could be easily halved, like a cake. He could not map a quarter to 15 minutes, and half an hour to 30 minutes, because his brain was never trained to do so. He first had to go through an individual paradigm change from digital representation of time to analog representation—and perception.

An individual paradigm shift can be brutal when the old paradigm was fundamental to a whole thought building—which is collapsing during the individual paradigm change due to a cornerstone of the thought building being destroyed.

People in eastern Germany that grew up in socialism had to go through the individual paradigm change that is the subject of the movie: Good Bye, Lenin! Resistance against such kind of fundamental change is very strong. Cultural change programs can help, but are often failing due to this strong resistance.

Community Paradigm Shift As a paradigm shift emerges it gets adopted by some members of the community of practitioners referred to as early adopters of the new paradigm. But it is only after a critical mass has adopted the new paradigm that one can speak of a community paradigm shift having occurred. The size of this critical mass depends on the kind of paradigm shift and the kind of relevant practitioner community. While the Kuhnian definition of paradigm shift defines a community paradigm shift as completed when the new paradigm is "universally recognized", a better definition may be gained by applying the diffusion of innovations theory from Everett Rogers (Rogers 1962) in a similar way as Geoffrey A. Moore in his landmark book "Crossing the Chasm" first published in1991 (Moore 1991). This may be achieved by substituting the term "product" by "paradigm", and defining a community paradigm shift as having occurred when the chasm has been crossed, and not only visionaries among the practitioners have adopted the new paradigm, but also pragmatists are starting to adopt the new paradigm. This is important for reaching a critical mass from where the adoption rate growth in the community of practitioners will be self-sustained.

3   Historic Paradigm Shifts in the Telco Industry

From the outset, the telco industry has struggled to cope with a number of deep paradigm shifts encountered so far during the digitization of the telco industry. And the speed of paradigm shift arrivals is only accelerating. To understand this deeper, let us look at recent history of the telco industry by selecting some paradigm shifts that have shaped it.

Telco digitization started in the late 1960s and early 1970s with the advent of early packet switching (ARPANET 1969 as precursor of the Internet) and early digital voice switching. The 1980s saw the advent of SS7 and ISDN in the telco space, Ethernet in the Enterprise space and Internet in the academic space. The 1990s brought the commercialization of the Internet whereby the Internet innovation diffused into the enterprise and Telco space, and the World Wide Web appeared as level-2 paradigm shift. Another Level-2 paradigm shift was LAN virtualization with IEEE 802.1 Q VLANs together with the emergence of Ethernet switching, which spread into the Telco space winning over ATM due to its superior underlying communication paradigm that is modelled after the communication at a table. The table paradigm of communications which Ethernet has inherited from its roots in Aloha net enables automatic address learning via broadcast and multicast frames—ATM had no chance because address assignment could not be as "brutally automated" as Ethernet's inherent method for zero touch automation of address assignment. The Internet brought about several major paradigm shifts, most importantly "from closed to open". The data networks of the time focused on closed user groups, while the Internet brought the idea that networks are for sharing and all endpoints in the public Internet shall be reachable (from today's perspective the root cause of many Internet security problems). Some paradigm change happened by accident due to ISPs and Telcos inability to introduce distance related charges for Internet traffic, not even volume based charges. This resulted in two paradigm shifts which I would summarize as "from bell-head to net-head". The first is: "from distance dependent charging" (voice) to "Distances don't matter" (flat rate tariffs). The second is: "from volume dependent charging of data (X.25)" to "data volumes don't matter" (due to flat rate tariffs). As a result of the Level-1 paradigm shift "Distances don't matter" a level-2 paradigm shift occurred that separated the voice control plane from the voice forwarding plane (soft switch and media-gateways), as implemented in the so called 3GPP Release 4 architecture of mobile networks. Subsequently a battle took place in 3GPP with the result that IMS was specified based on SIP, which may be perceived as a level-3 paradigm shift, superseding the soft switch based level-2 paradigm shift—leaving it as a transitory technology for interworking with legacy circuit switched voice that by now is on the way to being phased out. Along the way, the net-head paradigms started to take over control from the bell-head paradigms. Telcos and also Telco vendors struggled and implemented cultural change programs on their path to transforming the circuit switched legacy voice network (fixed and mobile) to an IP based NGN. This could be referred to as the "circuit-switched to packet-switched paradigm change" which was also powered by exponential growth of compute power, memory capacity per chip and transport capacity—as predicted for by Moore's law. The cost per byte of compute, storage and transport for inter-process communication dropped exponentially. The next paradigm shifts that came along were SDN and NFV. Telcos today are at various stages regarding the migration of their legacy networks. Many operators have not yet completed their migration to all-IP based so called next generation networks (NGNs) with IMS and VoLTE, and now face the next wave triggered by Software Defined Networking (SDN) and Network Function Virtualization (NFV). With 5G on the horizon, industry consensus has emerged that 5G networks will be based on SDN and NFV as key enablers of automation. Interestingly, one of the reasons is a paradigm shift that partly reverts a previous paradigm shift—so the industry went first from "distances matter" (for charging in legacy voice networks) to "distances don't matter" and now the new 5G paradigm appears to be "distances matter for latency", at least according to Carolin Chan, VP and GM, 5G Infrastructure Division, Network Platform Group (NPG) | Intel. She stated at the SDN NFV World conference 2017 (Chan 2017): "I would like to look at 5G not as a network built for a certain purpose, but rather look at 5G as built and architected for latency".

4   Paradigm Shifts Along the Path to Brutal Automation

In the following a number of paradigm shifts is examined that play a crucial role on the path to brutal automation. Some may be considered as historic already, but as they are still ongoing, they are listed in this chapter rather than in the previous one.

4.1   From Hardware to Software to Open Source Software

As Stanley Gibson wrote in a Computerworld article (June 9, 1989), The Software Industry was born in 1969 with IBM's decision to unbundle Software and Hardware (Gipson 1989). This has given way to the creation of independent software companies such as SAP (founded in 1972). Bill Gates and Paul Allen founded Microsoft in 1974. Due to IBM's 1981 decision to source the operating system for the IBM PC from Microsoft a whole new ecosystem of Software and Hardware vendors emerged—centered on a Microsoft owned operating system. Another operating system called UNIX started to escape from AT&T's Bell Laboratories in the early 1970s—in many different versions. In 1983 the free software activist Richard Stallman launched the GNU Project to create a Unix-like free operating system, and in 1989 Stallman pioneered the GNU public license—laying the legal foundation of the Open Source paradigm shift. The 21 year old student Linus Thorvald published in 1991 a first version of Linux—targeted at the Intel 80386 Hardware. Open Source was a rather disruptive paradigm shift for the software industry. Today, Linux has transformed the software industry. Fast Forward to the year 2011 when Marc Andreesen wrote: Software is eating the world. This statement stems from the article "Why Software is Eating the World" which was originally published in The Wall Street Journal on August 20, 2011. What happened in between? A large portion of software is based on open source—and running in the cloud. Open Source is about sharing development cost, and is about accessibility of software at low cost. However—cost is still in testing and support and in the combination with key pieces of some superior proprietary software. Not all is free and different business models apply, but in general, Open Source is itself a paradigm shift and it has accelerated the paradigm shift from hardware to software. After having transformed the software industry, the telco industry is next. At the 2017 SDN NFV World Conference in The Hague multiple open source projects demonstrated major progress and even breakthroughs, such as ONAP, P4 and CORD.

4.2   SDN: From Hardware Centric to Software Centric Paradigm

Networking has traditionally been defined in a hardware centric way, thus using a hardware centric paradigm. In part because of the classical network theory defining that a network consists of nodes (active network elements) and cables (passive network elements) connecting the nodes. In the network theory, which is a part of the mathematical graph theory, all networks can be abstracted to a level where they

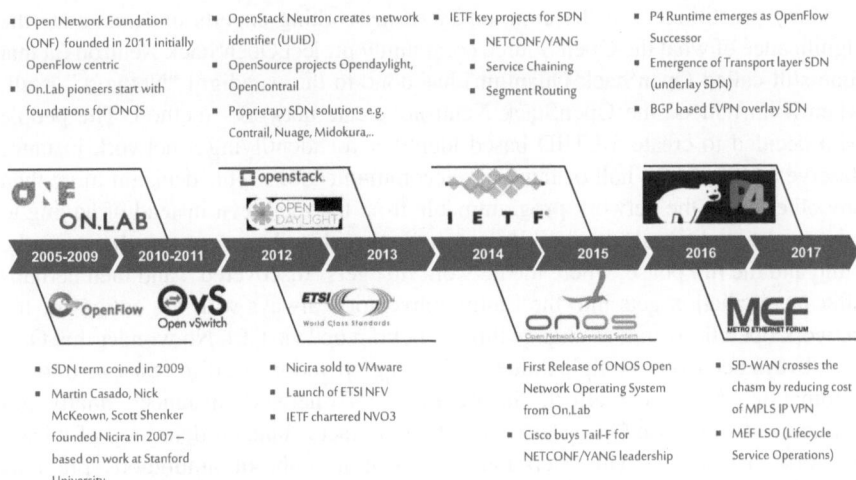

Fig. 1 History of SDN—selected milestones

consist of nodes and arcs. Arcs are an abstract representation of a physical medium such as a cable. Arcs typically show up as lines in a diagram with nodes connected by lines. A second reason for the hardware centric paradigm is the dominant role that hardware node vendors have played in the telecoms industry, leading to the network vendors being able to design their products—the nodes—in a proprietary way. It was only due to the groundbreaking work at Stanford University by Martin Casado, Nick McKeown and Scott Shenker around the year 2007 that OpenFlow was created as a byproduct when trying to open vendor proprietary routing and switching code to the academic community. The creation of OpenFlow marks the critical event in the history of SDN (see Fig. 1) that has let the software genie out of the bottle. Node hardware vendors have since reacted by offering various ways how to make the network more programmable for the purpose of automation. OpenFlow is by far not the only way—other so called southbound interfaces from an SDN-Controller have emerged quickly. But in all cases programmability means that a different community of practitioners takes over control: those who know how to program the network and those who know how to harness this programmability for network automation. Some of the methods are based on networking protocols and paradigms, such as BGP and NETCONF/YANG. Other approaches differ such as P4. The following diagram shows a short history of SDN. It shows that the relevant practitioner community is changing, not only because of individual paradigm changes performed by existing members, but increasingly also by members of the IT domain entering the networking domain, and of course by younger members joining who are familiar with newer software centric network paradigms. As a result, the community of practitioners changes. The Software centric paradigm slowly takes over control, favoring those who know how to program the network and those who know how to harness this programmability for network automation. Ultimately the objective is zero touch automation.

Many practitioners in the community of networking experts underestimate the significance of what the Open Source community project OpenStack Neutron (at that time still called OpenStack Quantum) has done to the paradigm "Network" itself, when it introduced the OpenStack Neutron "create network" method. The people who decided to create a UUID based identifier for identifying a network instance deserve a place in the hall of fame of telecommunications. This decision more than any other made the network programmable from the top down instead of having to rely on the traditional plan/build/run paradigm with a discovery step between the build and the run phase, where the network first gets "discovered", and then perhaps after reconciliation gets into the runtime inventory, always with the option for the network operations staff to change things "bottom up" via CLI. No wonder that OSS has always been very complex, due to allowing operations staff to resist zero touch automation. As a side effect, the definition of "network instance" finally got harmonized—because OpenStack has set the de-facto standard definition of what a network instance is. This step must be seen in light of ambiguous previous definitions, and in light of the networking industry having failed for over 40 years to create an industry standard uniform identifier for a "network instance". Most networking experts do not even know that such an identifier was missing—I guess again due to blinkers. It is a common feature of paradigm shifts that someone changes the ground rules of the game, such as the definition how the concept "network" is defined, or the creation of an identifier that makes it programmable throughout the whole lifecycle, but does not tell the community of networking experts about these rather fundamental changes. That paradigm shift allow IT-centric staff and their automation tools to take control of what was previously the exclusive realm of network node centric people using CLI commands and scripts for fragmented domain specific partial automation solutions. The OpenStack defined network instance is a central object at the northbound interface of SDN-controllers such as OpenDaylight, Contrail, OpenContrail or ONOS when speaking to an OpenStack based end to end Orchestrator sitting on top of them (northbound). Furthermore it is well aligned with MEF E-Line, E-LAN and E-Tree based "network as a service" (NaaS) definitions. MEF definitions even allow that one such "network as service" spans multiple operators. And MEF LSO (Lifecycle Service Orchestration) is all about zero touch brutal automation of such NaaS spanning multiple operators. The SDN automation has meanwhile reached the optical domain as well, with optical transport paths being setup "on demand", including across operator domains. And of course the IETF has chartered various key projects to support this journey to brutal automation, such as Segment Routing, Service Chaining, NVO3 to name just a few. Not to forget of course NETCONF/YANG and the Internet scale proven BGP protocol for automated discovery of network endpoint reachability—not just for layer3 endpoints (IP-addresses) but also for layer2 endpoints (MAC-addresses e.g. using EVPN). The P4 Runtime is perhaps the latest paradigm shift in SDN—applicable to programming switches and routers running at terabit/s speed—always at line-rate if the P4-Runtime program has compiled successfully.

4.3   NFV: From Hardware Appliance to VNF Software

In the same keynote speech where he coined the term "brutal automation" Arash Ashouriha also stated, that the original vision for Network Functions Virtualization (NFV), when it was launched 5 years ago at the SDN and OpenFlow World Conference in Darmstadt, was to learn from the web, the enterprise and the hyperscale companies, but that "the reality is, that we are still far away from the pure cloud native vision we had years ago". Onboarding new VNFs is still too cumbersome and license models have not yet been adapted sufficiently. ETSI NFV has reacted e.g. by setting up a working group looking after license models and also one that is investigating charging models for NFV—these both are a step in the direction of the vision outlined in "Future Telco Band I—Virtualization is transforming the telco industry". In the chapter titled "Innovative Billing Methods make new partnering models possible", it was stated that: "One could imagine the extension of today's online charging functions to include the simultaneous debiting and crediting of charges for the various parties involved in the performance of the service such as end customers, network operators, marketplace operators, and network application developers." Looking back to the history of NFV it must be noted that of course virtualization was built upon the pioneering work of VMware for x86 architecture hardware virtualization. The central role of the Open Source movement—in particular OpenStack as open source Cloud Software—becomes apparent (See Fig. 2). While NFV is now in a deployable state, some operators seem to hesitate and consider jumping over the level-1 paradigm shift (from real hardware to virtualized hardware) to a level-2 paradigm shift from virtualized hardware to virtualized software packages, also known as Containers. To work with containers, an operator needs a Container-Runtime such as from container pioneer Docker, or rkt (Open Source). In addition, a container orchestration solution is needed, such as Kubernetes (Open Source), Docker-Swarm or Mesosphere.

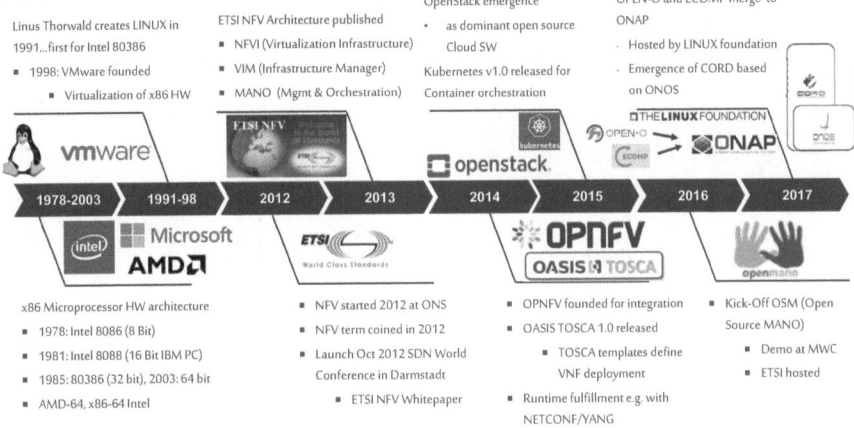

Fig. 2 History of NFV—selected milestones

An avant-garde is even already looking at a level-3 paradigm shift referred to as "server-less", which abstracts even from the need to orchestrate the deployment of the software package by reducing the software to pure functions, taking away from the developer the burden of having to think about deployment options. These more advanced deployment models using container or even serverless network functions are referred to as cloud native. NFV has been embracing container technology, but the reality today is that deployment for VM based VNFs is much more mature than deployment of containerized VNFs, and VM based VNFs will continue to play an important role for certain applications, even when the expected migration to cloud native will have advanced. Savvy operators engage therefore in multiple horse races.

Another area where NFV has progressed significantly is in the orchestration of VNFs, where both ETSI hosted OpenSource project OSM (OpenSourceMANO) and Linux-Foundation hosted ONAP have made significant progress, with the first ONAP release (Amsterdam release) released shortly after the conference in November 2017. Also the open source project CORD (Central Office re-architected) of the ONF has made major progress, not only for fixed network (Residential CORD) but also for mobile Networks (Mobile CORD) and Enterprise Networks (Enterprise CORD). CORD is leveraging Open Network Operating System (ONOS), which ON.LAB and ONF have built since the creation of ON.LAB in 2011 (see also Fig. 1). In 2017, ON.LAB and ONF have merged and a new ONF has been created, even more focused on creating code as part of open source projects such as CORD and ONOS.

4.4 Cloud Native: Monolithic VM to Containerized Micro-services

While NFV has matured to a certain degree, the next levels of paradigm shift already kicks in, which is referred to as "cloud native". Cloud native is centered on containerized applications. Containers are a paradigm shift in virtualization away from running applications in dedicated virtual machines (see Fig. 3). VMs are rather heavyweight, because each VM has its own instance of an Operating System (OS). As in most cases this operating system is Linux anyway, containers have been

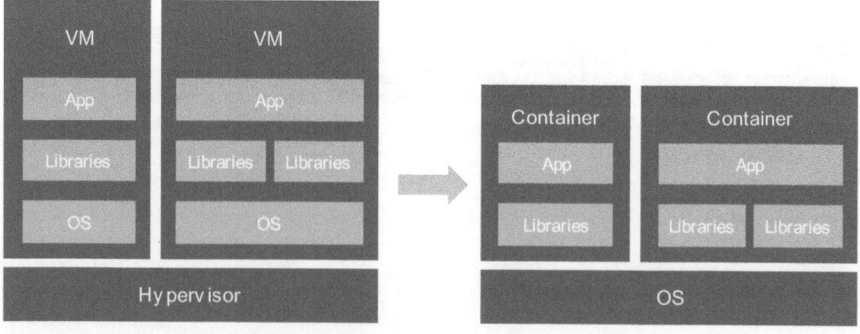

Fig. 3 Paradigm shift from VM to container

introduced initially by Docker as lightweight standalone executable packages of software that share with other applications the OS system, but still have their own environment such as their specific libraries.

For cloud native, previously rather monolithic network functions need to be decomposed into ideally re-usable software components referred to as micro-services, which interact via APIs, and which can be deployed as containers including built-in self-healing and self-scaling capabilities. This way, so called "carrier grade" networking with 99.999% availability ("five nines") can be built with components that do not need to have five nines availability. This paradigm shift has been best expressed in a diagram from a presentation that Marcus Brunner of Swisscom has held at the SDN NFV Conference 2017, contrasting a picture of an elephant against a picture showing a massive amount of ants. This contrast visualized quite nicely the paradigm shift from a monolithic VNF (one elephant), to cloud native (many ants). This paradigm shift expresses that it is possible to build reliable systems using non-reliable components. See also related aspects in the chapter on APIs (how the ants communicate) and RESTful (ants do not keep state, so another ant can takeover easily). This compares to monolithic applications derived from hardware appliances deployed as pair of redundant appliances, resulting in the need for a backup elephant and probably a complicated synchronization to reach the five nines availability target.

4.5   ONAP: From Legacy Processes to Zero Touch Automation

Why can ONAP be considered a paradigm change in itself? After all it is just an Open Source project aimed at the creation of an Open Network Automation Platform. Well, the scope and power of this is just unimaginable for telecom practitioners tangled in legacy paradigms, who from their experience believe that this is impossible. After all, telecom professionals have tried for decades to enhance time to market for new products—with very limited success.

ONAP promises with its first Release "Amsterdam" to be able to shorten that timeframe from months (sometimes years) to 15 min. This may be exaggerated where integration with legacy BSS/OSS is required, but in cases where the new service is ordered via Portal and where an integration with BSS charging is either simple (e.g. subscription based) or not necessary this may indeed be achievable. The ONAP Software Release "Amsterdam" has been released in Nov 2017. An impressive list of members representing over 55% of the worldwide subscriber base have aligned behind ONAP. The following diagram shows a high level view of architecture of the ONAP Amsterdam release (See Fig. 4).

The reasons for acceleration from months to minutes is that ONAP covers not only the Run-Time when orchestration happens but also the Design Time when products are created. ONAP leverages multiple paradigm shifts, including the model driven paradigm, the cloud native paradigm and the DevOps Paradigm. The Model driven paradigm allows to automate without hard coding. The Cloud Native paradigm brings many benefits including reliability built with unreliable components, that can be deployed more flexibly and with less constraints. The DevOps paradigm (see Fig. 5)

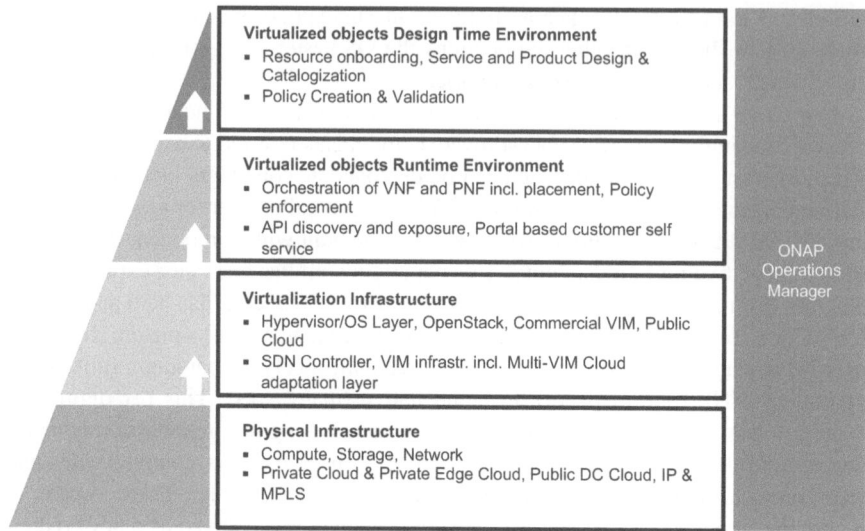

Fig. 4 ONAP Amsterdam architecture high level view

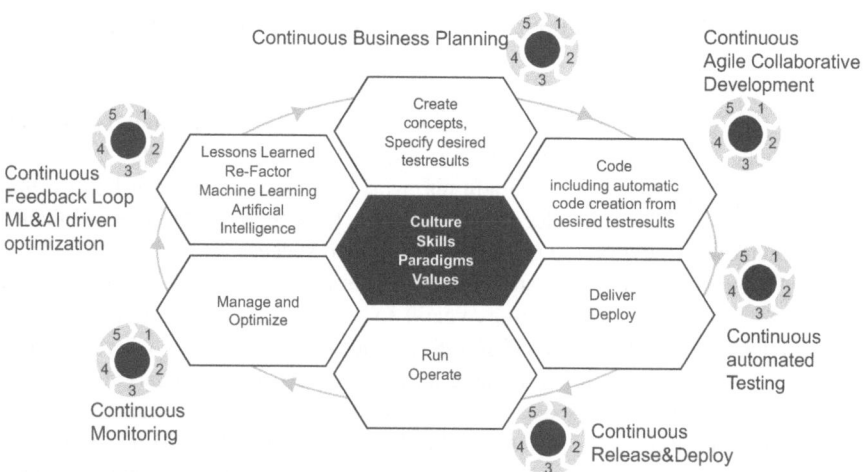

Fig. 5 DevOps paradigm

changes the process of how the telco operates, thereby making it more agile and transforming it into a closed loop process. In actual fact a closed loop process comprised of multiple closed loop processes—as indicated in the below diagram:

Some processes even get inverted for "brutal automation". For example the Code process may be transformed and thereby inverted from a process where testing including the verification of the test result is done at the end before launching a product to a process where the test result is used as starting point. Modern software methods recommend to specify the test result at the start of the coding—basically as output of the think process. A brutally automated coding sub-process could theoretically generate the code automatically from the specification of the test-results, and it could be automatically tested at the end of the Code sub-process. So far, this is futuristic but not impossible. In any case, an advice how brutal automation opportunities can be identified could be to look for the possibility of inverting processes, even if initially considered impossible. In closing his "brutal automation" keynote speech at SDN NFV World Congress Arash quoted a "wise man", who said: "How do you know what is possible, if you have not tried the impossible?". This is exactly the point—how do you know what is possible, if you have not looked at it with "paradigm shifted eyes"? Are you sure you cannot start a coding process with the specification of the test result? That is what students learn at universities today anyway. Looking how postpaid charging was inverted to prepaid charging gives a good example of such a brutal process inversion aiming at real time capability and automation.

4.6 From Brutal Automation with APIs to RESTful APIs in 5G

Some may think that Jeff Bezos was brutal, when he mandated the use of APIs with the threat to fire anyone who would not adopt this mandate. Some may think that this is the only way to succeed. Using APIs to expose was a critical step in the rise of Amazon to where it is today. Telcos have to realize that exposing their data is critical to success. But exposing via APIs is just a level-1 paradigm change. The practitioners have already moved on to a related level-2 paradigm shift that is referred to as RESTful APIs. RESTful is a paradigm created by Roy Fielding from an HTTP Object Model he proposed 1994 and refined to the REST architectural style in his dissertation in 2000. The REST paradigm mandates implementation via a stateless client server protocol, and de facto this is HTTP1.1 today, moving to HTTP 2 as natural evolution. Besides the stateless property, a so called HATEOAS paradigm is perhaps the most important aspect of RESTful architecture, as it allows a better decoupling of Client and Server domains. With HATEOAS (Hypermedia as the engine of state), the client navigates only based on URLs received from the server, easing the extensibility and adaptability of the API because such extensions and adaptions become possible with server side changes only. 3GPP has adopted the RESTful style paradigm for the 5G Core network.

4.7 From 4G to 5G: The Latency First Paradigm Shift

Carolin Chan, VP and GM 5G Infrastructure division at Intel stated at the conference (Chan 2017): "I would like to look at 5G not as a network built for a certain purpose, but rather look at 5G as built and architected for latency". This is an indication, that 5G will be architected based on a "latency paradigm", a paradigm shift away from a "distance and location does not matter" paradigm which has led to centralization of network functions in a few central locations of modern networks. The consequences of such paradigm shift are not yet fully understood. Certainly it means, that the 5G Core network will be much more distributed than the 4G Core network—at least those resources that are part of a network slice instance for Ultra-reliable Low Latency Communications (URLLC).

4.8   P4: From Virtualization to Anti-virtualization Paradigm Shift

The paradigm shift that P4—or more precisely P4 Runtime is based upon may be characterized as "from virtualization to anti-virtualization" or "from vendor data sheet defined to fully programmable Terabit/s switching and routing". Traditional approaches to x86 architecture CPU based routing have not been able to match the performance of high performance routers. OpenFlow based solutions have lacked true openness, as some vendor dependencies still remain, e.g. in the way actions must be specified. The keynote speech of Nick McKeown at the SDN NFV World Conference 2017 was pointing to the fact that now for the first time hardware chips are available, which do have the same or better performance at the same or better price then proprietary network processors. This is a proof that previous claims are false that full programmability would cost performance or be much more expensive or both. When Nick stated "it was as if we all had blinkers on" it was clear that he was talking about the biological aspects of an individual paradigm shift, where people simply cannot see what lies quite obvious in front of them, due to years of training that what lies in front of one's eyes is impossible and therefore invisible. P4 can be understood as anti-virtualization, because it does the opposite of virtualization. Virtualization comes from the Latin word "virtus", which has the original meaning of image, something that looks like being real but is not—e.g. an image reflected from flat water or from a mirror. Derived from that is that the picture has the same features and capabilities as the real physical original. Therefore, virtualization is about producing an image (e.g. a software image), that has the same capabilities and behaves like a physical original (say an Intel x386 processor that was the first hardware supporting Linux). So virtualization is about producing an image to a pre-existing hardware. Contrary to this, anti-virtualization is about producing a hardware to a pre-existing image, where the pre-existing image is the set of all successfully compiled P4-Runtime programs. Hardware vendors produce chips that are able to execute these programs. This reversal of direction—hardware exists prior to image versus image exists prior to hardware—that is the anti-virtualization paradigm change that comes with P4. Please see also diagram 6 below, that shows how virtualization and anti-virtualization paradigms can be merged together.

4.9   The Impact of ONF with ONOS, CORD and vOLTHA

The Open Networking Foundation (ONF) has been instrumental in starting the SDN revolution with OpenFlow. ON.LAB has started around the same time as ONF and created the open source project Open Network Operating System (ONOS). In 2017 the ONF has been merged with ON.LAB and increased the focus on Open Source. CORD started as ONOS use case and is now also an OpenSource project. ONOS is an open source SDN Controller that is integrated in the key open source project CORD (Central Office re-architected), which aims at providing one common network edge platform addressing all three domains of use: fixed/residential with R-CORD, mobile with M-CORD as well as enterprise E-CORD. Another important CORD project is virtual OLT Hardware Abstraction (vOLTHA). It targets the fixed/residential domain of use with yet another paradigm shift: Instead of having to abstract each vendors access node technology, vOLTHA aims to change the rules of the game by forcing the HW vendors to develop towards a pre-existing "image" referred to as "abstraction". This can be interpreted as an anti-virtualization paradigm shift—similar to P4. In the CORD workshop at the SDN NFV World Conference 2017 a CORD Introduction and Roadmap presentation was given that showed the CORD roadmap and vision. The following diagram (see Fig. 6) shows a Detecon interpretation of the CORD vision as presented at the Congress.

Detecon interprets this vision as the attempt of bringing together CORD with ONAP and CORD with P4 (and vOLTHA anyway). CORD is in all three domains of use focused on the network edge—under a top level orchestrator such as ONAP. The diagram shows in the three lower layers how the anti-virtualization paradigm shift on the right side (depicted as inverted pyramid) complements nicely with virtualization on the left side. P4 Runtime as fabric service allows high performance switching and routing, based on a paradigm shift where hardware has to follow the abstraction (the successfully compiled P4-Runtime image that exactly describes how the hardware has to behave—the vendor has to build the hardware accordingly). Similar to this is vOLTHA, which provides an abstraction for optical fiber access nodes, which vendors have to build their hardware to. Through vOLTHA the power position

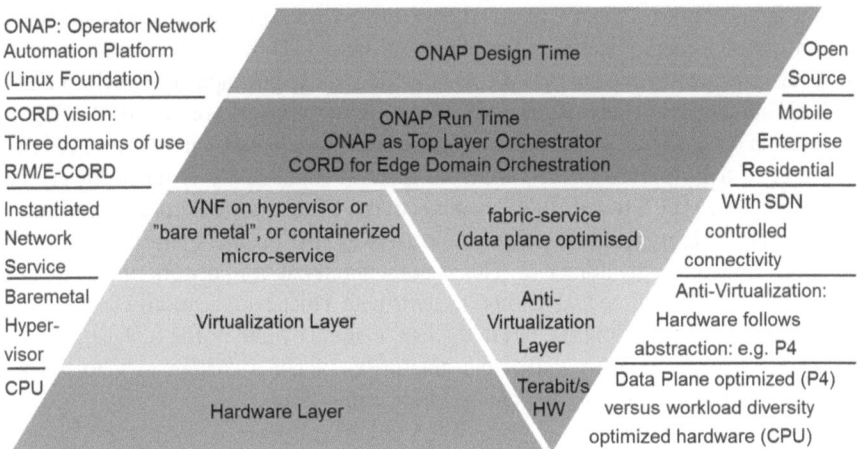

Fig. 6 Detecon interpretation of CORD vision

between software and hardware vendors gets inverted. It is not a vendor specific hardware element abstraction made after the vendor has created his design, rather it is a common abstraction created prior to the vendors having created the hardware design. Thus, the hardware vendors have to obey to the vOLTHA defined abstraction, not vice versa. In case of the CORD mobile domain of use an important aspect is the extension of the span of control into the radio access domain—the RAN, with Open Source xRAN project. This has brought major advances and shows how the RAN can be subject to network slicing—an important new concept that is key for building 5G networks in a cost efficient way while meeting vastly differentiated demands regarding quality, latency, security, scale, and target cost.

4.10   The Impact of SD-WAN (Shifting Power to the Enterprise)

SD-WAN stands for Software Defined WAN and at last year's SDN NFV World Conference Dan Pitt had pointed to the fact, that SD-WAN is not the same as WAN-SDN. WAN SDN is applying SDN in the Wide Area Network (WAN). SD-WAN is different, because it is leverages the fact that a virtualization infrastructure can be put into the CPE that gets delivered at customer premises. A virtualization infrastructure (e.g. x86 processor, storage and networking) inside an enterprise networking device that is either enterprise owned, or carrier owned and managed, or third party managed opens a path to Enterprise driven innovation—independent of network operators progress with SDN and NFV. This is a situation that is similar to the Voice over IP revolution. There is a network operator driven VoIP revolution—referred to as IMS and VoLTE, and there is an Enterprise driven VoIP innovation—referred to as unified communications. Similarly, there could be separate innovation paths for operator driven SDN and NFV and enterprise driven SDN and NFV—and of course savvy operators are jumping in and provide carrier managed SD-WAN solutions. Enterprise networking giant Cisco and Enterprise virtualization giant VMware have both acquired a market leader in the SD-WAN space. SD-WAN has already crossed the Chasm as it saves the enterprise cost of having to expand capacity of MPLS based IP-VPN network services. Rather, SD-WAN allows policy controlled steering of application specific traffic from the enterprise branch via public Internet connectivity. The cost of a Byte transported via public Internet connectivity is much lower than transported via MPLS based IP-VPN connectivity. With many enterprise applications increasingly residing in some public cloud anyway, the business case for SD-WAN solutions is increasingly attractive. This is not without risk for network operators, as it would not be the first time that disruption starts with a niche application. The paradigm shift behind this disruption risk lies in the fact, that the overall traffic orchestration for SD-WAN may be in the hand of the enterprise—again a change in who has the ultimate power and authority to control where traffic flows.

5   Additional Paradigm Shifts on the Path to Brutal Automation

The so far described paradigm shifts are not the only ones that must be watched and their impact considered when planning a path to zero touch automation—which may have to be a brutal automation. The following paradigm shifts may be equally important, or at least play an important role as well:

- From node to function
  - Operators staff are used to manages nodes with interfaces
  - Brutal automation brings about the disappearance of the node
  - Familiar interfaces like CLI disappear
- Procurement: from buyer market to seller market
  - Some breakthrough technologies tend to turn-around the position of seller and buyer—suddenly the procurement department is in the weak position, and the vendor is in the strong position as he can choose to which operator he delivers portions or all of his production capacity
  - Steve Jobs has exercised this well with buying the total production of 1 year of critical components, e.g. for the iPOD hard-disk.

---

6   Addressing the Skill Gap

One key take-away from the SDN NFV World Congress was the "skill gap", i.e. the inability to of telco operators to attract the kind of talent with deep software knowledge and the difficulty to re-train existing staff with the new skills.

Cultural change program Many operators have already implemented cultural change programs aiming at up-skilling the existing staff and preparing them for the changes that the paradigm changes described above bring about.

Up-Skill program Up-skilling existing staff is critical, as it is not sufficient to rely on new people who do not understand the technical, and business requirements and also not the processes that need to be transformed. Operations teams and also teams dealing with testing need to be upskilled for the new world. Also the process for product creation needs to be revisited for opportunities where new products can be designed using automated tools (that need to be created or may be available from ONAP). A massive push into Open Source cooperation is strongly recommended.

Recruiting program Recruiting programs need to consider these paradigm changes and need to anticipate the kind of skills that are required in the new world. Much more emphasis must be given to areas such as familiarity and experience with Open Source. Not only from a coding aspect, but also from legal and business aspects such as licensing, business model analysis and creation etc.

Greenfield spin-off Given the massive amount of parallel paradigm shifts it might be wise to consider the creation of a separate legal entity that can spearhead the innovation in much more agile ways than a traditional telco operator may be able to turn. It takes a long time to perform a right angle turn for a large tanker like a legacy rich telco operator, whereas a small boat can turn quickly—and should be able to fail fast—if it learns fast and finds the right path to brutal automation.

7   Summary: The Path to Brutal Automation Is Bumpy

The telco industry is undergoing fundamental change, triggered by multiple paradigm shifts. Coping with this multiplicity of parallel paradigm shifts is a huge challenge for network operators—it is brutal. This is particularly brutal for large operators that have a historically grown large set of legacy networks and an associated large set of IT-systems for OSS and BSS and—last but by far not least—an associated large staff trained in legacy paradigms. Paradigm changes that invert processes upside down (such as from test result at the end to test result at the start of the process) affect all the people in this process—this is brutal automation. And multiple processes get inverted or at least partially inverted due to the ongoing paradigm changes. Procurement has to learn to sit on the other side, when the market in certain critical areas transforms from a buyer market to a seller market- this is brutal as procurement has to leave the comfort zone, and go away from old style RFQ process and buyer market negotiation position. Also human resources has to undergo brutal change—maybe brutal automation by introducing new hiring approaches such as via organizing hackathon events. Transforming an operator to the new world of zero touch automation based on virtualization is a difficult task, not only technically, but perhaps more so with regard to the cultural change required not only in the own staff, but also in the relation with vendors and partners. Cultural change programs are required to support this massive change. And a strategy how to migrate from current mode of operation to future mode of operation. Perhaps it is required to live through multiple levels of paradigm change, e.g. by multiple generations as KPN has shown at the SDN NFV World Conference. This article has tried to look at the history behind these massive changes, and tried to shed some light on which paradigm changes are going on underneath the surface. Brutal Automation is brutal because many people and staff of telco operators will have to go through individual paradigm changes. The sheer amount and scope of the various paradigm changes described should enlighten senior management staff to better understand what this journey is all about, and the greatest challenge is not the technical transformation of the network. The greatest challenge is the transformation of the staff that has to go through all these paradigm changes. This is what makes the desired zero touch automation so brutal. The term "brutal automation" has been coined at the SDN NFV World Conference 2017. It captures very well the target—automation, and the impact to the network operator and the staff—it will be brutal. Radical change is waiting, and many individuals will have to "rewire their brains". Some will fail to do so, and for some it will be a painful experience—in both cases the word brutal captures the impact.

References

Andreesen, M. (2011, August 20). Why software is eating the world. *The Wall Street Journal*.

Ashouriha, A. (2017). Video recording of opening keynote speech at SDN NFV World Congress 2017. Last retrieved December 12, 2017, from https://www.layer123.com/sdn-webcast-mle123/

Chan, C. (2017). Video recording of speech "Enabling Next Generation Networks Today" at SDN NFV World Congress 2017. Last retrieved December 12, 2017, from https://www.layer123.com/sdn-webcast-mle123/

Gibson, S. (1989, June 19). Software industry born with IBM's unbundling. *Computerworld, 6*.

Gonsa, Chrestin, & Reith. (2014). Virtualization is transforming the telco industry. Future Telco, Detecon International GmbH.

Kuhn, T. S. (1962). *The structure of scientific revolutions*. University of Chicago Press, first edition published in 1962.

McKeown, N. (2017). Video recording of keynote speech at SDN NFV World Congress 2017. Last retrieved December 12, 2017, from https://www.layer123.com/sdn-webcast-mle123/

Moore, G. A. (1991). *Crossing the Chasm* (1st ed.). New York: HarperCollins.

Rogers, E. M. (1962). *The diffusion of innovations* (1st ed.). New York: Free Press of Glencoe.

Rowan, R., & Yegge, S. (2011). Last retrieved December 12, 2017, from https://plus.google.com/+RipRowan/posts/eVeouesvaVX

Edge Computing: The Third Major Step in the Evolution of Telco Networks

Christoph Goertz, Thomas Vits, and Claus Eßmann

1   Edge Computing in Context of Network Technologies

Describing telco networks to non-telco people is usually reasonably simplified as an ongoing evolution. For example in case of mobile networks from 2G to 3G to 4G LTE and currently towards 5G—it's always about more bandwidth, higher speed, i.e. a "better" data transfer. Almost everybody from the telco industry knows that actually some of these evolution steps were actually major stages with significant impact on process, organization and business with many challenges. Within the more than 100 years old capability to enable communication between humans over long distance, starting with telegraphy, evolving towards fixed and mobile telephony and finally to mobile broadband, there are up to now only two really disruptive steps in the context of telco networks: First, the circuit switched networks and second the migration to All-IP.

So, why will edge computing be the third major step in the evolution of telco networks? Compared to circuit switched and All-IP where the network services were communication centric and, more important, completely owned by the telco operator, edge computing will require to open the network for third party applications which will not be owned (and therefore controlled) by the telco operator. Furthermore, these applications will not be solely communication centric, but will differ from use case to case and therefore computation itself becomes a central service. However, many of these applications will have one thing in common, they require low latency compute. Telco networks will need to provide these compute capabilities and this is something completely different compared to what was done in the first two stages in the evolution of telco networks: Transmitting data requests from source to destination.

C. Goertz · T. Vits · C. Eßmann (✉)
Detecon International GmbH, Cologne, Germany
e-mail: Christoph.Goertz@detecon.com; Thomas.Vits@detecon.com; Claus.Essmann@detecon.com

© Springer International Publishing AG, part of Springer Nature 2019   131
P. Krüssel (ed.), *Future Telco*, Management for Professionals,
https://doi.org/10.1007/978-3-319-77724-5_11

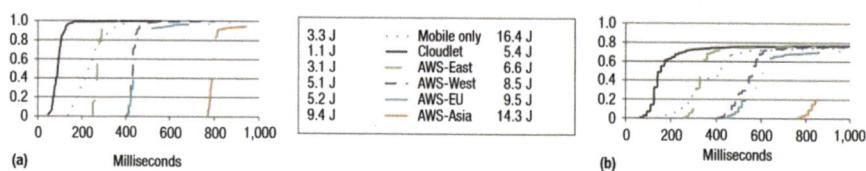

Fig. 1 Response time distribution and per-operation energy cost of an (**a**) augmented reality and (**b**) face recognition application on a mobile device

1.1   The Bidirectional Movement of Computation: Telco's Network Well Positioned for Edge Computing

The opportunities for telco operators for value creation in the field of edge computing results from a bidirectional movement of computation. On the one side, compute will move from cloud to the (telco) edge because the cloud is too far away (in terms of network topology and physical distance) to deliver low latency. One the other side compute will move from the device to the (telco) edge, because the device has only limited capabilities to deliver high performance compute. Implementing an telco edge starts by building the according infrastructure by placing data centers inside the operators network so that data from a client can be routed to such data centers ("cloudlets" Mahadev Satyanarayanan) via a direct and physically short path.

What will trigger these two movements? Future immersive applications like AR (Augmented Reality) or VR (Virtual Reality) will require real-time computing, i.e. a very low end-to-end latency in the order of a few milliseconds. High computation on a device is first limited by the form factor like size and weight, which is crucial for best customer experience—nobody wants to wear a helmet-like AR glass when actually one could look like a usual glass. Second, computation on a device is a cost factor—imagine a drone performing GPU-based image recognition of agricultural land, costs for the device but also initial development costs will decrease when computation is offloaded to a GPU located in Telco's Network. Third, the higher energy consumptions on the device due to computation, which will drain the battery, will affect the customer experience by a lower reach or heat dissipation. Imagine electric cars where one would like to avoid additional electric consumers like compute servers on board.

With the target of finding ways to decrease latency a lot of research work on edge computing was done in the recent years, a good summary is provided by (The Emergence of Edge Computing). Research by (Ha et al. 2013) shows the importance of cloudlets for low latency offload services, given for two applications (AR & VR) on the one hand a significant latency reduction compared to several Amazon Web Service (AWS) data center and on the other hand side a significant energy reduction, the results are provided in Fig. 1.

Based on the bidirectional movement of computation, telcos with their networks are well positioned for edge computing as they have their part of the network closer to the user than the public internet.

1.2   5G Will Need Edge Computing and Edge Computing Will Need 5G

Ultra low latency is one of the key attributes for future 5G networks, as it targets for 1 ms latency. This could be enabled i.e. by advanced technologies in the radio access part. But the full value of low latency for applications is only provided in case of end-to-end latency view, this means towards the server side and back, hereby is speed of light the limiting factor. For example, to enable 5 ms total end-to-end latency a cloudlet must be available in a distance of max 50 km.[1] Edge Computing is therefore a must for 5G, it is already part of the 3GPP 5G standardization (e.g. in 3GPP TS 23.501 V1.3.0).

Since edge computing is not restricted to a specific access technology, it can be implemented already before 5G. The obvious difference is the higher latency in the radio access part e.g. with 4G LTE ca. 20 ms, with LTE-adv ca. 10 ms, compared to the 1 ms target with 5G. Another major advantage of 5G will be provided with the concept of network slicing.

1.3   Building Edge Cloudlets Is More than SDN/NFV Based Network Virtualization

Software-Defined-Networking (SDN) and the related concept of Network Function Virtualization (NFV) are prominent technologies in the telco industry in the recent years which aim to enable the convergence of IT and telco network (often referred to as network virtualization). This network virtualization can be seen as a logical move within the All-IP migration of telco networks. The main driver is cost optimization and simplification of service delivery. Traditional network challenges can be solved by introducing of SDN/NFV networking architectures:

- Virtualized networking functions based on Commercial-Off-The-Shelf network hardware
- Continuous development and deployment through systematic testing on virtual infrastructure
- Introduction of increased network automation, virtualization and centralized intelligence
- Reduced cost through virtualized infrastructure and lower network over-head
- Flexibility through programmable network and elastic on-demand network architecture

---

[1]Light will travel around 200 km/ms in an optical fibre, so under perfect conditions, we would need a distance of 50 km for 0.5 ms round trip time. But to be realistic we are adding 3.5 ms on top coming from network cards, switches and network virtualization technologies to the 1 ms 5G access latency.

- SDN&NFV solutions promise to address operator issues by removing software and hardware dependencies thus improving network agility and operational efficiency.

All three technologies edge computing, 5G and SDN/NFV will emerge together—this is the vision. However, what about the reality?

Currently telco operators are facing major challenges on process and people with the introduction of SDN/NFV. This implies to break "Silo"-mindsets by reflecting cross-functional teams and expertise. New agile processes, tools and development methodologies as DevOps are required as well as a cultural mindset change with full cost control and flexible operations.

The SDN/NFV based network virtualization is limited to telcos own network functions—edge computing will require much more; this is a summary of the key questions:

- How to attract developers? There is a gap between telco edge networks and developers which is currently not covered.
- How to deal with the situation that exact application requirements for edge computing are today unknown? This is difficult to handle with current telco processes and mindset "Tell me your exact requirement first!"
- How ensure up-to-date infrastructure hardware? E.g. a GPU might to be changed after some months. How fast can telcos ensure this?
- How to deal with security concerns when opening the network infrastructure for third party applications? Current attitude is "My network is my kingdom, nobody is allowed to enter!"

The conclusion is here, that edge computing will have a huge impact on networks and its applications. It is important to understand that this is especially true for telco networks.

2 Edge Computing: An Opportunity for Operators?

All KPIs of a cellular network have evolved significantly from 2G to 4G networks despite of one: latency. While e.g. bandwidth was increased by more than factor $1000\times$, latency only decreased by the factor 30. However latency is one of the most important KPIs for computation, which have to be delivered in real time for many use cases as e.g. in the Internet of Things. So, how could the latency be decreased for the delivery of such computational results?

The answer to this question is not a new concept and is implemented already with Content Delivery Networks since years: bringing the computational node physically closer to the application or device where the results of the computation are needed to decrease the distance the signal has to travel. .

As the computation is nowadays often done in the cloud, the cloud itself, or a "piece" of the cloud has to be moved as close as possible to the requesting device/

application, it has to move to the "edge" of the network. This is called edge computing. As the network itself consists of different parts, there are different types edge computing, everyone characterized by its location inside the network.

2.1   Edge Computing in Different Flavors

2.1.1   Edge in the Public Internet

Edge computing can be of course provided by public cloud providers. But the access and aggregation network is in nearly all countries owned by telecom operators, no matter if it is fixed line or mobile network. The public cloud provider needs to take care about getting closer in the physical distance to the connecting point to the operators' access and aggregation network. This connecting point is the closest location a public cloud provider can reach to an application or device and would be called here the "edge" of the public internet. This is already done by Amazon with some AWS and content delivery locations. In this case, Amazon has setup an Elastic Cloud (called EC2) or content delivery location (Amazon Cloudfront) very close to the connecting points to operators' networks.

2.1.2   Operators Edge Computing

The Operator of fixed line or mobile networks has some more possibilities to get even closer to the application/device requesting the (real-time) computation. Operators are limited[2] i.e. by the balance between costs for setting up computation "cloudlets" and the required latency. If we see it as a black/white picture, a mobile operator could equip all base station with a cloudlet, which would result in a major invest for edge infrastructure as there are e.g. in Germany some 10,000 base stations, resulting in the same number of cloudlets. At the other extreme, the operator would setup only one or two edge cloudlets for the whole country, to ensure a specified latency, that will not significantly outperform the latency of the edge deployments of service providers in the public internet. The truth is somewhere in between and it depends on the requirement for (ultra-) low latency as well as the willingness to pay for ultra-low latency. To achieve a latency below 30 ms in a country from the size and geographical environment like Germany an operator would need around five cloudlets (dependent also on network topology), which have to be distributed in a smart way, having nearly the same distance to every point in the country.

2.1.3   Device Edge Computing aka Fog Computing

Another flavor of Edge Computing would be to have the computation on the device itself, while integrated the device (virtually) into the network and make it kind of a network node. In times of smartphones with powerful Gigahertz-clocked eight or ten core processors it seems to be a good idea to use these resources as efficient as

---

[2]They are not only limited by these factors, but also by technical and process restrictions, as they are not starting with a green field approach.

possible. But there are some impacts from the model of Edge Computing: even if mobile devices are really powerful, they are still limited in battery capacity: If such device capabilities will be used at their maximum level, the average smartphone has to be recharged within 4 h or less (assuming a battery capacity of 3500 mAh with system-on-chip Snapdragon 835).

But there could be also security impacts, as the security level of average smartphones is only a valid one until the next known security breach. The interesting on this approach is the fact, that it can be implemented without new hardware by just adding software to the incorporated devices.

2.2   General Trends in Edge Computing

Edge computing has already done a remarkable way forward: some years ago, nobody knew about this concept. Meanwhile every network and IT infrastructure vendor is offering equipment for Edge computing... but every vendor has a (slightly) different interpretation of the Edge topic.

But why Edge? It worked also before? Yes, on the first view this opinion seems to be valid, but due to the Internet of Things and its myriad of connected devices we have to expect the unexpected, meaning use cases we haven't thought of before. As Thomas J. Watson said in 1946 "There is only a demand for three computers in the world", now we have more computing power with a smartphone in our palm than one of these three computers with a size of a big room had.

As Edge Computing itself is already a trend, it is remarkable how the different players on the market are interpreting this trend and trying to drive it forward.

Most of the telco equipment providers are offering edge routers which are integrated in their strategy for 5G with SDN and NFV equipment.

The OEC (Open Edge Community) as one open initiative to drive the topic, is looking to EDGE Computing in a very pragmatic way: they see two megatrends today, mobile computing and cloud computing where Edge Computing is the missing link between these megatrends to speed up the latency as one of the last barriers for deploying real time use cases in the cloud.

But many equipment vendors are seeing Edge Computing only as part and an enhancement of their Content Delivery Network Strategy e.g. to optimize the picture quality according the needs of the demanding device for a video stream.

2.3   Specific Use Cases: Is Edge Really Needed?

If we talk about Edge Computing, we talk always about changes in the network, may it be in the public internet or in the operator's network. This will need some investments for operators they want to get back. This is leading to the question who will pay for the decreased latency which is only dependent on the use cases!

2.3.1   Use Case: Augmented Reality Maintenance

Augmented Reality is a use case where the reality will be extended by additional information, e.g. the front window of a car will be used to visualize the navigation information as an overlay to the reality, so no additional screen will be used for the navigation system as it is realized today.

In an Augmented Reality Maintenance use case the maintenance of a machine, robot or tool can be done remotely by an expert but needs to be mechanically executed on-site. One not necessarily experienced worker wears glasses which enrich the real picture with additional information. This augmented reality is sent to the displays of the expert who will guide the local colleague to maintain or repair the machine in the right way.

To achieve this remotely with a valid user experience the picture of the glasses has to be enriched in real time as well as the augmented picture has to be displayed in real time to the expert. This cannot be done with the usual deployment of the AR Maintenance solution in the cloud, as such a cloud deployment will provide a high latency and/or remarkable difference latency (no guaranteed maximum latency). With an Edge-based solution the usual latency would be decreased below the time of the human sense of reaction. But the Edge based solution could not only speed up the latency in general, due to the minimization of network hops and physical distance, the standard deviation from the lower average latency will also be minimized.

With this enhancement in latency from the deployment of the Edge concept, the user experience for the AR Maintenance will be increased significantly, changing the acceptance of users of the use case from "hardly usable" to "must have".

2.3.2   Use Case: Autonomous Cars with Artificial Intelligence

Autonomous Cars are advancing more and more: from supporting systems for the driver over semi-autonomous systems (your car is parking autonomously) to autonomous systems, driving your car without any support by a human driver. The last ones are not fully functional yet, in the moment the driver must be able to take over the control of the cars (e.g. by taking the steering wheel) in a few seconds. One major problem for autonomous cars is, that an implementation of a fixed algorithm, running locally on the IT systems of the car, can never cover all situations that can happen in today's traffic. A solution to this challenge could be an autonomous car, which is learning from new situations to react correctly. This would be an autonomous car with artificial intelligence. But where comes Edge computing into the game? If a car learns a new rule to react correctly to a new situation according to the local circumstances, the other cars which are physically close to this car should also get this information as fast as possible, as they could pass the same area a few seconds later. This transmission of the learning in real-time will not work with standard cloud implementations as the latency will be too high for cars directly following or in very close distance. The solution here would be again an Edge-based solution where the cars in close distance will get the information in real-time. In this case, Edge will not only enhance the user experience of this use case but will enable the use case as such, as the latencies with a usual cloud based implementation will not be sufficient.

2.4   Opportunity for the Operator?

The Edge concept has a clear opportunity for the operators, especially the operators edge approach, where the Edge cloudlet will move into the operators' premises. This will be an approach with advantages for both: the Edge cloudlet owner, in this case the operator and also for the user of the solution especially in a mobile use case, where the device, running the solution, is a mobile device with limited battery capabilities.

The Operators Edge is advanced over the Edge in the public internet, as the operators own the access and aggregation network and can bring the Edge cloudlet closer to the user which can make a difference in latency of 10 ms and more.

But, for a mobile use case the Operators' edge is also advanced over the device Edge, as the device Edge will use the computational resources of the device to speed up the latency for the computation. While the latency of such a (local) approach couldn't be better, the device Edge approach consumes a lot of energy and will eat up the battery of the mobile device within a few hours.

So, the Operators' Edge will bring the best of the two other approaches together giving the operators a big opportunity to get a foot in the door of low latency applications and solutions, which will be often from the Internet of Things domain. The question is only; will the operators take the chance this time (they "only" have to do it in the right way) or will they screw it up as they did many times before?

3   Edge Computing Delivered by Telcos: An Opportunity or Next Failure?

The rise—some people may call it "hype"—of edge computing opened for telcos a window of opportunity to transition from a data transfer service provider (aka "bit pipe") into a data transformation service provider. Many attempts have been done in the past to do so and many cash was burned to climb up the value chain—without any notable success.

Telco edge computing essentially combines two things: network control and compute capacities. The battle is on whether operators are able to add competitive compute capabilities and the corresponding interfaces to their network. If they fail again, e.g. by missing the window of opportunity, the established cloud players will be able to find a way to control the network in addition to the existing compute infrastructure they already have, e.g. by simply partnering with network operators and co-locating their infrastructure at the relevant locations within the network. As the cloud players are able to move faster thanks to a more agile and flexible mindset and financial assets this window for telco operators will be closing very quickly. If they don't act now they'll end up again as pure connectivity providers and chances are high that this may be the last big battle. It seems, not many operators already understood this opportunity well because it needs a bold decision, as deploying edge within a telco network is much more than just another telco service. It requires a change of mindset and attitude in many aspects.

3.1   Why Telcos Are Destined to Deploy an Edge Computing Infrastructure...

3.1.1   Physical Locations

Edge computing requires the deployment of a distributed compute infrastructure. Regardless of how far you distribute it (e.g. whether you go to the access nodes or to some aggregation sites) it first requires physical space which needs to be equipped with network access, power, climate control and security. As networks are built upon such locations, network operators do have access to that—either they still own it or, if they sold it, at least they have a long term contract for accessing those locations. This asset cannot be overestimated, as it is a big challenge to get ahold of such an infrastructure easily. It is one advantage telco operators have over global cloud players as they cannot expand so easily into this as it requires long negotiations with municipalities, building owners etc. When Amazon acquired Whole Foods in mid-2017 there was even some speculation that the actual purpose was to get those physical locations to run an edge computing infrastructure (Does Amazon's purchase of Whole Foods redefine "Edge Computing"?).

3.1.2   Network

Depending on the application using edge computing different kinds of connectivity from the application/device to the edge compute node are possible, ranging from simple Ethernet, WiFi or other private networks to mobile networks in licensed spectrum (and all flavors in between such as campus networks when you consider industrial edge computing use cases). For operator edge services, naturally the connectivity would be provided through the existing network of that operator.

3.1.3   Local Breakout

Network control is key for edge computing, as you need to ensure the local breakout of the relevant traffic to the edge compute nodes. Cloud service providers that do not have this control can only come as close to the user device as the internet exchange point (and still do have no control how the traffic is routed from the device to the internet exchange point and back). This is the reason why you typically find the cloud providers in the same physical locations as the big internet exchange points, e.g. for Germany in Frankfurt.

3.1.4   Quality of Service

Current networks are built and services sold following the best effort approach. There are many reasons for that but in fact, quality of service has not been successfully introduced into (mobile) networks (despite some very basic QoS such as prioritization of IMS/VoLTE traffic over normal data). While in principle edge computing service can be delivered also over best effort networks at least initially combining it with a specific QoS may be required at least for a range of use cases.

3.1.5   Network Optimization

Considering the closed-loop latency advantage telco edge computing provides over central cloud architectures the access network aspect plays an important role here. Consider that in LTE just the access latency is in the order of 10–20 ms. Optimizations here can be (selectively) deployed by the network operators only. Also specific improvements e.g. in the area of TTI may be required.

3.1.6   Network Virtualization

All telcos are moving towards SDN/NFV based networks. At least they claim to do so, the reality seems to be much behind the aspiration and claims. Nevertheless, telcos already spend a lot of thought (and invest) into it, some are already advanced, some are less. Telco edge computing fits conceptually well into the network virtualization concept (but requires some additional aspects to be considered as outlined below). In other words, telcos that do already have a well-defined/deployed network virtualization strategy can add edge computing into this strategy comfortably.

3.1.7   5G

Last but not least: 5G. It's not there yet and it's not clear how and when it will be there. It seems to become more and more obvious that some parts (like MIMO) will come much earlier than initially anticipated and some others may take much more time, so we will likely see a staggered rollout of 5G technologies. The point is: 5G requires some sort of edge computing to deliver up to its promised delivery of 1 ms latency, reason being the speed of light limit. In fiber the light signal takes roughly 1 ms for every 200 km. That means in order to consume let's say only 0.5 ms for the signal propagation delay end-to-end the distance between device and compute can be at maximum 50 km.

3.2   … and Why Telcos Should Not Even Think About Edge Computing

Telco operators have many crucial assets to deploy an edge computing service. However, the truth is to build a competitive service they need to be on par with existing cloud giants in any aspect to leverage their advantages.

3.2.1   Software R&D

Delivering a compute service means you need to understand and be able to manage the underlying stack and offer compelling services enriching that service. All that is pure software. Telco operators have no sufficient skills and mindset to understand software development as they are coming from a model where boxes that were built by vendors according to very detailed standards were deployed and operated by them. Actually many operators do face that reality in the current attempt to cloudify their networks. Unless telcos are able to transform into a software company edge computing (and many other things) won't be possible.

3.2.2 Developers, Developers, Developers!

In the end applications have to make use of edge computing services and applications are built by developers. Have you ever observed telco operators talking to developers? This rarely happens because even if a telco has offered some API they tried to force developers first to sign lengthy T&Cs before even talking. Another structural problem is the small footprint of operators being able to address only very limited geographical regions. The challenge is here how to address a global developer community respectively developers who would like to deploy their application on a global scale? Some may say standardization is the answer but again: operator standardization and the world of application development are running on two different time scales, which do not match.

3.2.3 Infrastructure Deployment and Upgrade Cycles at Cost and Speed

The telco edge infrastructure needs to be on par with the central cloud in order to leverage the advantage telcos have. That means they need to provide top-notch compute service as well as GPUs for graphic analysis and AI. Moreover, as (especially in the field of GPUs) the technology evolution is still going on at high speed the infrastructure needs to be constantly updated which not only requires very quick monetization of the infrastructure but also bears operational challenges new to telco operators.

3.2.4 Security

Can you imagine an operator to open his infrastructure to run arbitrary software on it, i.e. next to the telco core software? Every operator is scared about that and obviously this is a serious challenge. If this implies to build a completely separated infrastructure for edge computing only then telco operators would throw away deliberately a lot of the assets they have in their hands—not many would be surprised however if that would happen.

3.3   ... and How Telcos Could Do It: Conclusion

3.3.1   (Never) Mind the Gap!

Telcos need to be brave and focused to take the unique opportunity at hand. The challenges shown above are also opportunities to transform into future-ready modern software-driven agile organizations. This needs top-level management commitment but more importantly the understanding of the whole organization, of every single employee that the future of the traditional telco is doomed.

3.3.2   Forget About "Carrier-Grade" Quality

The terminology "carrier-grade" quality is the "made in Germany" quality seal proudly used by operators to describe a level of quality they think they are delivering. In fact, it refers to a five nine availability and fast fault recovery however this relates only little to the actual customer experience of the provided services. It's more (much, much more) important to continuously improve the services. That means also to lower the hurdles implied by standardization. This can be achieved by taking ownership (real ownership, not expressed by them means of things like vendor selection) of the software stack, either internally or in open source communities.

3.3.3 Don't Be Afraid! Go with the Flow

While security by design is important as in every other area, telco operators should not be afraid of edge computing as a security risk or risk of letting Trojan horses into their network. Security needs to be taken seriously, but if that is done, there is no problem to run the code for some edge computing service on the same physical CPU as your vEPC.

3.3.4 Edge Computing MVNO

The biggest obstacle to deploy edge computing within telcos is the state of mind. Unless the full organization is ready to move it won't be successful. It may take a long time to get an organization in this direction, so there are other ways to bridge that gap. Think for instance of an MVNO-like approach which—free from any heritage and organizational burdens—is sourcing the necessary network and compute services and selling this in an aggregated way to the application developers or consumers.

References

Does Amazon's purchase of Whole Foods redefine "Edge Computing"? – Dean Bubley's Disruptive Wireless: Thought-leading wireless industry analysis, http://disruptivewireless.blogspot.de/2017/06/does-amazons-purchase-of-whole-foods.html

Ha, K., et al. (2013). The impact of mobile multimedia applications on data center consolidation. In *Proceedings of IEEE International Conference on Cloud Engineering 2013 (IC2E 13)* (pp. 166–176).

Mahadev Satyanarayanan, Carnegie Mellon University 2009 first definition of cloudlet

The Emergence of Edge Computing – Mahadev Satyanarayanan, Carnegie Mellon University https://www.cs.cmu.edu/~satya/docdir/satya-edge2016.pdf

Overcoming VoLTE: Deployment Challenges

Thorsten Lotz and Krzysztof Korzunowicz

1 Business Drivers for VoLTE

With the emergence of mobile broadband and smartphones, subscribers increasingly started to use OTT applications like WhatsApp, LINE, Viber, etc. instead of services offered by MNOs. Revenue from mobile voice and messaging decline steadily, forcing MNOs to focus their investment on their network where growth is possible—mobile broadband—even to a greater extent.

To maximize revenues from their most precious asset—the radio frequency spectrum—to its fullest potential, MNOs are investing more efficient radio access technologies like LTE and 5G. As an undesired consequence, the rapid rollout of LTE networks and its increasing coverage have accelerated the erosion of revenues for mobile voice. Customers of OTT providers stand to benefit from the improved speed and low latency in LTE, bringing the experience in voice closer to what MNOs provide. In effect, MNOs are on the brink of becoming data pipe providers for third parties, who run their services on top of their network, without incurring any costs. Whereas they must maintain two expensive sets of infrastructures, one mainly for voice and the other for data, while complying with regulatory obligations of the country they reside in.

The major reason for MNOs to invest in VoLTE is the need to decommission the legacy 2/3G circuit switched domain to enable a full spectrum refarming to LTE and 5G for broadband data usage. On the feature level VoLTE services are similar to those in 2/3G and do not compete with OTT, but their goal is to leverage the biggest

T. Lotz (✉)
Detecon Asia Pacific Ltd., Bangkok, Thailand
e-mail: Thorsten.Lotz@detecon.com

K. Korzunowicz
Detecon International GmbH, Cologne, Germany
e-mail: Krzysztof.Korzunowicz@detecon.com

© Springer International Publishing AG, part of Springer Nature 2019
P. Krüssel (ed.), *Future Telco*, Management for Professionals,
https://doi.org/10.1007/978-3-319-77724-5_12

advantage the MNOs have over OTTs—seamless user experience between all access networks without disruption even in case of network congestion.

Central element of VoLTE is the IP Multimedia Subsystem (IMS), standardized by 3GPP. The IMS core together with Application Servers (AS) enable the service execution for native voice and messaging services in packet switched networks.

Originally IMS was designed access agnostic and it had a natural place in the core of the network as the voice production platform for multiple services, irrespective of access technology. Enhancements to enforce quality of service in mobile packet switched networks and interworking towards legacy 2/3G networks were developed much later. In terms of Fixed Mobile Convergence (FMC) strategy, IMS enables operators to converge LTE and Fixed VoIP offerings into the same core, causing substantial savings on operations. Beyond this, IMS allows MNOs to leverage infrastructure beyond their premises by enabling voice and text services via any Wi-Fi access. This functionality is called Voice over Wi-Fi (VoWi-Fi). Integration between MNOs cellular voice service and Wi-Fi calling is an emerging trend in the market as technology entrée barriers are low and user acceptance is high. With VoWi-Fi MNOs need to deploy less small cells to provide their services in buildings with poor coverage, as customers take on the task to build their own indoor networks.

2   Challenges in the Deployment of VoLTE

The deployment of VoLTE is most challenging task since the introduction of GSM 25 years ago. Complexity and technical barriers are high; implementation efforts, timeline and costs are often underestimated.

Of particular importance for VoLTE is the capability to allow voice call continuity, even when leaving LTE coverage and handover to 2/3G is required. This feature is called Single Radio Voice Call Continuity (SR-VCC). Though this feature sounds very useful it has several downsides:

- Complexity related to integration effort towards legacy 2/3G networks.
- Costs and time to implement both on network, as well as IT side, as it requires correlation of charging data records in VoLTE and 2/3G.
- User experience is in reality only "almost" seamless, as there are speech disruptions related to access technology swap every time when used.

Providing seamless handover between all networks is one of the most technically ambitious tasks. Especially in areas with poor LTE coverage, many inter-radio handovers are expected to occur. In the early deployment phase of VoLTE, users are likely to experience higher call drop rates and call setup failures and will see this as a degradation compared to service quality in 2/3G networks. Reaching maturity of SR-VCC takes time and requires additional effort. However, costs and effort spend in SR-VCC will become void, once LTE achieves the same or a bigger footprint as 2/3G.

For the same user experience as in legacy networks, operators tend to deploy the same fully blown feature set, e.g. ring back tone, call completion and multi SIM. This is referred to as "2/3G feature parity". Technical implementation challenges are not considered and often no calculation on the business case to do it is made—as features are dictated by marketing. This requires a complex and a costly integration of Intelligent Networks (IN), which is the service execution area in 2/3G. What is often forgotten is that once the LTE-only status of an MNO is achieved, a transfer of the service execution to the AS becomes the only sensible option for having one service delivery platform. The deployment of VoLTE should be seen as an opportunity to retire non-profitable and no longer needed features. Similar features like in 2/3G can easily be setup on AS layer, setting these up here right in the beginning reduces the effort once a full transition from IN to AS is required.

Other aspects impacting the costs and timeline for a VoLTE deployment are maturity of both the IMS solution and the handsets, as well as virtualization and interworking between IMS and AS components. In the past vendors have developed their solutions at the MNOs premises transferring a lot of the costs related to testing on the MNOs. Interworking between IMS components of different vendors has proven to be a challenge, as 3GPP standards have often been interpreted differently, causing components not to work with each other at initial phases of the deployment. High amount of testing and additional effort was required by the MNOs, who considered to deploy VoLTE early, causing delays until commercial launch of several years and additional costs.

MNOs taking on late the challenges that arise with VoLTE should leverage from lessons learned by choosing experienced vendors. Especially prior to selecting vendors for IMS and AS interoperability tests should be performed. Also taking the same vendor for the new VoLTE components has proven to benefit the timeline and costs of the deployment.

3   Reasons Against a Quick Deployment of VoLTE

In mature markets, large MNOs have deployed VoLTE trying to prove their technology leadership. The truth is that their customers have not even noticed, as their experience is at launch at best the same as in legacy, but it can be worse. A similar statement could be "others have it" that is why "I need it too".

The way many MNOs are often marketing the fact that they deployed VoLTE is amusing at best: typically a short statement in a local newspaper and a bit longer article in specialized press. The first one completely ignored due to almost identical user experience and the second reaching only a very limited group. The total outcome is having bragging rights on telco conventions. Hence, technology leadership nor "others have it and our customers will run away" should not be the reason behind VoLTE investment.

Beside a very complex deployment of VoLTE with interworking towards 2/3G networks, there are other options for mobile voice services in the future:

- Wait for LTE-2/3G coverage parity, turn on VoLTE and start planning decommissioning of 2/3G networks. This avoids a very costly IN and SR-VCC integration between VoLTE and 2/3G.
- Deciding never to upgrade to VoLTE, if revenues of voice are dropping very quickly. Let them die with 3G or partner with OTT providers to cover the left-overs of voice services.

When the above simplest solutions are not fit for the MNO, then several other aspects should be considered when making the decision. The right time to make the transition strikes a point of balance between:

- The costs of VoLTE, especially SR-VCC and IN integrations and changes in those costs related to maturity of VoLTE solutions,
- Need for spectrum refarming for broadband data usage,
- 2/3G end of life dates and possible savings on maintenance costs and
- Savings on having a single voice production for fixed and mobile, if applicable.

MNOs that decide to launch VoLTE late, benefit from innovations over the last years. Nowadays, a great number of VoLTE handsets is available from multiple vendors, with prices steadily declining, lowering the entry barriers especially in emerging markets, where device costs are of major concern. Out-of-the-box VoLTE solutions composing IMS core and AS will become available to greater extent, reducing the number of interfaces to be integrated.

4   Reasons for a Quick Deployment of VoLTE

Brownfield operators should notice that there is only one set of circumstances, which justifies a quick rollout of VoLTE, i.e. access networks are already or close to congestion and a relevant part of the spectrum is blocked by voice on circuit switched technology. Spectrum refarming to LTE and 5G is required to increase the service offering of mobile broadband. It is likely that such an operator could create a strong positive business case on VoLTE today.

For Greenfield MNOs the situation is different. This is because almost every smartphone is LTE-enabled, making it easy to embrace a large number of customers. Unfortunately, this does not imply VoLTE handset compatibility and subscribers will use OTTs for voice in data only networks. A deployment of VoLTE should therefore be considered in relation to the availability and costs for VoLTE handsets. Also waiting too long may also make it difficult of taking services back from OTTs.

5   Overcoming the Complexity of VoLTE by an Agile Like Deployment Approach

The introduction of VoLTE significantly differs from any deployment of fixed and mobile services and technologies in the past. For technologies like public switched telephone network (PSTN), 2G and even 3G the implementation of new features for voice services was limited by the speed of their development by standardization bodies, which was rather slow. Additionally, the nature of changes in technology, on the example of 2G to 3G, was only enhancing the previous solutions. Thus, the complexity of building the 3G voice solution with value added services (VAS), as we know it today, was spread over many innovation cycles and years.

A waterfall based VoLTE deployment assuming 2/3G parity and seamless hand-over between packet switched LTE and circuit switched 2/3G has proven to be very complex, as it requires creation of a new voice service delivery platform and its integration into many legacy systems. In effect, proceeding with a complete requirements-to-design mapping done in a single step, has turned out to be impossible. In previous technologies, changes were incremental and operators had time to learn how enhancements were to be integrated, whereas the knowledge on how to integrate the entire VoLTE solution needs to obtained first. Unfortunately, this is not something a person fully comprehends after reading design guidelines and standards. It takes months of trial and error to understand all its aspects in conjunction with its interworking towards legacy systems.

To streamline the introduction of VoLTE and to allow learning during the project, an agile like deployment approach of VoLTE is of advantage, which divides the complex task of deploying VoLTE into several steps. By applying this approach, the MNO only needs to have a basic understanding of his VoLTE scope and target architecture at the beginning of the project, while only in the steps, detailed requirements are defined and in a design, integration and testing phase the solution is build. Each step builds on top of previous steps concluding with a friendly user trial in production and a go to market decision. With this iterative approach, learning becomes an integral part of the process allowing constant improvement and hand-over of responsibilities from design, through build to run. The following steps are big enough to make them sub-projects but small enough to be fully understood during requirements-to-design mapping, resulting in a solid project and resource planning:

- The first step will enable basic VoLTE to VoLTE calling functionality.
- The second step will provide connectivity to other networks, i.e. to allow VoLTE call to and VoLTE to be called from the operator's 2/3G network, PSTN and fixed IMS as well as other operators.
- The third step will enable SR-VCC for seamless handover during voice calls from LTE to 2/3G, when moving out of LTE coverage. (This step can be omitted in case of LTE-2/3G network coverage parity.)
- The fourth step will enrich the user experience by enabling features known from 2/3G like conference, call waiting and introduce billing.

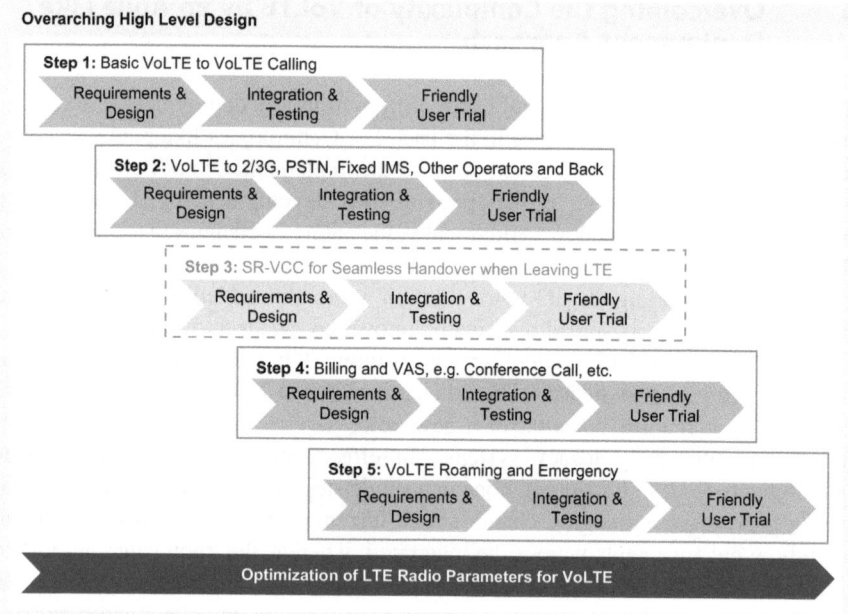

Fig. 1 Agile-like VoLTE deployment approach, comprising of five steps

- The fifth step focuses on the deployment of latest features standardized in 3GPP, which are VoLTE roaming and VoLTE emergency functionality with location information.

Due to fact that all steps conclude with a friendly user trial in production, radio fine-tuning can be started early in the deployment. This has shown to benefit the overall deployment timeline in reaching 2/3G-like quality of VoLTE services, whereas in a one shot scenario it requires additional effort and time that is seldom considered.

The figure shows the described agile like VoLTE deployment approach (Fig. 1).

6   How Does Virtualization Fit to the Picture of VoLTE?

IMS is one of the main systems impacted by the idea of Network Function Virtualization (NFV) and both IMS and AS can be introduced as virtualized (today) or containerized (soon) solutions. This enables MNOs to benefit from:

- Using commercial off-the-shelf hardware and simple expansions,
- Reduced time to market, by having less specialized elements of the network, making the purchasing process simpler and less risky

- Simplified operation, by automated re-configuration and moving network workloads onto spare capacity and
- Optimizing the network configuration and/or topology in near real time.

In a virtualized setup, the entire VoLTE solution may be cheaper when considered individually, especially in a small-scale deployment. The reason being that in a joint hardware and software based deployment, the capacity could be multiple times too big for the needs of a small MNO, while a big chunk of the cost will be this dedicated hardware. On the other hand, the price of a license based only virtualized solution will scale according to usage. Additionally, savings on having multiple test and production systems on a single platform can be substantial, if done right.

All the above is theory. Practice has shown that virtualized software also brings challenges, that are not obvious to an MNO unexperienced with cloudified and purely software based solutions.

The biggest issue is the hardware platform itself. Even though the off-the-shelf sounds easy and cheap, the reality could not be farther from the truth. There are multiple companies in the world that spent a very considerable effort into building their own "telco clouds". Very few were as successful as they expected to be. Most were not able to do it by themselves and ended up spending considerable amounts on support of external companies and training of their own people. Which is not a problem when virtualization becomes a strategic initiative, but becomes an issue when it is only planned to be done for IMS and AS.

Another issue that is critical for success of virtualization, is that the software that the cloud must carry cannot be written like it was for huge hardware equipment. Especially redundancy and deployment mechanisms must be completely software based, which changes how many things will be measured and executed. That work needs to be done both in the MNO as well as on vendor sides.

6.1   Utilizing Public Clouds for Telco Services

One could say that the MNO should outsource the cloud problem completely and use what is available on the market. There are many public cloud providers and some of them even able to introduce what was considered "telco specific"—hardware acceleration. The idea seems good, unfortunately there are many limitations. The availability of the public cloud is not the same across the world, also very often particularities of telecommunications law might be a blocker of using the public cloud. Still, the public cloud option should at least be considered if the limitations are not present and capabilities, skills and approaches should be understood and sensibly copied from them as IT has gone through this cloudification process a while back already and it is easier to find people with related skillset.

6.2   Building a Telco Cloud

There is of course the possibility that the telco creates its own cloud, for its internal purposes. Although that is costly, for multiple reasons. First one is the inherent daily variation of usage, that makes the cloud as big as the peak usage, while it is used on average a lot less. Which means, that the economy of scale is less pervasive but there are other "costs".

One of the key learnings of big strategic virtualization initiatives is that it is not enough to just "get a couple of programmers". It requires competence in areas of: data center networking, virtualization/containerization, continuous integration/delivery/deployment, very fine-grained monitoring of everything and much stricter software-like requirements management enabling automated testing. Those skills are not easy to find and build-up. And that is just the technology side. The topic requires a fundamental shift of the company values, methods and organization. Especially, telco management needs to understand that the divide and control methodology that could be easily applied to standardized networks is just not effective with clouds. It is critical to think end-to-end. That is indeed a learning to be made, and a big obstacle at the beginning.

6.3   Virtualized IMS: Cheaper or ...?

The goal of having a "cheaper" IMS will not be easy to achieve based on the cloud. The general return on investment is big, but not when considering it as an IMS only initiative. Operations and testing automation that can be done will have a huge impact on the cost side, but only when spread out across all of the services and not without a huge investment at the beginning of the cloud journey.

What is even more business impactful is that it is possible to have and IMS that is built better, more customer oriented, with a friendlier payment model, and easier to manage, maintain and change. That can be achieved by giving chances to new entrants into the NFV market, allowing to create some ferment in the rather stable telecom vendor industry. Cloudification will require a lot of work both on the MNO as well as those new suppliers, but it will be done as software development, where the needed investment is not as big as it used to be, when telco equipment was uniquely huge hardware pieces with complex tightly coupled software on top. Production of such complex products required a huge risk to be taken and that set the pricing levels. Thus, one may say that the general goal of the IMS in the cloud is to make risk for every participant of the market smaller.

7   Conclusion

This article presented mistakes and challenges during the deployment of VoLTE in the past. As one of the most important lessons learned it was highlighted, that technology leadership should not be the driver to invest in VoLTE. The deployment of VoLTE should only be driven by the need of spectrum refarming for LTE and 5G and the need to decommission 2/3G networks. If revenues of voice are dropping very quickly, it can also be a rational to never invest in VoLTE and become a pure broadband data provider. MNOs deciding to invest in VoLTE late benefit from mature technology, greater interoperability, lower costs for VoLTE handsets and more experienced vendors. It was pointed out, that reaching 2/3G-LTE coverage parity prior to deploying VoLTE benefits the business case, as no spending in a complex and interim solution for seamless handover from LTE to 2/3G (SR-VCC) is required. Finally, this article concluded with a new, agile like VoLTE deployment approach. With this approach, learning becomes an integral part of the deployment by delivering little by little of the entire VoLTE solution and streamlines the timeline as cumbersome radio optimization can be started early. The article also showed consideration points for virtualization, that should allow for a plan to be made related to IMS virtualization as part of a fundamental, value-increasing transformation to a cloud based provider.

Part III

Innovation

Successful Navigation Through Digital Transformation Using an Innovation Radar

Jan Heinemann, Oliver Platzen, and Carolina Schiefer

1 Digital Disruption and Resulting Challenges for Telecom Operators

As fast-moving technological progress is affecting everyday life of businesses and their customers, it bears chances and risks that drive digital change. Everything is enabled by technology—rapid advancements heavily fuel the pace of change in customer demands, market roles and value creation. Digital change drives new paradigms in product and business model innovation across industries. During the last years, as indicated in the Fig. 1, companies in most industries moved from product- to product- and service-based business models or were disrupted by new market entrants focusing on the service-component (debundling of traditional value chains).

Accordingly, the telecom industry continues to be shaped by advances in technology which foster new applications and digital services such as artificial intelligence (AI), microservices or augmented reality (AR), new devices but also network improvements with regard to 5G or edge computing. While technological improvements inherit new business potential for operators, they also force them to transform the IT and network landscape. Simultaneously, telecom operators are facing one of the most challenging times in history as revenues for traditional services are dwindling and new competitors are continuously entering the market.

Not only will over-the-top (OTT) service providers continue to steal valuable revenue from telecom operators, indicated by the decline in Average Revenue Per

J. Heinemann · C. Schiefer (✉)
Detecon (Schweiz) AG, Zurich, Switzerland
e-mail: jan.heinemann@detecon.com; carolina.schiefer@detecon.com

O. Platzen
Detecon Inc. (USA), San Francisco, CA, USA
e-mail: oliver.platzen@detecon.com

© Springer International Publishing AG, part of Springer Nature 2019
P. Krüssel (ed.), *Future Telco*, Management for Professionals,
https://doi.org/10.1007/978-3-319-77724-5_13

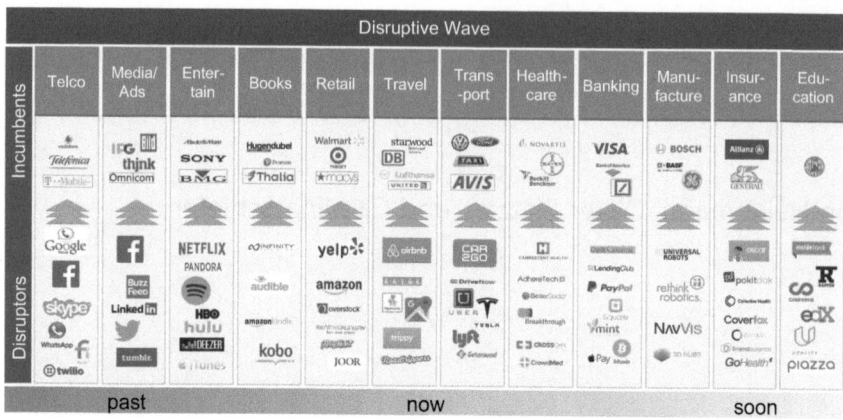

Fig. 1 Disruptive wave of digital transformation for industries

User (ARPU) for mobile services of Western Europe telecom operators from 28.93% in 2007 to 17.94% in 2016 (Analysys Mason Limited 2017). But also consumer habits and requirements are rapidly changing. This is supported by the expected growth of mobile data traffic to 49 exabytes per month by 2021, a sevenfold increase over 2016 (Cisco 2016). Furthermore, telecom operators are under pressure to develop platforms that support high levels of connectivity and enable Internet of Things (IoT) solutions as well as 5G networks ultimately facilitating IoT—which are both also targeted by OTTs such as Apple or Facebook.

Ultimately, telecom operators must elaborate on new alternative revenue streams and transform their business from connectivity providers towards data centric operators, offering additional content heavy services like machine-to-machine (M2M), health or media. This implies that the telecom industry's revenue no longer will grow by offering fixed or mobile connectivity, but new products and services based on data traffic and value-added content services for both, B2B and B2C customers, will fuel the future. The key for telecom operators is thus collaboration, as the past has shown, that they often do not have the assets and capabilities to successfully provide data centric offerings alone.

2   Technology and Startup Radar: From Technology Intelligence to Strategic Actions

These fast-moving dynamics lead to complexity and uncertainty about the future and force all industries to constantly innovate and adapt to their environment for success and survival. This implies a fundamental change of a telecom operator's positioning vis-à-vis OTT players and new customer requirements. This especially refers to a transformation into software oriented and agile organizations with continuous

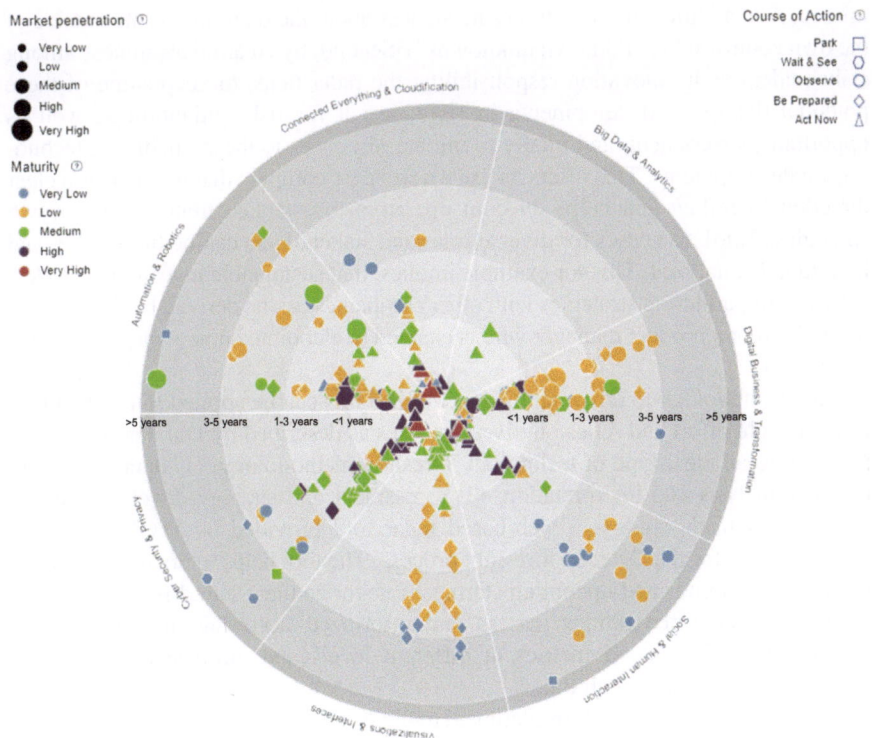

Fig. 2 Exemplary radar layout

deployment and customer proposition along the value chain of web-scales[1] based on the implementation of DevOps teams with end-to-end responsibility, and an autonomous standing within the organization with highly automated processes supported by artificial intelligence. Therefore, companies need to establish adequate capabilities for technology foresight in order to turn threats coming from such changes into opportunities and to use information advantage for sustainable business success. This requires a corporate foresight mechanism that gathers in-depth information and then enables users and stakeholders to distinguish signal and noise via intelligent clustering and rating for decision making or developing an innovation strategy.

For coordinating foresight efforts, a technology and startup radar is an ideal tool and methodology to identify, assess, and communicate technological developments and trends within an organization's environment (for exemplary radar layout see Fig. 2). First, it raises awareness about technologies early in time that are likely to be of strategic importance for organizations ensuring that market opportunities are not

---

[1]Web-scale, originally introduced by Gartner, refers to cloud services firms such as Google, Amazon, Rackspace, Netflix, Facebook, etc. (http://blogs.gartner.com/cameron_haight/2013/05/16/enter-web-scale-it/)

missed. And additionally, it enforces awareness about the ongoing events outside of the own control sphere ("known unknowns"). Second, by creating awareness among stakeholders with innovation responsibility, the radar helps to keep some of these potential disruptive developments in sight, use it for risk mitigation as well as opportunity assessment, and finally to initiate responses to these upcoming technological developments. This offers its users a strategic compass that guarantees a clear direction to follow and helps telecom operators to establish trend based growth strategies. Third, use cases for diverse roles and stakeholders can be thoroughly and individually analyzed. This for example implies, that sustainable investments, M&A activities or partnering strategies with other companies can be derived by the strategy department or a product manager who is enabled to elaborate on new inspirations for its product.

The methodology to arrive at this level of insights can be applied to any technology foresight effort and relies on diverse phases as described by Durst et al. (2015), referring to an ideal-type of technology foresight methodology: (1) data collection, (2) data analysis and interpretation, (3) assessment and strategy development. To support this methodology, a web-based radar tool provided by ITONICS is the software basis for the Detecon foresight process. The tool helps to document current technological trends and upcoming startups, categorize them into different maturity stages and relevance ratings, assess and prioritized according to market- and customer-specific defined metrics at different levels and ultimately analyze the corporate environment at all times.

Data collection starts by formulating a strategic question for determining the scope of the foresight project. This might lead to a specific use case, for which a solution needs to be elaborated, a holistic understanding of industry-specific technologies or even a concrete partnering activity for which potential startups need to be screened and assessed. Having this scope in mind, data needs to be gathered from different sources and appropriately stored. This implies a concept of fully integrated, connected and customized content plans in order to address an organizations specific content needs. Key for data is the access to a network of scouts in innovation hubs worldwide, the access to databases and the experts, who can verify and validate assumptions behind a technology's or startup's promise in order to finally decide which trends are relevant for the scoping—or not. Following the identified needs of its users, the radar is scanning for developments in seven technological areas: connected everything and cloudification, big data and analytics, automation and robotics, cyber security and privacy, visualization and interfaces, digital business and transformation, social and human interaction.

The *data analysis and interpretation* phase includes the prioritization, categorization and filtering towards a short-listing with in-depth technology profiles as well as an assessment regarding diverse tailored criteria. Using the ITONICS tool, trends and technologies are shown in an intuitive radar layout, which can be navigated quickly with the help of different filter and assessment options. Moreover, trends and startups are distinguished concerning their maturity level and market penetration and visualized accordingly. Changes and opportunities, that Detecon Research considers to be currently interesting for telecom operators or other industries and to which organizations should pay attention to and consider using in strategic decision making, thus can be thoroughly analyzed. It reflects the idiosyncratic opinion of Detecon's senior experts based on day-to-day work and broad secondary research. Thereby, the radar tool allows telecom operators to combine and store the in-depth findings from all scanning activities on a central platform and to process and evaluate them individually. Besides that, the tool can be used for a collaborative evaluation of relevant trends and technologies by different stakeholders.

Assessment and strategy development will depend on the standpoint taken by the respective stakeholder. Using the data at hand by framing the evaluated factor within a solid model of the future, will help organizations to define growth- and value-oriented strategies based on a deep understanding of future markets including customer needs, the competitor landscape, and technological changes and disruptions.

Based on the customer-specific trend, technology and startup evaluation, the technology and startup radar offers businesses a holistic overview on the organization's environment, which is updated on a regularly basis. Thus organizations are able keep track of all relevant technologies and trends and highlight new market opportunities and threats for businesses. This ultimately helps telecom operators and other industries to track technological improvements in production, or other core processes with regard to internal efficiency and to elaborate revenues streams via a new, digital product- and service portfolio for meeting new requirements coming from digital disruption.

3   Taking a Closer Look at Technology and Startup Radar Examples

The Detecon technology and startup radar offers multiple views and answers to predefined use cases that can be tailored to the specific needs and questions that the user is interested in: (a) depending on the value chain, and specific process steps in focus, and (b) depending on the specific function or department within the company.

Accordingly, the tailored radar can serve a wide range of needs on different levels depending on the desired outcome and interests in focus, e.g.:

- Perspective: internal (industry) trends versus global (socio-economic) trends
- Timing: initiation versus periodic guidance for innovation tracking
- Usage: management dashboard versus spying glass (competitive analysis)
- Cause-effect-relation: impact on innovation versus impact by innovation
- Development: M&A versus partner pool

In either way, there are two different types of an innovation radar, depending if the focus is more on technology and related trends, or rather on a startup landscape with regard to these trends and a corresponding detection of potential future market leaders.

First, the technology radar displays the latest innovative technologies including potential new services and products but also social or organizational topics. Some guiding questions are: What is happening in the market? Which technology could substitute our business model? Which innovation could help reduce costs or increase sales? Second, the startup radar displays relevant startups in the global entrepreneurial scene. Some guiding questions for this radar are: Are there potential new competitors? Which companies could we merge with or acquire? Are there companies in the market we could partner with?

3.1 Technology Intelligence: The Technology Radar

After having looked at the radar concept in general, let's take a specific example for one trend sector in the radar (see Fig. 3). AI is currently driving broad industry interest and public discussions, fueled by huge investments: according to Tracxn (2017), AI investment in 2015/2016 constituted around 25% of all trending technology investments, such as AR/VR, wearables or smart cars. A more detailed view on the trend sector AI enables a first break down of this topic, showing all the important concepts around AI: those that are relevant today, such as robotic process automation

Fig. 3 Exemplary radar view on AI (excerpt)

(RPA) and proprietary data, and also those that will become relevant in the coming years, such as context recognition or brain-to-machine interface. RPA is a very prominent current example, also sometimes described as the preface for many AI implementations, which is also discussed in more detail in chapter "Technology Transformation: Reinvent the Network and Change the Way of Working, Now!". All of this underlines one of the strengths of the radar concept: providing transparency on general and industry-specific trends and applications in a complex field such as AI to fuel the business innovation funnel.

IDC (2016) predicts that "75% of enterprises and independent software vendors will include cognitive/AI functionality in at least one application by 2018." Thus it is also important which application or technology might be relevant for individual industries or functions, such as AI-based credit score rating applications or AI in claims handling for the financial industry. For telecom operators various AI application fields need to be considered—from network and production and internal processes (e.g. advanced analytics for planning and financial control) to sales, customer service (e.g. predictive servicing, digital assistants) and products.

According to a report from Forrester (2017), 35% of companies still do not have enough information about AI to develop strategies or use cases around it. Supporting this, Gartner (2017) reports that most organizations are still in the early stage of AI adoption (59% of organizations are still gathering knowledge and investigating the technology). Trend charts in the radar offer further background on each trend for a more detailed assessment. So looking into AI, a basic definition is the 'understanding and application of human intelligence processes ("thinking and learning") by machines, especially computer systems'. But more importantly this view provides an in-depth analysis of AI. It is particularly interesting to look at the components of this trend. AI is a collection of technologies and concepts, all aiming at mimicking the human brain, i.e. cognitive capabilities such as reasoning and decision making or planning and strategizing as well as technology capabilities such as natural language processing or computer vision.

While benefits of AI are fairly easy to understand, such as enabling the analysis and recommendation of vast amounts of data that humans are not capable of (e.g. IBM Watson in healthcare), there are multiple complex challenges. These are a consequence of AI's impact on society and workforce, stemming of course from its ambition itself to replicate human intelligence. While ethical discussions dominate public discourse about the topic, there are also simple practical issues that come with data quality (such as inherent data bias) as well as that this trend is being currently driven a lot by private companies or startups, that are looking at solving specific business challenges rather than more agnostic platforms.

Figure 4 provides a detailed description and evaluation of the respective trend as well as links to related trends, studies, statistics and companies. Another crucial element for the full assessment of the trend beside the evaluation and analysis of underlying concepts and technologies will be described in the next section, where we look at relevant businesses and business opportunities that come out of this trend.

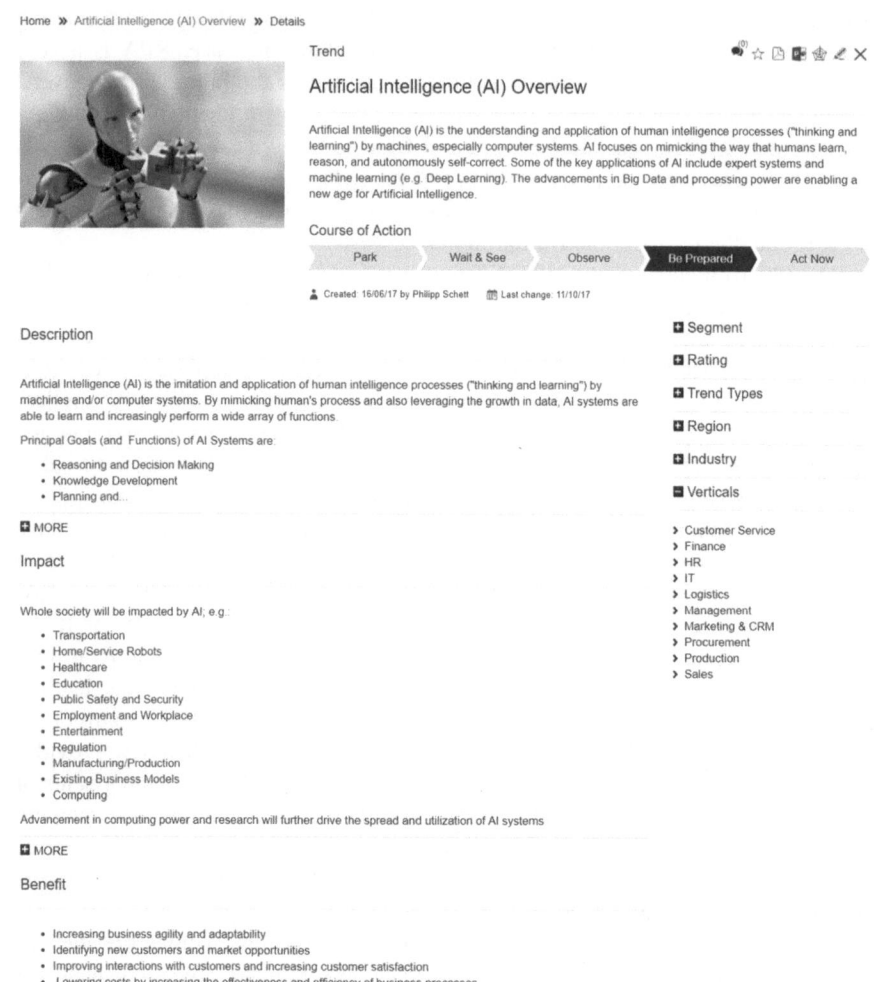

Fig. 4 Detailed view on AI

3.2   Startup Intelligence: The Opportunity Radar

During the past years, the landscape in the AI startup ecosystem has dramatically changed. The number of deals with involvement of AI startups increased by more than 400% during the last five years, from close to 160 in 2012 to ca. 700 in 2016 (CB Insight 2017). Funding during this time increased from almost USD600 million in 2012 to above USD5000 million in 2016 (CB Insight 2017).

Given this very dynamic and constantly changing startup environment, it urges the need for market participants to constantly stay ahead on recent opportunities in the AI market. Opportunities in the radar can be structured according to a wide range of market defining criteria: by industry, by segments, by topic (e.g., AI), by ten different verticals (e.g., finance or procurement), or by the course of action proposed (park, wait and see, observe, be prepared, or act now). In addition, further criteria can help to specifically narrow down the characteristic of the startups to be scoped, such as: The customer focus of the market (B2B, B2C, B2B2C, other), year of foundation, funding stage, number of employees, number of users, regions the startup is active in, number of Twitter, Facebook, or LinkedIn followers, as well as the total funding (in USD) and the home country.

Based on these different filter criteria at hand, the tool offers a wide range of options to further structure and analyze opportunities within the focused area. Hence, further prioritizations along a handful of criteria can help the user to tailor down the analysis, with the result of a unique list of opportunities according to the predefined criteria and needs.

In order to further present the analytical power that the use of the startup radar can create, three different AI related fields will be employed as an example: (1) machine learning, (2) video recognition, and (3) self-driving cars.

First, in the field of *machine learning*, two different startups are picked from the radar, as they show variations in their current market perspectives: Graphcore is a UK-based startup that was founded in 2016. The company is building a new type of processors that shall be able to accelerate the processing of machine learning applications. In contrast, ElementAI, founded 2016 in Canada, is a leading applied AI company, offering a full suite of AI-related products and service applications, helping clients to unlock the value of their data. For example, their customized AI models can help improve their clients' fraud detection, predictive maintenance services, or implement dynamic pricing models.

The difference between both companies is not only between a hardware and an AI service provider. Moreover, Graphcore is still very early in maturity with low recognition in social media, whereas ElementAI is already mature in their time-to-market and with very high media recognition. Hence, depending on the strategic focus of the user, the next strategic actions can differ significantly.

Second, Affectiva and Carifai, both focusing on *video recognition*, show a disparity between their current maturities. Affectiva, which shows a higher maturity, was founded in 2009 in the US, and develops products that detect people's emotional states via video recognition. Clarifai, a US-based startup founded in 2013, offers advanced image and video recognition technology that uses AI to summarize videos or recognize specific activities. Even though both products are already relatively evolved in their development, different levels of risk aversion might lead to different reactions when it comes to further strategic analysis.

Finally, the third example is the US startup Embark, founded in 2016. Embark develops AI and *machine learning* software that enables self-driving commercial trucks for the long haul stretches. The software itself cannot be categorized as very unique, nor is it very mature yet. The company shows currently very low media attention. However, given it is embedded in a very solid network of experts and researches and has a strong team of employees, Embark still might have the potential to become an interesting company to partner with.

These three examples offer a glance at how the radar can support its users to stay ahead in a constantly changing environment, with the help of a structured overview of relevant developments in the startup ecosystem. Further, it allows for tailored answers, fitted to the users' predefined use case or focus, e.g. in terms of topic, risk-aversion (maturity of the startup), or just in search for a hidden champion. Based on the first data analysis and interpretation, the user is well prepared for the next stage to further assess and develop a strategy.

4   What Can an Innovation Radar Do for Your Company?

When companies are facing a constantly changing and disruptive environment, as outlined above with the example of telecom operators, the radar methodology provides valuable guidance in order to lead and actively shape the innovation landscape. The radar's foresight mechanism provides transparency within the technology as well as startup ecosystem, and enables creative discourse and action about change, adaptation and mitigation at any corporate level. However, the key for a successful and impactful use of the radar is an intelligent integration into the internal innovation and partner management processes, in order to unlock the full value from identification and deduction up to strategic actions and implementations.

To demonstrate some of the characteristics and benefits of the radar with a concrete example case, the trend category AI was chosen. The example provided a deeper look into functionalities and structure of the radar, including related technological trends, as well as an introduction to specific startups in the AI ecosystem. These deep dives and selected examples demonstrated what value a radar tool and methodology can provide. A company cannot only gain insights into trend developments, but can also use the analytical base to further structure the gained insights into strategic action. Hence, and in slight adaptation to a well-known quote, it is 'not what a trend radar can do for your company, but what your company needs the trend radar to find answers for'.

References

Analysys Mason Limited. (2017). Telecoms Market Matrix – Western Europe.

CB Insight. (2017). The 2016 AI recap: Startups see record high in deals and funding. Accessed October 02, 2017, from https://www.cbinsights.com/research/artificial-intelligence-startup-funding/

Cisco. (2016). Cisco visual networking index: Global mobile data traffic forecast update, 2016–2021.

Durst, C., Durst, M., Kolonko, T., Neef, A., & Greif, F. (2015). A holistic approach to strategic foresight: A foresight support system for the German Federal Armed Forces. *Technological Forecasting and Social Change, 97,* 91–104.

Forrester Research. (2017). AI: The next Generation of Marketing.

Gartner Research. (2017). Cultivate enterprise intelligence with AI. Accessed October 19, 2017, from https://www.gartner.com/smarterwithgartner/cultivate-enterprise-intelligence-with-ai/

IDC Research. (2016). *IDC FutureScape: Worldwide analytics, cognitive/AI, and Big Data 2017 PredictionsMichael Quinn Patton (2002) qualitative research and evaluation methods* (3rd ed.). Newbury Park, CA: Sage Publications.

Tracxn Technologies. (2017). https://tracxn.com/. Startup discovery data base.

The Virtual Representation of the World is Emerging

Gregory Lukowski, Andreas Rauch, and Thomas Rosendahl

1 What Is the Virtual Representation of the World and Why Should We Care?

Within 5 years, there will be over 21 billion connected machines, cars, buildings and other forms of smart devices in the world, all developed to collect, analyze and share all kinds of data. In total, by 2020, our accumulated digital universe of data will grow to around 44 trillion gigabytes. This development is accompanied by the increasing sophistication of data processing technologies. Just think of recent advancements in AI, high performance- or quantum computing. One of the most recent examples being Google's AI algorithm beating a professional player at the Chinese board game Go, which is considered the world's most complex board game. As a result, the world is currently at the dawn of a technology-driven revolution which is fueled by a combination of the explosion in data production and data processing technologies. The combined effect of those two forces is changing the way we operate every "object" and manifests itself in the form of digital twins and eventually the emergence of the future internet.

So what are digital twins? And why are companies so excited about them? A digital twin is a virtual dynamic model which includes everything that is known about an object (see Fig. 1). In other words, digital twins are exact replica of their physical counterparts that change with the current environment in real time to help companies (and people for that matter) monitor, test, treat and maintain any number of systems. They are continuously getting richer with data by integrating real-time location data, temperature data, energy consumption and other relevant data. Digital

G. Lukowski · A. Rauch (✉)
Detecon International GmbH, Cologne, Germany
e-mail: Gregory.Lukowski@detecon.com; Andreas.Rauch@detecon.com

T. Rosendahl
Deutsche Telekom AG, Bonn, Germany
e-mail: T.Rosendahl@telekom.com

© Springer International Publishing AG, part of Springer Nature 2019
P. Krüssel (ed.), *Future Telco*, Management for Professionals,
https://doi.org/10.1007/978-3-319-77724-5_14

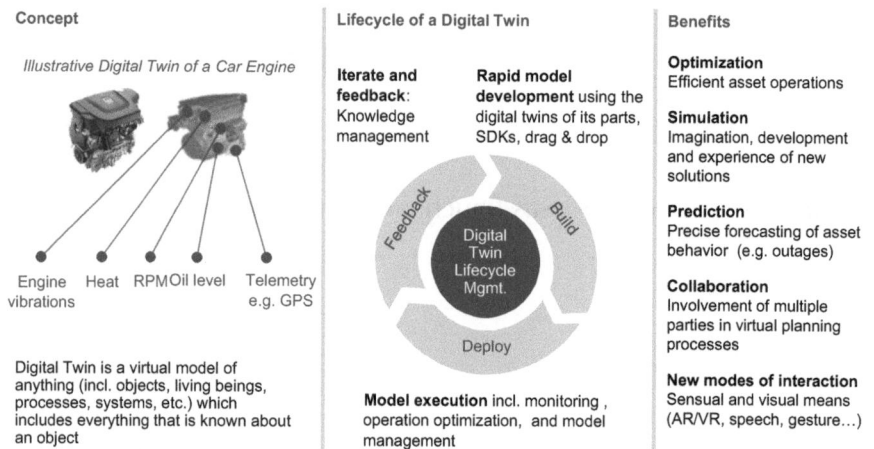

Fig. 1 What is a digital twin?

twins are emerging everywhere across every industry, in business and in consumer markets because of two factors: their ability to integrate large amounts of static, real-time, structured and unstructured data and to combine this data with advanced data processing methods such as AI, machine learning or high-performance computing. Digital twins are already adding tremendous benefits in terms of efficiency gains and innovation. For example, GE aviation is realizing cost savings of 125 million dollars through operating digital twins of jet engines alone. In a similar vein, Lufthansa Technik is working towards digitally modelling every single customer aircraft throughout the entire lifecycle as part of a digital aircraft twin. That way they can perform predictive maintenance effectively across their entire fleet. Moreover, Siemens has built a fully automated auto factory which not only produces physical cars but also their digital twins including all relevant data. These twins are continuously fed with real-time data once the car has left the factory halls reducing time to market from 30 to 16 months.

As a result of this development, the next generation of the internet, i.e. the virtual representation of the world, is slowly emerging. While the "old" communication internet revolutionized the way information is shared by connecting global computer networks, the next generation of the internet will revolutionize the use of intelligence in the form of algorithms, advanced analytics, AI and autonomous systems by connecting all physical things—i.e. objects, people and organizations—via their digital twins. In that way, the future internet is creating a continuous stream of big data to everyone and everything who is connected to it.

Importantly, in the next generation of the internet, all digital twins within the network can reason and make decisions with each other rather than operating in isolation (Rifikin 2014). Why is this interesting? Think of humans having digital twins. To understand how the body works and improve its health, your digital twin must model more than the physical parts of the body. It must model how your entire body works together. This model includes not just organs, bones and other parts but

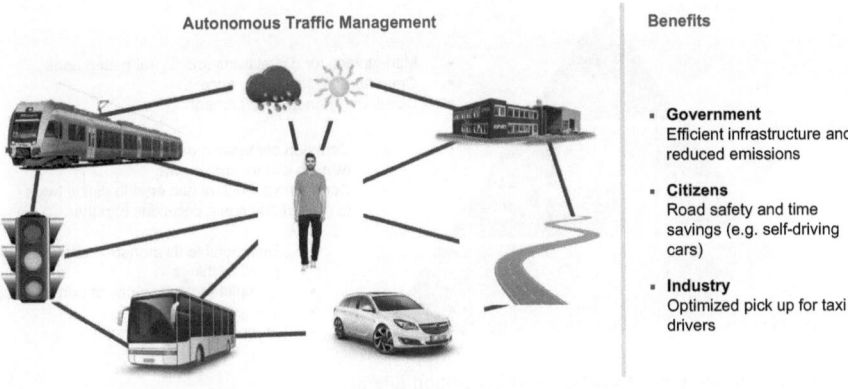

Fig. 2 The emerging virtual representation of the world—e.g.: Autonomous traffic management

also describes processes like blood flow, metabolism and the interactions among organs. Leg pain may be caused not by a leg problem but rather by a pinched nerve in the spine. A standalone model of a leg won't help you diagnose that problem. And if you only gave medication for the leg pain, without finding a treatment for the root problem, you condition would not improve.

Consequently, by enabling digital twins of different types to "speak" and make decision with each other, a whole new level of collaboration, simulation, prediction and optimization can be reached by performing simulations and predictions on the data of all interconnected digital twins. As a result, intelligent infrastructures such as smart cities, smart supply chains, smart energy etc., all of which depend on managing, orchestrating and optimizing complex systems, can be much more easily planned, build and run and thus become a lot more widespread. Moreover, in contrast to the vertically integrated industries that ruled the twentieth century economy, the future internet will enable horizontally integrated networks where new types of businesses, prosumers and social enterprises can plug in and take advantage of its distributed architecture (Rifikin 2014).

Imagine if we could access and connect the digital twins of all streets, cars, drivers, passengers, pedestrians, traffic, whether conditions, personal itineraries etc. (see Fig. 2) to increase productivity by means of optimized supply-and demand matching, predictive maintenance and simulation? Quite likely, such a system would considerably ease the transition towards autonomous traffic management based on self-driving vehicles by eliminating most of today's barriers, such as insufficient car-to-car communications, safety hazard recognition, legal issues, and unclear whether and road conditions. It would also promote the move from ownership- to access based- societies resulting in better resource conservation and social cohesion.

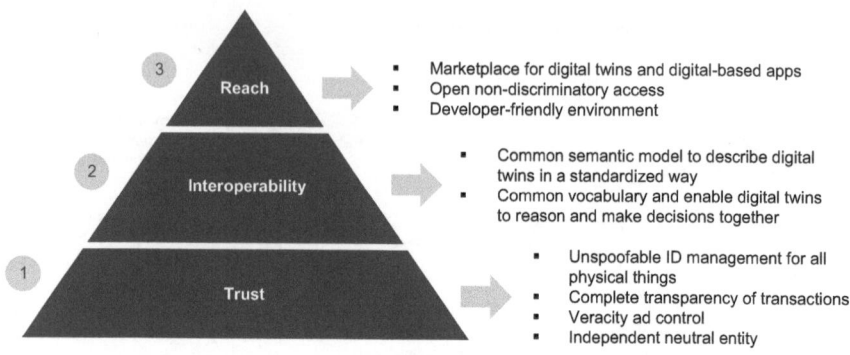

Marketplace for digital twins and digital-based apps
Open non-discriminatory access
Developer-friendly environment

Common semantic model to describe digital twins in a standardized way
Common vocabulary and enable digital twins to reason and make decisions together

Unspoofable ID management for all physical things
Complete transparency of transactions
Veracity ad control
Independent neutral entity

Fig. 3 The building blocks of the next generation internet

2   What Are the Main Challenges to Build the Virtual Representation of the World?

As stated above, the virtual representation of the world is emerging. However, until now its speed of development is still very much at a crawl. The reason why it has not taken off so far is that there are currently three major impediments which first need to be overcome: Lack of trust, lack of interoperability and lack of reach (see Fig. 3).

First, consider trust. A study of McKinsey (2016) promised that internet of things (IoT) markets would generate up to $11.1 trillion US of potential economic impact including consumer surplus by 2025. Still, because mounting cyberattacks have increased the awareness that most IoT devices are for one still trivial to exploit and at the same time multiplying entry points for hackers and cybercriminals to critical business processes and other highly sensitive data, many companies are currently reconsidering their ambitious growth strategies if not suspending them altogether. In fact, a recent study by Bitkom (2016) found that security concerns are the biggest obstacle in the industry when it comes to implementing IoT projects. Next to providing end-to-end (E2E) security to connected devices, data privacy and sovereignty of digital twin owners and users need to be guaranteed to build trust among participants of the future internet. More specifically, digital twin users must only use records that respect their access rights while digital twin owners must stay in control of their data at all times (Otto et al. 2016). Of course, being able to enforce data property rights is inseparably tied with the ability to provide, manage and authenticate unique and tamper-proof identities for all digital twins, digital twin owners and users. Although commercial solutions in the form of identity and access management platforms such as OneLogin, IAM cloud or Covisint are steadily gaining popularity, each of these players is pushing its own solution at the cost of being able to manage identities laterally across platforms. Still, the basic technologies in the form of standards like OAuth 2.0 or JSON Web tokens for lateral identity management do exist. What is missing is a neutral and trusted entity which takes over the performance of identification, authentication and authorization services.

Secondly, existing digital twins have appeared disparately, i.e. digital twins of different types, domains or industries are most likely not able to connect and thus talk to each other unless they are designed by the same company. The reason behind this is that the big manufacturing companies like Siemens and GE are trying to protect their market shares as they are scrambling to manage the transformation from makers of machines to fully digital businesses while at the same time the big IT firms such as Google, IBM or SAP are themselves trying to control the virtual part of manufacturing by establishing their own standards. Still, the expectation that ad-hoc networks of smart devices and services can be constantly formed and re-formed to manifest transient value systems is driving the need for broad agreement on how such devices interoperate and understand each other. Shared metadata, i.e. data that describes the content ant context of data, is a core part of the solution. Again, the basic technology, i.e. vocabularies and domain-specific ontologies for metadata, are already existing and are being developed further (Bauer et al. 2016). Examples are the resource description framework (RDF) which provides globally unique identifiers for metadata, the web ontology language OWL which can be used to express semantic models or linked open vocabularies (LOVs) which facilitates the sharing of metadata. Best practices for ontology engineering, publication, discovery, reuse and mapping are also available through the semantic and linked data initiatives. Realizing semantic interoperability at scale, however, will require collaboration and coordination across standard organizations, consortia, alliances, and open source projects (Schaffers et al. 2013). The need for a shared roadmap and commitment to work together is also self-evident.

Thirdly, until today digital twins have appeared (at least for the most part) selectively, i.e. favoring complex objects such as machines and plants in high-value asset industries. Thus, we are far off the complete virtual representation of the real world. The reality is still mostly digital terra incognita. Especially small and medium sized enterprises (SMEs) and normal citizens have little control over their data. To be able to scale the future internet needs to leverage the success factors of the old communication internet: openness, distributed architecture, and peer-to-peer collaboration. By leveraging these same factors the future internet will be able to allow anyone, anywhere, and at any time the opportunity to search, access and use its digital twin data, semantic models and application program interfaces (APIs) to create new applications and business models. Like the old internet the future internet cannot be owned by anyone (Rifikin 2014). Companies can only be providers (e.g. of communication infrastructure) and facilitators (e.g. for sharing semantic models). At the same time, web companies, non-profits and public-private partnerships will inhabit the future internet and coordinate its content. The future internet itself will be a virtual public square where anyone who pays for a connection will gain admission. The steps towards making the next-generation internet reality are summarized in Fig. 4.

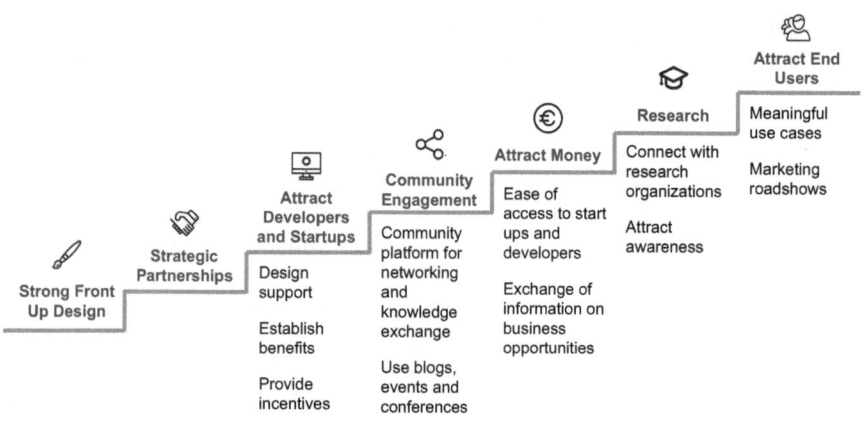

Fig. 4 How to make the next-generation internet reality

3 Why Is It Important for Telcos to Be Part of the Virtual Representation of the World?

The telecommunication business is currently under siege. Indeed, telco companies have been in a comfortable situation for the past 30 years, having enjoyed huge margins on their connectivity business with no real competition. However, in the last 10 years, digitization has increasingly disrupted every industry including telecommunication with companies such as Skype, Whatsapp or Facebook. These companies leveraged the Voice over IP group of technologies to eat into the comfortable margins that telcos were making on mainly international calls and roaming fees. These new players have created a lot of pressure on this industry and telcos need now to reinvent themselves as their traditional telecommunication business appears to be becoming obsolete.

Moreover, not only is connectivity increasingly becoming a commodity, but also the client relationship is being stolen from telcos by the leading digital companies. Indeed, using Facebook's or Google's services is mostly free of charge, very convenient and often complemented by many other services that telecommunications companies are for the most part unable to offer at similar prices (e.g. face chat, movie and series delivery). Thus, providing connectivity is no longer enough to be financially successful. This is why many telcos have started moving up in the value chain (i.e. closer to the customer) which can be seen, for example, in their Merger and Acquisition activities. For example, Verizon has made acquisitions in the mobility industry such as Hughes Telematics in 2012 or Fleetmatics in 2016. Telstra has made no less than 10 acquisition since 2013 in the health industry alone. This is certainly a good option, but there appears to emerge an alternative which is much closer to telcos' core capabilities.

With the emergence of the virtual representation of the world which has been described in chapters one and two, we see a great opportunity for telcos to reinvent their business models and reinvigorate growth. More specifically, we suggest that telcos need to play a key role in connecting all physical things and their digital twins. This growth is already happening elsewhere. For example, IoT platforms are flourishing all over the world. According to IoT Analytics (2015) their number has risen to more than 450. This makes sense because in order to be able to manage and profit from the data collected by connected devices of all kinds, these devices need to be connected to an IoT platform. In a similar vein, digital twins are appearing in silos within various industries. To enable cross-industry services, these digital twins need to be connected to each other. We believe that telcos should be this player which connects all these digital objects.

Telcos are in a very good position to drive the development of the virtual representation of the world for three reasons: neutrality, security, and connectivity. Firstly, connected objects and digital twins are developed in many industries which telcos are not really a part of. However, companies in these industries are frequently biased towards players in the same industry when it comes to sharing data, combining assets let alone build a platform that will connect all their digital products. Therefore, telcos are in a great position to play a central role in orchestrating the development of such a solution because they are, for the most part, neutral to intra-industry conflicts of interests and, importantly, much less dependent of the use cases. It is already apparent that industries highly impacted by IoT will naturally drive the development of the first use cases. Secondly, because of the development of the PLWAN networks (e.g. LoRa, LoRaWAN, Ultra Narrow band, etc.) and the advance of IPv6, IoT is growing at an exponential pace and is bringing along massive security challenges. As mentioned above, connecting all these devices presents hostile hackers with unknown opportunities to usurp the virtual representation of the world. Multinationals that are well positioned in the security market, such as many telcos are therefore in a prime position to provide the necessary E2E-security for the new internet. Finally, telcos have been successfully connecting people for several decades. Their vision of connecting all the people could be easily enlarged to encompass everything that constitutes our world: people, processes, physical objects, digital models.

4   What Are the Implications of This Future Internet for Telco Companies?

To develop the virtual representation of the world and overcome the three challenges mentioned above, telcos will have to focus on two key actions: reinforcing or developing new capabilities and developing an ecosystem by building new partnerships.

First, consider the necessary capabilities. Most telcos are incumbent companies whose current structures, skills and way of working not yet fit to play a dominant role in developing the virtual representation of the world. In our view, telcos should excel in the following five areas: IoT E2E-security, scalable solution, ecosystem-play, interoperability and semantics modeling (see Table 1).

Table 1 Telco capabilities for the virtual representation of the world

	Security: in the IoT world, security concerns are present in the entire value chain: from protecting physical hardware over providing authentication and authorization services to ensuring E2E-secured connectivity. Without deep knowledge about all these elements, a proper solution that should reach a critical mass of users cannot be developed
	Scalability: every manager must has already heard of terms such as "Agile methodology" or "MVP". We believe that this is not enough and that every manager must have experience with them. Such experience is the base to understand the thinking of developers and others who can build scalable solutions
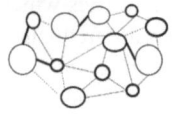	Ecosystem-play: we believe that an open and sharing mentality, together with a global perspective is more and more necessary to make partnerships successful. Furthermore, if several companies align on a common mission and vision, they will be much more likely to develop a solution which will be quickly adopted by the market. From a tech point of view, by employing the container-as-a-service model, telcos can foster cooperation
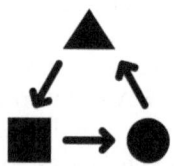	Interoperability and semantic modeling: We recommend that telcos engage in collaboration and coordination across standard organizations, consortia, alliances, and open source projects. To do so, we recommend the development of shared roadmaps and commitment to work together

Telcos and industrial incumbents have existed for a very long time and are therefore naturally very different from the new digital companies. Moreover, they are not in a position to push their solutions in the market in the same way that Google, Apple, Facebook, or Amazons does. Thus, for the time being, no one in Europe is in a position to develop and commercialize a solution at the necessary scale alone. Finding the right partners and establishing collaborative relations based on fair incentivation for each will be two fundamental factors to succeed. Furthermore, the business environment has never changed so fast and has never been so fragmented. This fosters alliances and ecosystem of companies to share similar visions, beliefs and complementary solutions.

Regarding this partnering activity, we see four core types of partners for telcos: leaders, industry partners, technology partners and organizational bodies.

Core Partner Beside building and operating the solution, the founding partners will orchestrate this new ecosystem. This means they will decide on the type of the cooperation (i.e. joint venture, consortium, European funded or semi-private-public), the governance model, the prioritization of use cases and the partnering structure. These few companies should have a complementary set of skills as well as a sufficient reach in several different markets.

Industry Partners The industry leaders will definitely drive the prioritization of use cases as they will directly benefit from the existence of this solution. The general building blocks of the solution should be naturally independent of the use cases and these should be used to confirm the general architecture of the final solution.

Technical Partners Technological companies will provide the core building blocks of the target solution which should be based on a container-as-a-service architectural solution. This will help to develop quickly a working solution as tech provider solutions should be able to plug-and-play in this system.

Organization Bodies This solution could definitely help all these organizational bodies. Being a government, the European Union (EU) or standard organizations, to just mention a few, such an internet layer would enable companies to manage and control access to their data. It is therefore very important to engage with them as they could support such an initiative.

References

Bauer, M., Davies, J., Girod-Genet, M., et al. (2016). *Semantic interoperability for the web of things.* Available via Research Gate. Accessed November 27, 2017, from https://www.researchgate.net/publication/307122744_Semantic_Interoperability_for_the_Web_of_Things

Bitkom. (2016). *Industrie 4.0 – das unbekannte Wesen.* Bitkom-research. Accessed November 27, 2017, from https://www.bitkom-research.de/Industrie-40

IoT Analytics. (2015). *IoT platforms – The central backbone of the internet of things.* Available via: IoT Ana. Accessed 27 November 2017, from http://iot-analytics.com/wp/wp-content/uploads/2016/01/White-paper-IoT-platforms-The-central-backbone-for-the-Internet-of-Things-Nov-2015-vfi5.pdf

McKinsey. (2016). *The internet of things: Mapping the value behind the Hype.* McKinsey Global Institute. Accessed November 27, 2017, from https://www.mckinsey.com/business-functions/digital-mckinsey/our-insights/the-internet-of-things-the-value-of-digitizing-the-physical-world

Otto, B., Jürjens, J., Schon, J., et al. (2016). *Industrial data space – Digital sovereignty over data.* Accessed November 27, 2017, from https://www.fraunhofer.de/content/dam/zv/en/fields-of-research/industrial-data-space/whitepaper-industrial-data-space-eng.pdf. Fraunhofer-Gesellschaft zur Förderung der Angewandten Forschung e.V

Rifikin, J. (2014). *The zero marginal cost society.* New York, NY: Palgrave Macmillan.

Schaffers H, Grace P, Sällström A. (2013) *Fire strategy towards 2020 – Shaping FIRE's future in the internet ecosystem.* Available via: Amplifire. Accessed November 27, 2017, from https://www.ict-fire.eu/wp-content/uploads/AmpliFIRE_2013_White_Paper_on_FIRE_Strategy.pdf

Low Power Wide Area Networks: The Game Changer for the Internet of Things

Daniel Weber, Christian Schilling, and Frank Wisselink

1 LPWA Technologies Unlock the Internet of Things

The "Internet of Things" (IoT) has been around for many years before the term has become well-known throughout different industries. ITU defines the IoT as: "a global infrastructure for the information society, enabling advanced services by interconnecting (physical and virtual) things based on existing and evolving interoperable information and communication technologies" (ITU 2012). However, until recently, IoT use cases did show a modest growth only.

Promising IoT use cases, for instance tracking, are focusing on a mobile application. Thus, they need small sensors and hence can only be equipped with small batteries. Other major IoT application area, e.g. metering, sensors have to be installed under the ground. Providing connectivity to these major IoT use cases (c.f. Analysys Mason 2016) has been a challenge:

On the one hand, connectivity based on cellular technologies offered too much bandwidth for many IoT applications. On the other hand, the associated power consumption was too high. This caused repetitive recharging for tracking solutions or requiring "clunky" form factors limiting the development of tracking use cases. Moreover, the associated connectivity cost for these oversized connectivity solutions hampered valid business models from being developed. For sensors installed underground, the installation of hubs with dedicated local area communications were required in order to provide connectivity underground making installation and device handling complex and the total solution expensive. Summarized, new connectivity technologies were required to power the Internet of Things.

D. Weber
Detecon (Schweiz) AG, Zürich, Switzerland
e-mail: daniel.weber@detecon.com

C. Schilling (✉) · F. Wisselink
Detecon International GmbH, Cologne, Germany
e-mail: christian.schilling@detecon.com; frank.wisselink@detecon.com

© Springer International Publishing AG, part of Springer Nature 2019
P. Krüssel (ed.), *Future Telco*, Management for Professionals,
https://doi.org/10.1007/978-3-319-77724-5_15

Low Power Wide Area (LPWA) technologies solve many of the issues above. They are currently quite rightly so considered as one of "the hottest innovation areas in telecoms" (Markkanen 2016). Low energy consumption, deep indoor penetration for a low cost at limited data rates characterize these types of technology. These characteristics may be seen as essential for enabling the IoT. And even better, they have the potential to enable a new playground for innovators, whose use cases have yet to be invented.

Hence, LPWA technologies have the potential for exponential growth not only because they may enable use cases in terms of having a positive impact on their business case (Chua 2017), but also open new possibilities which are unlocked by its unique characteristics.

Since the LPWA opportunity is still fresh yet quickly arising, operators must act now and find the right approach for grasping the full benefits of them. They cannot just wait for the market to identify "the right" technology, as the entry barriers into the market will be too high at this point (MacKenzie 2017). But therefore, a clear vision about the different technologies and developments is needed. In the following, different LPWA technologies and implications for the Telecommunications industry will be thoroughly analyzed. In addition, possible courses of action will be given.

2   The Strategic Choice for LPWA Is Between a Proprietary or a Standardized Approach

In order to understand the business implications of LPWA technologies for Telcos, it is essential to understand which technology options are available and how they differ from each other. Currently, there are two main groups of technologies, which typically fall into the LPWA domain: one following a proprietary approach, while the other one is following a standardized approach.

While both groups focus on the same key characteristics (low cost, low energy consumption, deep indoor penetration, low data rates) they may be distinguished by their spectrum usage and technology standardization (see Table 1). The proprietary approach and the standardized approach are described in the following section.

2.1   The Proprietary Approach Implies the Use of Unlicensed Spectrum and Non-standardized Technologies, and Can Offer Fast Time to Market

The first group of proprietary LPWA technologies utilizes *unlicensed spectrum*. That is, this group follows an open approach. Examples are Sigfox or LoRa. Since there are no licenses to be acquired and/or frequencies to be managed this enables a fast time-to-market and therefore quick realization of use cases. Due to the open approach, there is also a large ecosystem readily available around these technologies, which enables a head start within the LPWA market. Part of this ecosystem are also

Table 1 Different approaches to LPWA technologies

	Proprietary approach	Standardized approach
Spectrum usage	Unlicensed spectrum • Fast time-to-market • Risk of future congestions in the network • Potential difficulties to provide standards on the Quality of Service • Mostly dedicated, geographically limited networks	Licensed spectrum • Radio spectrum managed by Telco operators • High level of Quality of Service • Guaranteed Service Level Agreements
Technology standardization	Non-standardized technology • Fast time-to-market • Open approach allows for a large ecosystem • International, nationwide usage restricted	Standardized technology • Takes some time to develop standards • International coverage: devices can work with different vendors • Ecosystem with powerful players (Telcos, network suppliers, etc.)

players from outside the Telecommunications industry to enter and compete in this market by building their own dedicated networks.

However, while technologies utilizing unlicensed spectrum may be the first ones to be rolled out in a large scale, they have one major downside: their sustainability. While unlicensed spectrum may work well today, there is a risk of it becoming more congested over the next few years, also by the increased number of IoT applications and therefore rising level of interferences. This is especially critical for use cases, which require high Quality of Service (QoS). These type of use cases cannot rely on the data being transmitted just anytime, because they often refer to safety-critical applications, such as smoke detectors, where every second counts. In addition, these LPWA technologies have limitations in their downlink capabilities. This is important especially for soft- or firmware updates.

Another drawback for technologies operating in unlicensed spectrum are often dedicated and functioning in a limited geographical area only without any link to the currently operated mobile networks. For an operator this implies that he has to operate a dedicated infrastructure for LPWA in parallel of its existing mobile infrastructure. This can lead to higher operational cost in the long run.

2.2 The Standardized Approach Leads to Quality of Service and Long Term Benefits of Scale but Requires a Longer Time to Market

The standardized approach utilizes *licensed spectrum*. This resource is available for registered Telcos only, who have to acquire the spectrum at a high cost. Compared to unlicensed spectrum, which is free of charge, the usage of licensed spectrum is therefore expensive. However, in utilizing this spectrum, Telcos may guarantee a certain level for QoS. Furthermore, it mitigates the risk of congestion in the network, not only because the spectrum is manageable, but also because the type of devices and their respective behavior may be managed to a further extent. This means a certainty in network availability which can be valuable for Telcos and their clients. In addition, Telcos can offer Service Level Agreements (SLA), which customers are willing to pay for. Last but not least, the restrictions for the downlink capabilities are not so strict—the Telco itself may manage the restrictions.

The usage of licensed spectrum therefore defines their Unique Selling Point (USP) against its Non-Telco competitors in the LPWA market.

The big advantage of standardization of the technology combined with licensed spectrum is that it also eases risks of congestion. In addition, other problems in the network by preventing unauthorized usage behaviors or non-conformant devices not in conformance with the standard can be avoided. This again enables a higher Quality of Service (QoS), but also to guarantee SLAs for the customer. Technology standardization also ensures that devices will be operable in networks of different vendors in multiple countries.

One of the most prominent standardized LPWA technologies for Telcos currently is Narrowband-IoT (NB-IoT), which particularly addresses the massive IoT market. This is why we will therefore refer to it in the following.[1]

The standardized approach to LPWA, however, also has a downside: it is much more time-consuming compared to the proprietary approach, as there are many obstacles yet to be overcome. Besides license considerations, network components such as the radio access and core network have to be set up first and according to the 3GPP standard. This may give the impression that standardized LPWA technologies have already lost the race against its proprietary competitors, before it has even started. Nevertheless, there are far more factors which must be included in a thorough analysis, which is conducted in the following.

2.3 The Standardized Approach Seems to Be the Sustainable Option for a Telco

As stated before, the utilization of LPWA technologies will be key for Telcos to participate in the Internet of Things. It is now up to them to decide, which approach for technology they want to pursue.

Both major groups of technologies have their advantages and disadvantages. A strategic decision, however, is more about *time-to-market versus sustainability* and the resulting implications. While it may seem that the standardized approach lags behind its proprietary opponent, this may not necessarily be true, considering that standardization and utilization of licensed spectrum is more sustainable and a true asset for Telcos. Furthermore, the standardized approach, while having to face some obstacles still, may seem less risky, as the spectrum is manageable and the devices are standardized. Therefore, their behavior is much more predictable. DT Group

---

[1]NB-IoT is standardized by the 3GPP.

however showed that a quick roll-out of standardized approach is feasible in a comparable short time frame (Deutsche Telekom 2017a).

Some Telcos[2] have chosen to hedge the risk by pursuing a hybrid strategy and provide both, standardized and unstandardized technologies (MacKenzie 2017). However, this approach bears the risk that there will be one legacy system at the end, which must still be managed by the Telcos, even if there is only a small customer group still using it. This risk is especially critical when considering the nature of LPWA use cases, which often require devices to be in the field for many years.

It is currently hard to determine, which of the technologies will succeed in the long run. It may even be that every technology will find its niche: it is much more likely, that 3GPP standardized technologies will serve use cases, which require nationwide and uniform connectivity within multiple countries, more efficiently than non-standardized technologies, which are more competitive in dedicated and limited areas (Markkanen 2016).

However, Telcos must now decide for an approach towards LPWA they want to pursue. Not participating in the race at all may not seem feasible though, as a lot of potential in the massive IoT market not only by offering connectivity, but also on-top services will be lost. This again, is very important for the monetization of the LPWA opportunity, as we will see in the following. At the end it will be a choice dependent on which use cases the respective Telco wants to serve with LPWA technology, as well as cost considerations and ecosystem support (MacKenzie 2017).

Furthermore, Telcos always have to consider that this time, compared to developments in earlier mobile cellular technologies, they are not the only player in the market. This time, there are players from other industries pushing into the market, as they can utilize proprietary LPWA technologies to build up their own, dedicated networks.

3   LPWA Technologies Enable Business Models for Mass Applications and Are Fundamentally Different from Incumbent Ones

LPWA technologies have many benefits enabling mass applications of IoT, yet only under the condition of limited data-use. Since most business models of Telco's rely on data use growth, LWPA requires a fundamental reassessment of the operational and business models. Low prices and mass markets set rigid constraints on operational costs. Introducing LPWA in a Telco is only feasible by a rigorous focus on the simplification of the product, pricing scheme and associated operational processes.

Another crucial aspect is the positioning of LPWA in relation to a Telco's existing Machine-to-Machine (M2M) Portfolio. In order to prevent unwanted price driven migration of "data hungry" use cases to LPWA, careful positioning and sales

---

[2]E.g. Swisscom, Sprint.

governance is required in order to combat price erosion and operational cost increase. Otherwise, precious network resources are wasted and operational cost will rise due to extensive spectrum usage.

Prospective business models require an active development of the market and building of the associated eco systems, which is only successful by investing in business and market development. Moreover, LPWA requires alternative models for monetization and Telcos must consider specificities of LPWA use cases. In order to realize the full potential of the technology: revenues are driven by the sheer volume of devices as well as the opportunity to offer services on top of connectivity.

The next chapter will focus on these use cases enabling application of IoT.

3.1 "Batteries Not Required": Current Use Cases and Field Trials Indicate That LPWA Is a Game Changer

What would be the impact if sensors would operate autonomously for a lifetime?

In order to exploit the LPWA opportunity, Telcos must not only bear in mind the key characteristics of the technologies themselves, but also the specificities of the use cases in order to be successful. As stated in the beginning, LPWA is developed for applications, which require only low amounts of data and offer long battery lifetime and deep indoor penetration, all at low cost.

Hence, LPWA is a game changer for different types of existing use cases: The ones which struggle to have a positive business case due to high connectivity costs in the past using classical M2M communication or, for example, for exchanging batteries in the field (which becomes relieved by the characteristics of long battery lifetime).

An example indicating that a game changer is materializing here, are the sensors NOWI develops (BTG 2017) in partnership with T-Mobile in the Netherlands. NOWI developed a technology to harvest energy from 4G networks. Although the harvested amount of energy is limited, the reduced low energy use of NB-IoT (T-Mobile Netherlands has already rolled out NB-IoT in the Netherlands) allows sensors to operate for a lifetime. This enables use cases which would not have been feasible before.

The first use case is a temperature sensor which can be integrated in the tarmac of a road. It can stay there for a lifetime and monitors the road surface temperature. It assists to determine when to grid a road to preventing it getting icy. This saves gritting cost. Moreover, it saves installation of wires in the road to power the sensors and to provide connectivity.

It is expected that the technology will find its application in the consumer market in the near future as well. Connected smart watches and fitness bands with no need to get powered externally or by battery change are feasible in the future.

Another type of use case are the ones which have not, up until now, been feasible due to technological restrictions, i.e. the smart metering device in the basement without connection due to limited indoor penetration of existing networks or power supply (Keysight Technologies 2017).

In the area of Smart Cities, there are even more use cases, where LPWA is at least a technological boost. Smart Parking is one of them. Smart Parking is no new application and has the potential to reduce many challenges of modern cities. In urban areas, the search for free parking lots is causing traffic and hence has an impact on the carbon emission, noise pollution and overall the quality of life of the city's citizens. The fundamental functionality of smart parking solutions is relatively easy: a device, usually set in the ground, determines the vacancy of its parking lot and provides two potential information, taken or free. This works for both, on-street parking lots and car parks. LPWA fits perfectly for smart parking, since the application needs a very limited amount of data transferred and a long battery-lifetime (Deutsche Telekom 2017b).

Another example from the area of Smart Cities is Smart Lighting. "Intelligent" street lights being able to receive information with regards to daily or weekly times for sunrise and -set, bad weather coming along with heavy rain or fog having an impact on the general brightness, spend less energy and therefore save money for municipalities. LPWA is of benefit, since street lights often have no additional power source, which can be used to supply energy to the devices. Furthermore, the high number of street lights per city and the cost for connectivity and modules might have been a barrier for the missing breakthrough of Smart Lighting.

To name another use case where LPWA has a positive impact is Smart Waste Management, with no external power supply and partially trash bins being in the basement of houses and no coverage.

A use case not coming from the Smart City area, but with high demand and attraction from the market is the tracking of devices. The logistics sector is very interested in the possibility of tracking their containers and hence being able to know where their shipments are at which time and when they will either be delivered to their clients or reaching their sites. Large logistic companies are shipping millions of units each day and would benefit a lot from the knowledge of the current position of them. Besides containers and large shipments, the tracking of high value goods is interesting for business-to-business customers as well. Theft on construction sites is an existing issue for construction companies and enterprises selling and leasing the machinery and equipment. But also the emerging market of shared bikes and motor bikes is seeking for such solutions and can be addressed.

Moreover, LPWA technologies have or will have in the future network based location services. Since a GPS uses much power this network based location service enables tracking in the consumer space.

There are various examples for end customers seeking to protect themselves against theft and put tracking devices on their valued goods like bicycles, motor bikes or expensive prams. If we think about watches with LPWA tracking devices for children and elderly people or collars for pets, we can see that there is also an emotional product in the making. It creates additional chances to approach end customers directly through up-selling on their existing contract and create value for parents or pet owners which results in higher customer satisfaction and less churn.

However, even though these use cases may bear a lot of potential, it is especially the vast range of use cases, which do not even exist today and may not even be explored, which makes up the true potential and the fascination of LPWA technologies. One example is the leasing of machines. Construction equipment for instance is at the moment leased for a fixed price, depending mainly on the lease term. A new approach could be to track the actual use of the device and charge clients accordingly. Frequent use and hence a higher degradation during the same leasing term is of disadvantage for the lessor and could now be priced higher than the less frequent use during the same period. We will see more of such examples in the future and find out how LPWA as enabler for new applications contributes to making all our lives more convenient.

3.2   Sustainable Business Models Are Feasible If Telcos Monetize More Than Just Connectivity

There are many game changing use cases, but what is the right strategy to monetize them? In the following, the unique position of the Telco within the LPWA ecosystem will be evaluated and why Telcos must utilize it.

There are basically two aspects, which must be taken into consideration, when finding ways to monetize the LPWA opportunity as a Telco:

The first aspect is generating revenues with "typical" Telco business, which is selling pure connectivity. This is feasible, especially when considering the massive IoT use cases not feasible before, which LPWA technologies have the potential to unlock. Here, Telcos must capture a large portion of this market in order to create the benefit of scale especially because the margins here are low. This is due to the fact that one major requirement for customers of LPWA is low costs.

In order to obtain a sustainable business model, acquiring scale and offering the most competitive prices only is not enough and leads as described before to unwanted price erosion and churn from the existing M2M Portfolio. Other models are necessary to generate revenue beyond connectivity. This is why a second source of monetization needs to be considered. In the following, the focus will be on how Telcos can build up a comprehensive offering (Newton-Smith 2014) with additional services.

Here, it is important for Telcos to utilize and build powerful ecosystems in order to grasp the full benefits of the LPWA potential. The ecosystem in IoT and LPWA in particular is closely connected to a layer architecture (see Fig. 1):

At the very beginning, there is the sensor, which may measure different data, e.g. temperature, moisture, and light. The data is then sent via a communication module (in this case a LPWA module) to the communication network. Within the layer architecture, horizontals (e.g. cloud platforms, application platforms) and verticals (e.g. Smart City) and the actual application for the end user can be built upon connectivity:

In the past, the communication network has been the core domain of the Telcos. They therefore find themselves at the very center of that ecosystem. They have

Fig. 1 Generating revenues
beyond connectivity

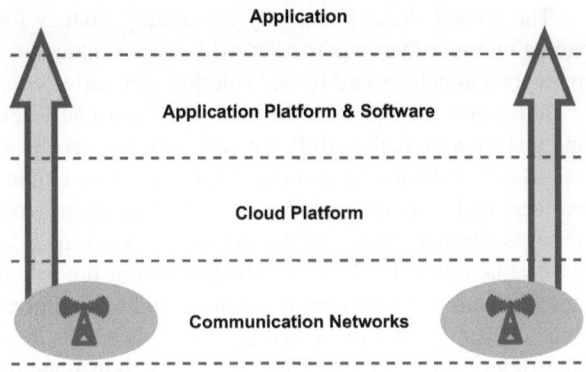

interactions and ties with all the other players, regardless if it is the module provider, partner from the cloud layer or even the end user device and application. The latter is the most important one, when it comes to monetization of an LPWA offering.

This position offers Telco operators a great chance to generate revenues beyond connectivity and cover a larger part of the value chain. Depending on i.e. the individual market position and size, operators can follow different strategies, which will be discussed in the following:

The first strategy is to focus on selected verticals, such as Smart Cities, logistics, or utilities. The needs of the end customer in each vertical can be very different and by building expertise in that area, Telcos have the opportunity to become a trusted partner for their clients advising and offering not only connectivity but end-to-end solutions (Newton-Smith). The decision on which verticals to focus on depends on various factors. Often Telcos already have a certain knowledge in some use cases and verticals such as Smart Home, which gives them a natural advantage. More important though is the individual market and client structure, deciding on the commercial potential and access to the relevant players.

The second strategy is to not only focus on one vertical, but build up and become the leader of a powerful ecosystem. The operator uses its central position within the ecosystem, having various partners to deliver end-to-end solutions. Partners can be everything between a start-up and powerful, experienced corporations, who have been leaders in their industries for decades. Every partner may provide different advantages. For example, the development cycles of start-ups and thereby the time from a "sketch" to an actual solution is often shorter than the ones a large Telco usually can provide. The advantage here is that the Telco and the start-up can gain early experiences with a fresh technology. However, it is also important to have experienced, industry leading corporations aboard, as this is most important to understand the needs of the specific industries and their expectations towards this new technology.

The second strategy is a very demanding strategy for rather larger Telcos being active on more than one market, with a certain amount of power to attract the best partners and deliver end-to-end solutions for various verticals and customers.

In Europe, for example, we have seen Deutsche Telekom and Vodafone[3] following this strategy with both being very active in accelerating the market introduction of NB-IoT and building and expanding the ecosystem including experts, technology partners and solution providers. For Telcos, the approach of bridging makers and markets during the network developing process is quite new, but very promising. On the technology side, the early involvement of the solution partners and the mutual development of prototypes generates insights the operators would usually only be able to identify after the actual go live. This results in a smoother roll out with less uncertainty and more expertise and ability to respond to client's questions after the go live.

4   The Game Around LPWA Has Just Started and Telcos Have a Chance to Change Their Game Extensively

Regardless which strategy a Telco provider follows, the evolvement of LPWA offers an additional and promising opportunity. A powerful example here is the evolvement of standardized LPWA technologies: 3GPP has completed the standardization of Narrowband IoT in 2016 (3GPP 2016) which initiated and fueled the NB-IoT ecosystem. Since then a new market has emerged with module manufacturer developing the first products and end solution providers moving to NB-IoT connections. The Telcos, again, at the very center of that development, are playing an essential part with developing the new network. As a consequence of that they had and still have a unique chance to not only help shape the market but really lead it. By this, they become an orchestrator of the new ecosystem.

However, Telcos must not rely on their powerful position within the ecosystem, which is also a result of past developments. This is especially true when the LPWA ecosystem becomes increasingly challenged by competitors from outside the industry. In this setting, Telcos must push forward and defend their valuable position in order to be successful.

References

3GPP. (2016). *Standardization of NB-IoT completed.* Accessed October 18, 2017, from http://www.3gpp.org/news-events/3gpp-news/1785-nb_iot_complete

Analysys Mason. (2016). *LPWA networks for IoT: Worldwide trends and forecasts, 2015–2025.* Analysys Mason.

---

[3]NB-IoT Prototyping Hub, Warp NB-IoT of Deutsche Telekom and NB-IoT Open Lab from Vodafone.

BTG. (2017). *Nooit meer batterijen: IoT- sensoren van NOWI halen stroom uit 4G.* Accessed October 16, 2017, from http://www.btg.org/2017/08/09/nooit-meer-batterijen-iot-sensoren-nowi-halen-stroom-4g/

Chua, G. (2017). *Market insight: CSPs should explore creative approaches to monetizing LPWA.* Gartner Inc.

Deutsche Telekom. Accessed October 06, 2017, from https://warpaccelerator.com/nb-iot/

Deutsche Telekom. (2017a). *Deutsche Telekom rolls out NarrowBand-IoT Network across Europe,* Press Release Deutsche Telekom AG 23.2.2017.

Deutsche Telekom. (2017b). *Einfacher Parken mit "Park and Joy".* Accessed October 16, 2017, from https://www.telekom.com/de/konzern/details/einfacher-parken-mit-park-and-joy-502194

ITU. (2012). ITU Recommendation ITU-T Y.2060 2012.

Keysight Technologies. (2017). *Narrowband IoT (NB-IoT): Cellular technology for the Hyperconnected IoT.* Accessed October 04, 2017, from http://literature.cdn.keysight.com/litweb/pdf/5992-2360EN.pdf

MacKenzie, M. (2017). *Preparing for LPWA: How operators are maximising their chances of success.* Research Strategy Report, Analysys Mason 2017.

Markkanen, A. (2016). *LPWA outlook update: The new 3GPP standards strengthen the MNOs' hand.* Research Note, Machina Research 2016.

Newton-Smith, C. (2014). *Monetizing the internet of things – Challenges and opportunities for telco operators.* Redknee, Looking Beyond. Accessed October 06, 2017, from https://www.redknee.com/about_us/news_events/newsletter/featured_articles/july_2014_feature_3

Becoming Faster and More Direct: Telco Carrier Product Management Turns Towards Customers, Partners and Colleagues

Interview with Thomas Heilen, Head of Product Management Consumer at Telekom Deutschland GmbH

Thomas Heilen

How do you go about developing, positioning, and launching your products in times of change and disruption? What capabilities must telecommunications carriers acquire to support their chosen business models? Thomas Heilen, for many years head of Deutsche Telekom's German Product Management, has a wealth of profound knowledge and an impressive track record. In the following interview, he shares his viewpoints and experience related to some of the most hotly debated topics in the telco industry today: the potential advantages of integrated carriers, the role of network wholebuy, 5G and traffic structuring versus unlimited data rates. He outlines key aspects of products and services that will be offered in the future, speculates on whether they will focus on networks, customer experience or content, and considers how they can enhance differentiation among carriers. The discussion continues with a look at the motivation behind an approach related to households and users and what benefits it can generate. Our interview closes with a prediction about the working and operating mode of telco carrier product management of the future, including external (more in-depth integration of partners and customers) and internal (getting away from Tayloristic setups) aspects.

Do integrated network operators enjoy a production advantage over carriers who have only one network domain available? Can integrated carriers turn this advantage into a competitive advantage for customers? Is FMC a pronounced customer wish?

T. Heilen: Data traffic in our networks is growing steadily, and strong growth is forecast to continue. The resultant challenge is clearly discernible, especially in

T. Heilen (✉)
Telekom Deutschland GmbH, Bonn, Germany
e-mail: thomas.heilen@telekom.de

densely populated urban areas. The challenge and the principle for a solution are comparable to those of urban development, where "densification" is the key word for the process of creating the required capacities. The situation for us is similar; physical laws ultimately demand an increasingly dense network as utilization loads rise. Such a dense network is comprised of smaller and smaller (micro and pico) mobile network cells. This small-scale structure must in turn be connected—and so the structure of the mobile network in the end becomes increasingly similar to that of the fixed network. The "production", as you call it, becomes increasingly convergent. Naturally, as a fixed network operator, I would like to fall back on its possibilities for densification.

The consideration of fixed and mobile networks from a common perspective is a central one for us as an integrated provider—from the portfolio view as well as from the production standpoint. After all, our customers want convergence. Of course, none of them call it that . . . For our customers, the advantages of integrated use are decisive: more convenience from a single contact, a cost advantage as part of an overall package, and the integration of attractive services. One example of convenience: Do you, the customer, really want to have to worry all the time about how your data traffic is handled? When people go to a party, asking the host for the WiFi password is standard practice—it does not have to be that way.

These advantages are not merely hypothetical; they are concrete and relevant to the customers, so we can derive competitive advantages from them. Our sales success, the highly positive satisfaction values, and even the feedback from our customers about the content of MagentaEINS are clear confirmation that our convergence approach is the right way to go.

What role will wholebuy and wholesale of fixed and mobile network services play in the future? Or to state the question differently, are cooperation on the network side in the area of coverage and capacity increasingly important for offering services that convince customers?

T. Heilen: Wholesale is a major cornerstone of the market structure in Germany. Regulations obligate us to offer our services to others as well. Looking at it from the other direction, it is becoming more and more interesting for us as well to find ways to close gaps other than by expanding our own network. When we talk about the fixed network, it can in some cases make better sense—from both the provider as well as the economic perspective—to purchase the access. We know how expensive fixed network expansion is, and there is little point from an overall economic viewpoint in having multiple providers laying parallel optic fiber cables.

At this time, wholebuy is not really an option for mobile networks. Our clear goal here is to offer the best mobile network. We also see ourselves as pioneers for the network of the future, for 5G. We are banking on complete control at all levels of the network infrastructure and want to remain the leaders in area coverage and quality as well.

How important is it in this context to control the network? Do you see opportunities for quality-assured services, or will the best-effort principle in service performance still be sufficient in the future?

T.Heilen: As we are a premium provider, we offer high-quality services—and customers honor this by paying rates that are above market average. This objective alone shows why we need a network whose quality and performance we can control and influence. Ultimately, IP-only traffic offers only best effort. Best effort would be acceptable for all solely in a hypothetical world in which unlimited capacities are available because they would then be sufficient for the most highly diversified requirements and all customers.

In the real world, there must always be a balanced differentiation because our standard of premium quality is not restricted to a small niche, but includes the entire market. One example: Instead of investing in vectoring and fiber to the curb, we could have put our resources exclusively into laying FTTH optic fiber to residences. In that case, the bandwidths in these residences would have been several times higher—but then only a fraction of the households would have benefited from an improvement in their Internet access.

The importance of performance features fluctuates over time. The lifecycle of the mobile network generations has followed a typical course. At the beginning, the major emphasis is on differentiation. Intense marketing campaigns arouse attention and generate demand, but other questions are decisive and differentiate in competition as well. Who is the first to master completely the technology? Who is the first to have the right devices in adequate quantities? What services provide a tangible experience of the new advantages? At the end of a mobile network generation, these topics are strictly hygiene factors. So it is necessary to provide more complex services to a large number of users—and to invest in these services.

Topics such as edge computing, low latency, and 5G are being heavily promoted by Telekom. In view of these new functionalities, how important will partnering be in the future, i.e., better coordination between telecommunications providers and various industries? How important, how new will partnering and co-innovation be as future focal points in product management?

T. Heilen: Consider our corporation's partnership with Zeiss that we announced at the Mobile World Congress in 2017. It clearly illustrates two of your points at the same time: one is the necessity of developing concrete solutions as partners, the other is the concrete benefit that is tied to the capabilities of new technologies such as 5G.

Generally, for many future applications, the edge computing technology concept is decisive. This is what makes it possible to take the computing performance required for applications and operating systems out of tiny devices and to locate them instead in cloud elements forming part of low latency networks. This is a decisive factor if future devices—like wearables—are to become light, affordable, pleasant to wear with attractive form factors, and supplied with almost unlimited compute power.

As far as development in partnerships goes, we will move ahead jointly with our partners to determine the borders of the respective application areas, for example. Approaches like open developer platforms can help here greatly. These joint efforts are an enormous aid to us as we concretize our specific development for the most widely diverse concepts such as edge computing. This is the only way to define and answer concrete questions, e.g., about size, features and distribution of cloudlets (the physical storage units in mobile networks) or regarding the concrete form of latency.

Connectivity is becoming increasingly indispensable for digital life, both personal and business. 5G offers the chance to create tailored configurations for various use cases; that is important for us so that we can continue to make efficient use of limited network resources as well as for customers and users because they also have differing use cases and related requirements.

When we look at 5G use cases, we currently see a broad range of new opportunities, especially in commercial and industrial demand: (industrial) IoT and autonomous driving are only two of innumerable examples. In the consumer sector, customers will profit in the future from higher bandwidths and lower latency—these factors will make themselves noticeable in a positive sense in video consumption. We do not yet, however, see the revolutionary "new" 5G use case—will it be VR/AR or will it be something else? We certainly cannot claim to be AR/VR specialists, so we are cooperating with other companies, including startups, so that we can become better acquainted with technical enablers as well as assess the market.

What do you think will happen to the weighting between customer experience, the performance capability of the network, and service and content offers in the future? Will carriers remain (active) in all these fields? Where will they be able to set themselves apart from the rest?

T. Heilen: We are active in all of the fields you mention and will certainly remain so. The network(s) is/are the beating heart of our value generation. At the same time, we see that our service platforms (take Qivicon for smart homes as an example) also cover growing customer needs. We are well positioned in this respect, in part through our differentiating approach following strict comprehensive European data protection rules.

Content has gained substantially in significance. Exclusive content enhances the value of our fixed network access offering, of course. We started here with top sports events and have now expanded it by adding exclusive series. Our premium standard is also reflected in the selection of the content; The Handmaid's Tale has received 8 Emmy Awards and is one of the best series in 2017.

Despite all the innovations—one important point is, and will remain, voice telephony. To be sure, many (OTT) providers with enormous numbers of users have integrated voice as a service in their apps over the past 2 years. But we see that traditional voice telephony—and the phone number—are still used happily and intensely. This is also the point where the limitations for new services become visible; for instance, in their dependency on a certain app or operating system for both, the calling and the receiving party. In many cases, the traditional method of dialing a phone number and pushing the green button is for many customers the simpler, more secure, and more reliable method. We are convinced that our voice portfolio will remain highly relevant for a long time to come and are investing accordingly in its quality and performance capability.

As far as customer experience is concerned, we always strongly assessed customer satisfaction in the past according to their experience while using the network. This part of the customer experience is certainly fundamental—but we also want to improve substantially in the purchase and service experience. The customer notices (and by the way, so do we) that we are still working with processes and systems from the last millennium. We will do a lot of catching up here; companies like Amazon have set clear standards in this respect, and these standards are being used more and more as the benchmarks for our services as well. That is especially true because we are a technology company; everything on our sites must work with one click as well.

We regard a "simple" and appealing digital customer experience—throughout the customer's entire lifecycle—as a key competitive advantage. We need this advantage as well because the rising demand for higher bandwidths will stimulate the market and prompt customers to examine the offered services and products more actively than before.

From contract partner and connection owner in the fixed network to better penetration of households: Telekom Deutschland has addressed the households on the market with special products and services under the name MagentaEINS. How successful is Telekom Deutschland with these offers on the market? What are the major success factors? What role does household penetration play as a core element of future strategy?

T. Heilen: The cutthroat competition in our industry makes the household perspective increasingly significant. The place where people live together is often the site of closely interwoven economic interests. Regardless of whether I am a network operator, insurance company, or retailer: if I can generate a crossover benefit for this household group, I have an understandable and attractive logic—one that we are already offering in the form of today's MagentaEINS package.

Besides the aforementioned economic motives for the households, the knowledge of the individual user in the household is moving more and more into the spotlight. Our view of customers and users is historically dominated by terms such as "connections, SIM cards, and contract partners." If we know/recognize the specific individual more precisely, both sides profit. For instance, many customers want personalized address and specific content recommendations. At the same time, the routines for the necessary identification in the service case are shortened, reducing our costs and enabling us to solve the issue more quickly.

We cooperate with our customers to develop new channels here. A very important point: we will not do anything here in excessive haste because T stands for "trusted"—securing the trust of our customers is our top priority. We must also keep in mind that we European telcos are constrained by significantly stricter privacy regulations than the companies we compete against (in the USA, for example).

Is security an increasingly important criterion for the consumer/business segment? What aspects of the general topic of security to you consider to be relevant for the future (e.g., privacy, data integrity, protection from cyber-attacks)? Is there a

willingness to pay for this, or do the customers more or less expect a premium
provider to include these attributes in its services as a matter of course?

T. Heilen: In this point as well: a part of our premium promise is the protection of our customers from certain risks. We have very tangible actions in this respect—from our offer of antivirus programs for mobile and fixed networks to software for the protection of children to the Cyber Defense and Security Operation Center (SOC).

An explicit willingness to pay for preventive care and protection from risks during the use of digital services, however, is still slow to develop, much like the principle for insurance: something has to happen to my neighbors before I decide to buy insurance myself. Malware/ransomware on smartphones, incorrectly delivered messages—these events are at this time (still) rare, thank goodness. I assume that risks will continue to increase and will materialize more and more frequently. The willingness to pay will not increase until there are incidents with a direct impact on people.

National initiatives such as Verimi demonstrate that a business model is seen from the perspective of the need for trustworthy and simple, cross-provider management of personal data. We see that Germans in particular are significantly more aware of the risks in using and storing their personal data than people in other regions such as in the Scandinavian countries.

I still see dynamics in the field of tension between convenience, privacy regulation, and digital business models.

Telekom Deutschland has introduced a zero-rating product in mobile service to the
market called StreamOn. How well has it been accepted by customers? What
regulatory hurdles must it overcome? How do you assess them? Isn't this product
at its heart an entry into the world of unlimited service rates that has been
common practice in fixed networks for some time now and that is perceived as
a major hurdle to the monetization of data services? What will follow it? What
opportunities for differentiation with pricing are still available to the carriers?

T. Heilen: The limitations on data use volume for mobile networks is a major concern for many users today. The second sentence the host of a house party hears is the question about the WiFi password. Even a poor WiFi connection is regarded as better—because it's cheaper. At the moment, virtually the only form of differentiation among providers of mobile services is the data volume—more volume up to unlimited is a key subject in the industry.

A completely unlimited data rate is really difficult to imagine at this time. For one thing, the networks differ in their coverage and performance capability. Moreover, our experience with unlimited voice calls shows that this ultimately leads to undesirable side effects of significant scope. We would frequently see more or less stationary use with abnormal use volume—which would be to the detriment of other users in the same cell.

I believe that traffic shaping such as that of the StreamOn product is an outstanding approach as it reconciles the intense use of data on the one hand and network quality on the other to the satisfaction of all parties. That is also how our customers see it; at this time, the StreamOn product is an outstanding differentiation feature that has been welcomed by customers and partners.

How important are partnerships (or will they be) at the service level with OTT providers? Are there differing opinions about cooperation with the Big 5 (Amazon, Google, Apple, Facebook, and Microsoft) on the one hand and with the smaller, more specialized providers such as Deezer, Spotify, etc. on the other? Will we be able to see ecosystems, perhaps aimed at cooperative ventures with carriers?

T. Heilen: We at Telekom began to reach out to partnerships very early. I would remind you of the partnership with Apple and the exclusive marketing of the iPhone 10 years ago. Or think about Spotify, with whom we introduced an exclusive "volume" partnership to the market in 2012—at a time when Spotify was far from being as well-known as it is today.

We observe certain cycles here as well. The example of Spotify demonstrates that partnerships undergo changes over the course of time; an exclusive partnership that originally supported our differentiation can become a more open model with multiple providers. A number of music streaming providers participate in our service StreamOn today.

We will continue to search for partners, especially in those areas outside our core competence, including, and increasingly, in the TV content sector. You can see this today in our partnerships with Sky and Netflix. The premium concept is authoritative here as well; we look for the partners with whom we can offer to our customers the highest quality in content and experience.

Major simplifications can certainly be imagined from the customers' perspective in this respect. Today, every single partner still asks you to register as a customer—sometimes in greater, sometimes in lesser detail—and to provide your personal data. The "possession" of the customer is viewed as a key value for every company. An exciting development in this respect is seen in the current cross-industry initiatives, such as the data alliance Verimi, that are striking out on new paths. There is active competition with the single sign-on procedures used at Facebook and Google, but with data security and privacy in conformity with EU law.

As far as ecosystems are concerned, I see the driving forces here at the global level, including players from Asia such as Alibaba or Tencent, and not only the competition between the EU and the USA. The next few exciting years will reveal what role telecommunication operators will play at the end of the day. Besides the significantly higher number of users in absolute figures, the scope and quality of the data will play a decisive role. WeChat offers many different functions that enable users to spend the major part of their digital lives within this single sphere. Telecommunication providers operating within the European regulatory framework have in comparison a significantly smaller treasure trove of data.

New cooperation models, shorter innovation cycles demand greater flexibility and agility in marketing and the product management of the carriers as well. What skills and capabilities in product management will be required in the future? How should work be organized in this setting? How will carriers successfully disengage from the processes and legacy worlds that have grown up historically and that are frequently sluggish and slow?

T. Heilen: Digitalization, innovation, speed, and "fast fail"—we cannot optimally model any of these on the Tayloristic principle that is here still often anchored in the organization and the target system (one position is concerned with technology, another with IT, the third thinks about the market).

We are also striving to gain the strength of startups and other companies that approach customers more quickly and agilely and monetize them faster. These companies are ultimately an object lesson for us that we must set up end to end. Teams that cover all aspects of development of a service or product are one example. This leads to substantially more devotion to the customers and to the services and products we offer to them; the technician thinks and feels along with the colleague from marketing—and both along with the customers above all and the experience that they will have with us. This takes us to more product- or portfolio-specific setups in our organization and processes that bring together an entire range of disciplines.

Can an entire corporate group be organized along these lines? Google and others like it show that this is possible. But we are not starting out with a wish list on a green meadow, but must adequately deal with our legacy—these are the systems and models that represent work and content today and, as a minimum, tomorrow as well. Naturally, we want to be closer to the market and faster in realization. A modular IT setup is a great help here—for the simpler integration of wholebuy components, for example.

We have long since begun introducing new methods and concepts such as Design Thinking and agile development models into our projects and systems, step by step. The results so far are highly encouraging. Above all, we see a cultural transformation in the direction of greater orientation to the customer, demonstrated (among other ways) in significantly simpler processes for our customers.

This becomes especially clear among the younger employees—they want to work on sustainable, crossover experiences that are tangible for customers. But the same thing can be said for everyone: It is just more fun when you can see your success faster and more directly.

Part IV
Partnering

Success with Partnering: Results of an International Study

Christian Krämer

1 Executive Summary

The main findings of the study confirm the following hypotheses:

1. Partnering has become established as a strategic factor in telco companies and will play an increasingly vital role for corporate development in the middle term (the next 3 years) as well.
2. Partnering is employed especially as a growth driver. Further digital added value is expected to stimulate revenue development, prevent the loss of customers (churn) in the core business, and improve profitability. Sales and innovation partnerships play the dominant role for raising revenues; from the perspective of profitability, service partnerships are also of interest.
3. The contribution of partnering to profits and revenues is still at a conservative level. For Example, the contribution to revenues is no more than 3–5% for 32% of companies.
4. However, 92% operate their cooperative ventures at a positive return on sales!
5. We believe that there is still unmined potential in partner management that could be exploited through partnering to generate contributions to revenues of more than 20% of total revenues (achieved at 36% of the respondents) or the benchmark of 3–5% return on revenues.
6. The examination of the instrumental partnering mix that is used reveals that the partner management maturity of the telco industry still has room for improvement. The possible methods and tools are used only by half or two-thirds, depending on the sector. There are frequently still methodological gaps.
7. Success factor research reveals that companies with strengths in the factors partner strategy, partner selection, partner integration, and partner development

C. Krämer (✉)
Detecon International GmbH, Cologne, Germany
e-mail: Christian.Kraemer@detecon.com

© Springer International Publishing AG, part of Springer Nature 2019
P. Krüssel (ed.), *Future Telco*, Management for Professionals,
https://doi.org/10.1007/978-3-319-77724-5_17

tend to post company profits or contributions to revenue that are twice as high as the corresponding figures in companies with less emphasis on these success factors.

Further professionalization of partner management from the strategy to the realization can enable telco companies to increase the contribution to revenues from cooperation. The greatest lever for success is not only the capability to partner, but above all the speed of cooperation End-2-End.

2 Motivation for the Study and Hypotheses

Three developments inspired the decision to conduct this international study:

1. The dramatic growth in partnerships in the telco industry over the last 3 years
2. The broad and diversified range of partnerships that are distributed and utilized across the entire added-value chain to enhance the services provided by telcos
3. The increase in the announcements of partnerships in the business press and the investor relations community. But what is the decisive difference between the examples of success and the cases in which partnerships fail and that become public now and again?

One striking point is the growth in the number of partnerships between telcos and OTTs. 943 OTT/Telco partnerships have been counted by Ovum Tracker (2017) in the period 2010–2016, as outlined in the next following "OTT-Partnering" article. There is also a sustained rise in platform partnerships between telcos, outfitters, and machine-2-machine partner companies. Telecommunications carriers began in 2009 to invest specifically in M2M and Internet of Things platforms. Eighty percent of the platforms decisive for the partner networks—IT systems and cloud applications above all—have been acquired by the telecommunications companies in the last 4 years. While 39 such partnerships for the build-up of ecosystems for the Internet of Things were noted in 2012, the dynamics almost tripled in 2015, when 87 cooperative ventures of this type were established (see Ovum 2015). These platform partnerships are intended to achieve a broad range of links to partners and market penetration within the framework of smart business networks.

The diversity of cooperative ventures is not restricted to innovation partnerships only. We also see a strong rise in sales and marketing partnerships. The distribution partnership of ProSiebenSat.1 and the cable network operator Unity Media Kabel BW for the purpose of broadcasting regional advertising is an example that falls under this heading. National television broadcasters want to use regional advertising to attract above all new customers with regional market coverage such as furniture stores or breweries. Cable or network operators make the access infrastructure for addressing the target groups available. The network operators want to monetarize further their network services by means of revenue sharing from advertising income (See Horizont net 2015).

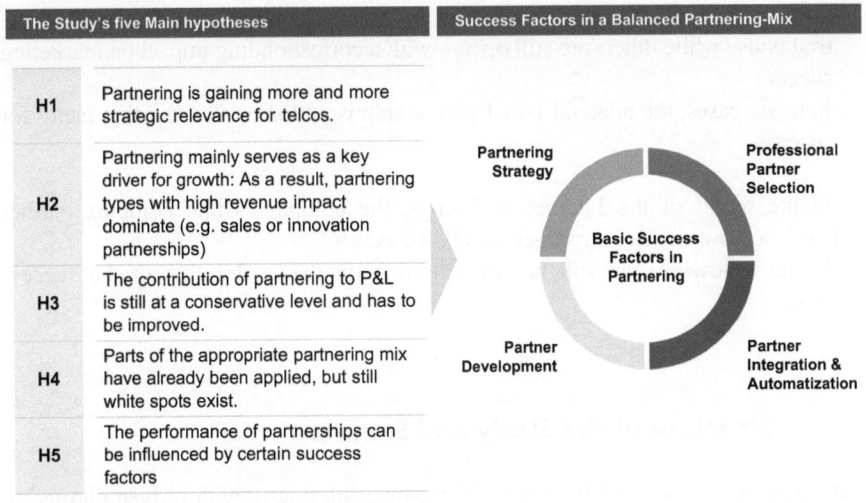

The Study's five Main hypotheses	
H1	Partnering is gaining more and more strategic relevance for telcos.
H2	Partnering mainly serves as a key driver for growth: As a result, partnering types with high revenue impact dominate (e.g. sales or innovation partnerships)
H3	The contribution of partnering to P&L is still at a conservative level and has to be improved.
H4	Parts of the appropriate partnering mix have already been applied, but still white spots exist.
H5	The performance of partnerships can be influenced by certain success factors

Fig. 1 Hypotheses and guidelines of the partnering study

In addition, the competition of a partnership changes from individual companies to competition of value chains (see Krämer et al. 2012). The steering of alliances gains additional significance. The overall issue here is the handling of smart business networks, i.e., trans-industry alliances comprising multiple partners from many different industries and technology contexts (Der Münchener Kreis, 2012). One example here is the ICT alliance built around the greenfield company Ngena (see www.ngena.net. Ngena is a global alliance of service providers based on a sharing economy concept. Cisco is the technology partner. Alliance partners on the service provider side are CenturyLink, Deutsche Telekom, Reliance Jio, and SK Telecom. The alliance partners provide (local) network services while Ngena represents the "hub". Ngena pulls together the connections to the benefit of business customers worldwide. Additional service providers are expected to join the alliance as partners gradually.

3 Our Hypotheses

We have developed the following hypotheses from our market observations and project experience, and they act as the red thread through the study (see Fig. 1).

1. The study assumes that partnering will be assigned an increasingly strategic task. While traditionally organic growth and M&A were at the forefront, partnering is gaining key significance for the corporate development of telco companies.
2. Partnering acts primarily as a growth driver. However, we know from our projects that partnering is a relatively new corporate function. It is suspected that the return on partnering is still conservative, but it must increase as the level of maturity rises.

3. Some of the required partnering tools and methods are being used or are in the trial phase while others are still open—with a corresponding impact on marketing success.
4. In many cases, the bilateral 1-to-1 partnership is no longer the deciding factor for success.

In the world of the Internet of Things, the companies that complex partner networks over a number of stages will be successful.

In the following, we will be concentrating on the exploration of the success factors.

4   Structure of the Study and Sample

The study was conducted by means of personal interviews with proven partnering professionals to obtain the desired depth of detail and confidentiality. Telecommunications companies and associated digital companies were the target group.

Over the course of the in-depth interviews, 25 companies were interviewed—half of them in Germany, half of them abroad. The picture in terms of the size of the interviewed companies was balanced: 38% have fewer than 20,000 employees, 33% have between 20,000 and 100,000 employees, and 29% have more than 100,000 employees. This clearly shows that both market leaders (incumbents) as well as market followers are represented in the sample. In terms of revenue, the focus was between 1 billion euros and 20 billion euros in 2015. The surveyed companies were, for the most part, internationally active.

Interview partners were the heads of partnering departments, CxOs and vice-presidents from the areas strategy/marketing/product and technology at telco network providers who had a close relationship to cooperative ventures. Twenty-nine percent of the study participants held the position of head of partnering. Seventeen percent of the participants were CxOs, i.e., they were members of top management. Twenty-five percent of the respondents were heads of a business unit and were involved in cooperative ventures because of this position.

The interviews were carried out mainly in 2016. They were conducted by a team of Detecon consultants from the Global Knowledge Community "Strategy/Innovation/Products".

The standardized questionnaire encompassed the entire partnering life cycle and explored success factors of cooperative ventures at the corporate level and at the level of strategic business units—in accordance with an underlying reference framework (see Krämer and Newen 2015).

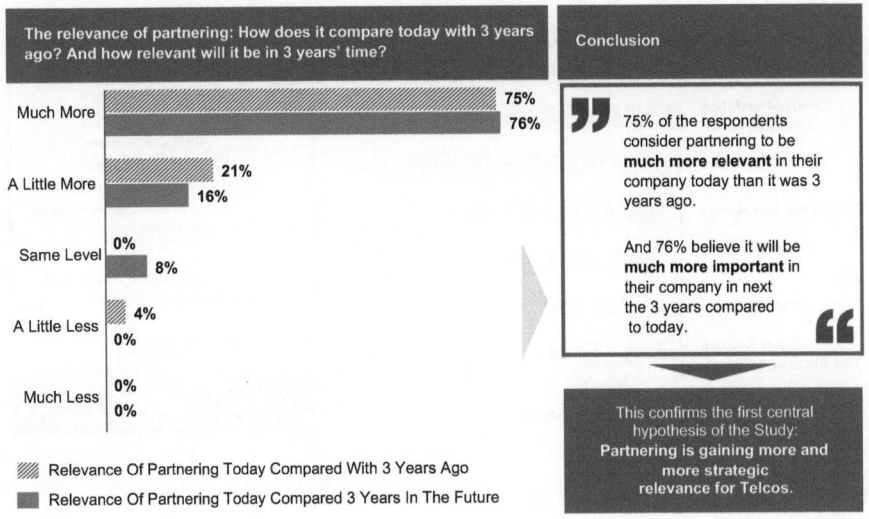

Fig. 2 Significance of partnering

5 Partnering, a Weapon for Growth

5.1 Strategic Relevance for Telcos

The thesis that partnering has matured into an important component of corporate strategy was impressively confirmed. As Fig. 2 illustrates, 75% of the participants confirmed that partnering had gained "very significantly in importance" and 21% noted that it had achieved "rather more relevance" in corporate strategy than 3 years ago. Looking ahead, the participants expect this trend to accelerate further as 92% assume that the significance of partnering in corporate strategy will grow strongly or to some degree over the next 3 years.

5.2 Principle Driving Forces

The desire to increase revenues is at the top of the list and is rated as very or fairly significant by 84%, see Fig. 3. This motive is followed by the goal of portfolio expansion. Sixty-four percent of the questioned telcos rate portfolio expansion and portfolio differentiation by means of partnering as *highly significant*. Sixteen percent still see it as *fairly* significant.

The effort to accelerate the time to market is given by 80% as another important primary reason for entering into cooperative ventures. Instead of the protracted development of their own products or services, companies now look more often

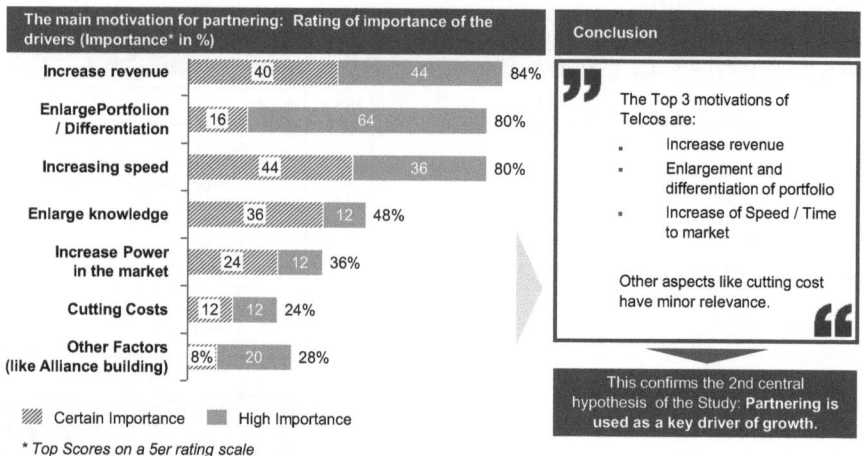

Fig. 3 Primary motives for partnering

for services that "are already functioning on the market". They want to cut back on the costs of their own development work.

Other driving forces include the desire to broaden knowledge through cooperative ventures or to improve the power base, e.g., in the sense of an opposite pole. These objectives, however, are rated as significant by only 48% (knowledge) or 36% (power) of the participants. It is surprising that the motive of reducing costs is mentioned as significant by only 24% of the respondents. This underscores the character of the present wave of cooperation as an offensive action.

5.3   Barriers

Forces that inhibit the development of cooperative ventures have contrary effects. In response to the open question about which stumbling blocks are especially distressing, participants mentioned many different barriers found within the company as well as on the outside. The greatest stumbling blocks outside the company are "problems with the partner". They include differing cultures, a lack of agreement in the partnership goals or the promised services, communications bottlenecks, and/or overlaps that foster competition (e.g., in the offered portfolio).

The majority of the responses concerned a large number of in-company "foot-draggers" that can be roughly grouped under the headings of "target and strategy conflicts", "inadequate authorizations", or "lack of resources".

The stumbling blocks mentioned here must be taken seriously. Unresolved partnership conflicts at BlackBerry in 2014 led to the discontinuation of the partnership with T-Mobile, significantly reducing BlackBerry's corporate value US (See Total Telecom, 2 April 2014). BlackBerry's relationship with T-Mobile US degenerated when it was unable to come to an agreement about device strategies.

In the end, T-Mobile recommended that its BlackBerry users upgrade to an iPhone. BlackBerry terminated the license agreement and suffered a dramatic drop in revenues in 2014.

The necessity to remove stumbling blocks means first and foremost a high standard of excellence for partner management so that the company's internal goals can be balanced against the partner's interests. There is additionally a certain need for resources to enable such steps as the automation of cooperation processes.

5.4   Economic Performance

When we look at the current revenue contribution made by partnering to strategic business units, it is currently between 3 and 5% for one-third of the respondents. In 3 years, only 12% of the respondents will still be striving to achieve this revenue level. Instead, 80% are pushing for higher revenue contributions. Twenty-seven percent of the interviewees want to raise the revenue contribution to the range of 10 to 20% (currently achieved by only 9%); E.g. 44% see partnering's revenue share at a level higher than 20% in 3 years (currently already the case for 36%).

When we look at the current contribution to profit (measured here as return on sales), partnering is already profitable for more than 90% of the respondents. The focus is on 3–5% return on sales, and that is reported by 48% of the survey participants. However, more than half still have profit of less than 2%: namely, 44% with low return on sales of 0–2% or even 8% of the companies that do not achieve any profit or even operate at a loss.

The profit position of partnering must be improved for the majority. A return of 3–5% should be set as an initial benchmark here and has proved to be feasible.

6   Excellence in Partner Management

6.1   Partnering Types

The partnering strategy charts the course enabling companies to achieve corporate objectives through the employment of partners and in which fields they want to press ahead with collaborative initiatives. Key starting point here is the decision regarding the added-value areas in which the company wants to achieve its goals with cooperative ventures. As our secondary research revealed, the four project types of innovation partnerships, sales and marketing partnerships, service cooperative ventures, and purchasing cooperative ventures represent the essential partnership fields for telecommunications companies.

This partnering study sharpens the focus of the image. An explicit prioritization emerges: sales and innovation partnerships are the top of the line.

In their observation of added value, 80% of the interviewed companies indicated that the sales and marketing area is supported the most by cooperative ventures. The innovation area is supported by cooperative ventures almost as heavily (76% of the responses).

The result showing sales, marketing, and innovation partnerships in first place confirms the central hypothesis that telcos utilize cooperative activities to achieve growth and that they prefer types of cooperation that have a strong impact on revenue.

The service sector (services and infrastructure close to the customer) is strengthened through partnerships for 36% and has a midfield position. The purchasing sector lags behind with 24% taking advantage of cooperative activities.

6.2   Partnering Strategy

Only 40% of the companies have a fully defined partnering strategy that is backed up by concrete plans for realization. The majority have a partnering strategy that takes the form of a "mission statement", but without any specific plans for execution. And one-third of the respondents manage cooperative activities in the form of individual initiatives without the framework of a corporate strategy.

There is definitely potential for improvement in the degree of hardness of strategy formulation.

But what strategic action building blocks do telecommunications companies use to profile their cooperation course?

Binding partnering governance featuring clear rules and decision-making processes in the company and with respect to the partners is of high significance for 76% of the respondents. The telcos' desire to maintain control over partner activities stands out here. This reverts to the special features of cooperative ventures; after all, they are based on collaboration with partners who are economically independent. Nevertheless, any impulses to take control should be reined in.

Moreover, 72% of the respondents check a key feature of cooperative initiatives: Does the cooperation make a sustained contribution to the added value of the company?

Third place is held by a managed partner portfolio, i.e., clear strategies determining which partners should be fostered, maintained, or discontinued.

But, coordinated business plans as well as defined partner models with clear "advantage promises (benefits)" range in the middle of the field and are mentioned as relevant by 48%. The problems with partners described before not surprising. It appears that in half of the cases there are no binding and precise agreements and that too little value is placed on the incentive character for the partner. Common understanding, anchored in concrete form in an agreed business plan, along with defined ratings of benefits could significantly reduce the potential for conflict.

Complex configuration models such as joint ventures and the creation of alliances are of only slight relevance. This is still in contrast to the ambitious business models for the Internet of Things, for smart metering, or smart home, which are founded on cooperation with a large number of added-value partners.

There is still potential for improvement and a need to catch up in comparison with other industries such as airlines, which have been optimizing the flight networks through cooperative associations for decades ("Star Alliance", for instance).

6.3 Partner Assessment

"You know that a good partner uses his elbows only to link arms" (Werner Hadulla). But how do you find this partner with the "best fit" for linking arms and persuade him to do so?

Sixty-eight percent of the respondents explained that they follow a systematic approach for the selection of cooperation partners. Sixteen percent stated that they seldom or never filter a partner company systematically and specifically. And another 16% improvise during the selection process.

This is evidence that the famous golf course partnering is still taking place, but nowadays it has only marginal importance or is a product of broader coincidence. The inherent risk here must not be taken lightly.

But what methods and procedures do telcos use during the partner selection phase?

The method of choice is the market and competition analysis, mentioned by 72% of the respondents as important. Conversely, 12% of the respondents described precisely this method as unimportant.

The range of answers about the use of structured selection processes for finding partners is similarly broad (e.g., using gradated selection processes with partner boards and scoring models). This is an important method that is used by 56% of the surveyed companies. On the other hand, only 4% explicitly classified it as unimportant, underscoring the low impact of the "golf course effect".

Surprisingly, due diligence and trend scouting are regarded as relevant by only 36% and 40%, respectively. This is in contrast to the mission of the popular innovation and sales partnerships that rely heavily on market appeal and the trendy degree of novelty in the partner solution. The due diligence method also has no more than niche character, and appears to be reserved for larger-scale partnerships with capital impact.

In summary, it can be concluded that the method of market and competition research familiar from the strategy and marketing field provides the most support for the selection process. The utilization of cooperation-specific methods such as systematic selection processes, structured partner approaches, and trend scouting is important for half of the respondents (more or less), but not for the other half. All is all, there is still potential for enhancing the accuracy of partner selection.

6.4 Partner Integration and Time to Market

When speaking of partnerships, the "time to market" is the length of time it takes to develop a partner from the stage "selected" into operating effectiveness on the

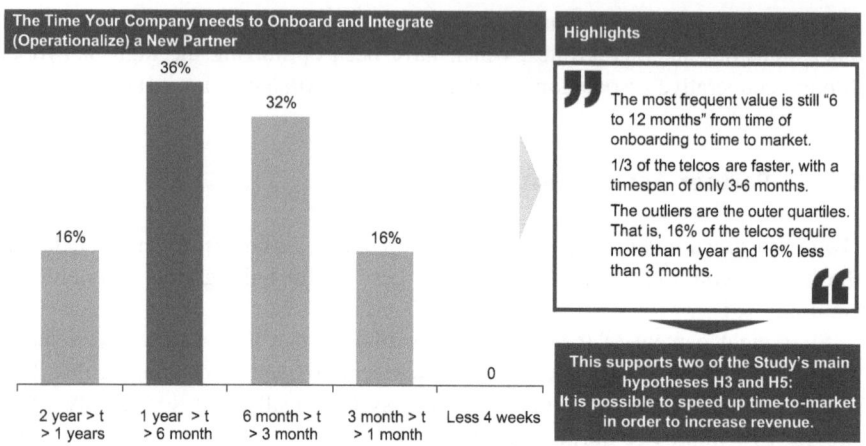

* t =: Time span from „partner selected" to „partnership now operational in the market"

Fig. 4 Time to market

market. The "time to market of a partnership" is an important performance indicator because telecommunications companies frequently enter into cooperative undertakings in search of speed and prefer the partnerships to protracted development on their own. In the case of a solution partnership for business customers (e.g., a CRM Software as a Service (SaaS)), this would be the length of time required by the cooperation partners to conclude the cooperation agreement and to offer the SaaS service in the desired cooperation format to the business customers on the market.

The feedback from the interviews is charted in Fig. 4. The most frequent time period from onboarding to market launch is less than 1 year, but more than 6 months. This was the period given by 36% of the companies, closely followed by 32% of the companies that require between 3 and 6 months for this. The two outliers of this "normal" range each covered 16% of the respondents. The time to market for the "fast" outliers was between 1 and 3 months. The market operationalization for the 16% "slow" outliers was between 1 and 2 years.

In view of the growing competition on the cooperation market and the upscaling of partnerships, rapid integration times are essential. The aim will tend more toward a benchmark of about 3 months than the current time of 6–12 months.

When asked about the automation of the cooperation processes, only 4% of the respondents reported that their processes for providing the services in cooperation ran "automatically and smoothly". The majority (56% of the respondents) reported interface issues and processed the cooperation product to the final customer with high levels of manual work.

The rapid procedural and automatic connection of the partners is consequently a key to success. Deutsche Telekom has recognized the importance of this key factor and uses a so-called "power strip" (See Onlinekosten.de 2014) with IT support to pursue a policy of fast "Connect and Disconnect" of partners. The power strip stands for a new Internet services platform that Telekom partners with their services can simply dock onto and feed their applications straight into the fast Telekom network.

6.5   Partner Development

The value of a partnership is in no small degree dependent on the partner's engagement. In sales partnerships, it can be measured on the sales figures that are posted. Accordingly, 72% of the surveyed telcos answered that they actively manage their partners.

If you probe more deeply and ask about the instruments they use for this purpose, one method stands out. Sixty-four percent of the respondents steer their partners according to principles of management by objectives—i.e., by setting targets. In contrast, the other methods for steering a partner program that can be considered, especially incentives or the provision of support tools (e.g., partner portals), are hardly used at all (by about one-third of the respondents and, in some cases, even fewer).

The determination that a partner-specific development/career path is offered by only one out of ten companies (12%, to be precise) is significant.

In other words, the investment in partner relationships is a differentiation factor that is rarely used. A recent study (see AMI Partners 2016) reveals that there is profitability potential lying dormant here. During a partner satisfaction measurement with cloud service providers, IT partners confirmed that Microsoft offered the best support tools. Support is related to the entire partner company and includes marketing programs, product training programs, productivity and developer software, and certification programs. This degree of support is in correlation with corporate success because Microsoft Gold Partners achieve the highest margins for cloud solutions ahead of Google, Amazon, and Salesforce.

7   A Look at Top Performers and Success Factors

What sets companies that are successful in partnering apart from companies that are less successful?

We accordingly broke down the sample into successful and less successful companies. Our initial hypothesis was that the successful companies would set themselves apart from the less successful enterprises by their greater commitment to the four partnering management dimensions.

The empirical results An inspiring, mutually coordinated, holistic partner strategy with a sustained degree of hardness drives success potential. Extended tests showed that top performers are differentiated in the area of strategy by convincing benefit models (revenue sharing, cost sharing) and also create special funds for partnering. The investments in the partner activity are rewarded by higher revenue and return values.

Another clear success factor is <u>partner integration</u>. Along with standardized and automated processes, it lays the foundation for upscaling the cooperative venture and profitable growth.

Top performers in revenue scored a higher index in <u>partner development</u>. Above-average engagement in partner development pays for itself in higher partnering revenues as a share of total revenue. In the extended analysis, top performers stood out from less successful telcos because of their offer of a partner-specific career path. Moreover, they guide their partners more sustainably with "management by objectives"—guiding the joint operation to success.

Top performers set themselves apart through their <u>professional partner selection</u>! As a result, they recognize partner risks at an early stage, and their choice falls on the best-fit candidate to produce good mutual potential for success.

The **four instrumental factors of are** fostered by general strategic factors that also have an impact on success. They include:

1. **The geographic orientation**: Companies with a **regional market expansion** (i.e., that operate on at least two national markets at the same time) perform at a higher level than operators of national cooperative ventures. Economies of scale, greater learning effects, and the simpler degression of partner investments play an economic role here.
2. **Service, maintenance, and infrastructure partnerships** are more frequently found at top performer companies. This category includes platform partnerships in which telcos are placed at the center as well as outsourcing partnerships. This business model is coupled directly to concrete revenue flows.
3. **Purchasing partnerships**: Even though purchasing cooperatives are far from widespread, they are overproportionately represented in companies that disclose above-average returns.
4. The **organization of the partnering function**: Top performers skillfully combine centralized and decentralized partner organization.

---

8   Perspective: A Vision for High Return on Partnering

Conclusion and main statement of this research: **Partnering success can be programmed**.

1. The major success factors for accomplishing this are the central factors of the partnering mix: a binding strategy, professional partner selection, speed in onboarding and rapid operationalization of the partners, automated partnering processes, plus the depiction of development prospects and the fostering of the acquired partners.

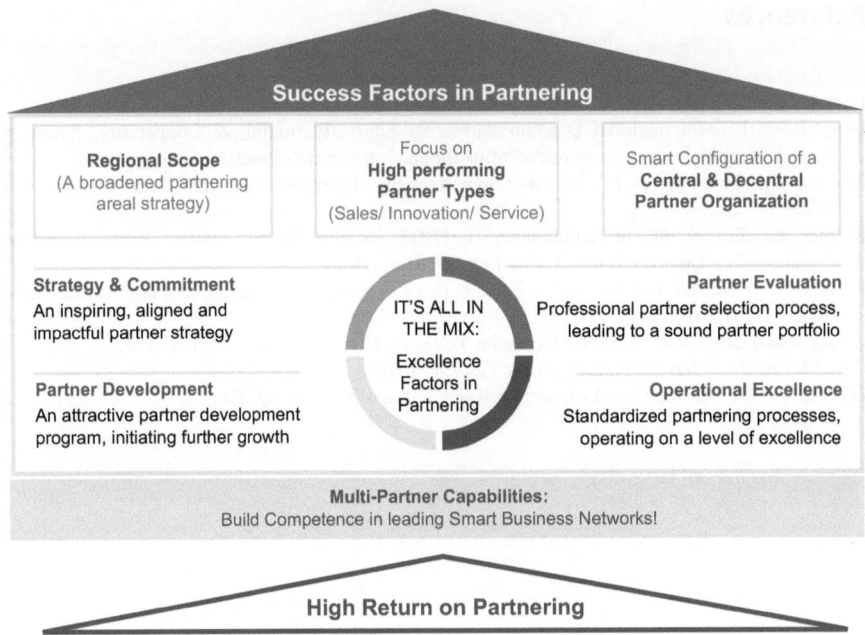

Fig. 5 Vision for high return on partnering

2. General factors such as regional areal strategy, a combined centralized/ decentralized partnering organization, and setting of focal points for the chosen partner types enhance the success of these partner instruments.

If you want to check on the complete details of the study please refer to our Executive study paper "International Detecon Partnering Study (2017)".

Even though there is no "one size fits all" model on the dynamic telecommunications market, the study describes a vision that is illustrated in Fig. 5.

In conclusion, the above investigated principles have a favorable effect that increases the return on partnering.

And there is still potential for improvement Although we identify high-performing action fields in the success factor research, they are frequently not used. We believe that dormant potential for success can be mined by a professional partner selection, fast onboarding, and a rapid operationalization of the partners in the processes, and special potential for differentiation is waiting here.

What ultimately counts is not that you (at some point) are cooperating with someone, but what you make of the cooperation and that you rapidly achieve initial as well as sustained success.

References

Ami Partners. (2016). *Channel partner profitability study*. New York.

Der Münchener Kreis. (2012). *Vortragsband smart business networks*. München.

Detecon. (2017). International Detecon Partnering Study, Partnering & Cooperation Research Project, link: www.detecon.com/de/Publikationen/internatinal-detecon-partnering-study

Horizont.net. (2015, March 25). Distribution partnership of ProSiebenSat.1 and Unity Media Kabel BW.

Krämer, C., Zenner, D., & Schuhbauer, B. (2012). Smarte Grenzgänger – Next Generation Collaboration, Detecon Telco 2032 – Landing Page, Bonn.

Krämer, C., & Newen, K. (2015). Das Beste in den Partnern hervorbringen, Detecon Management Report 2/2015, Cologne.

Onlinekosten.de. (2014, March 6). Deutsche Telekom, Höttges und die Steckerleiste.

OVUM. (2017). Telco-OTT Partnerships Tracker, 2016.

OVUM. (2015). Machine-to-Machine and IOT Contracts Tracker 4Q, 2015.

OTT Partnering with Telcos: On the Rise

A Situational Partner Management Approach

Christian Krämer and Riem Jalajel

1 OTT Partnering on the Advance

1.1 Emergence of OTTs

In today's world, the services and contents offered by over-the-top (OTT) companies like WhatsApp, Netflix, or Spotify have become a part of the everyday lives of millions of people all around the globe. Quickly accepted by tech-savvy millennials—the generation born between 1980 and 2000 and the first to grow up with smartphones—the number of OTT players is constantly growing.

As is their impact on the telco market. When Skype, WhatsApp, and other OTTs first appeared on the communications market, operators felt threatened and unsure about how to react. They pursued several strategies, ranging from denial to emulation, in their efforts to deal with the new competition. The results were mixed, at best slowing down the growth of these services. Now we see a turnaround in strategy that can be summed up in a single word: "cooperation". The hype surrounding partnerships with OTTs generated by this strategic about-face is substantial. Telcos have come to perceive linking with the creativity and efficiency of the global OTT/app market to be more promising than opposing it.

In the past six years, 943 partnerships of this type have been established worldwide. Their numbers have grown significantly with the advent of content providers such as Spotify, HBO Go, or Netflix and have doubled since 2014. Indeed, 61% of these partnerships were concluded with content provider partners in 2016 alone, as shown in Fig. 1 (see Ovum Partnerships Tracker 2017).

C. Krämer (✉) · R. Jalajel
Detecon International GmbH, Cologne, Germany
e-mail: Christian.Kraemer@detecon.com; Riem.Jalajel@detecon.com

© Springer International Publishing AG, part of Springer Nature 2019
P. Krüssel (ed.), *Future Telco*, Management for Professionals,
https://doi.org/10.1007/978-3-319-77724-5_18

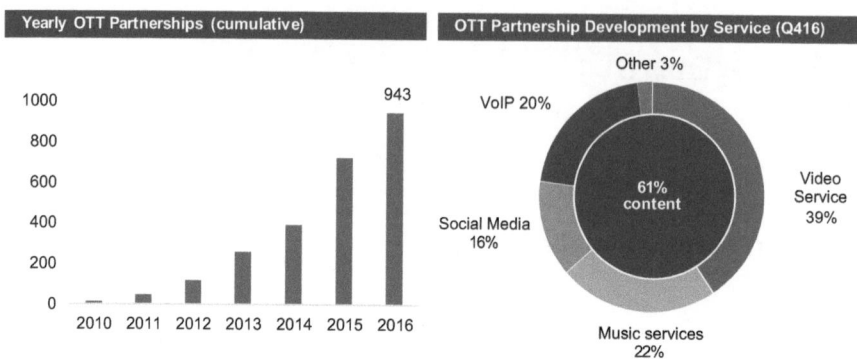

Fig. 1 Development of OTT partnerships and their functional scope (see Ovum Partnerships Tracker 2017)

1.2   Snapshots of OTT Partnerships

According to Detecon research, WhatsApp, Facebook, and Instagram have outgrown eight major telecommunication operators, including Verizon, Vodafone, and Deutsche Telekom in terms of subscribers (OTTs: 2.457 billion active users compared to 2.110 billion total mobile subscribers on the operator side).

Alongside this development cooperation models between telcos and OTTs have evolved from partnerships focusing on sales and innovation to more strategic partnerships and joint ventures. Facebook, for example, has joined broadband provider Surf in Kenya in a partnership to provide Express Wi-Fi that will give access to affordable Wi-Fi hotspots to customers and has already launched more than 600 hotspots (see Mumo 2017). In March 2017, Netflix announced a strategic partnership with Vodafone that will make streaming service available to Vodafones mobile customers (see ET Telecom 2017).

1.3   Benefits and Risks of OTT/Telco Cooperation

As the examples partly illustrate, both parties benefit when engaging in a partnership. Operators form partnerships with OTT players to strengthen customer loyalty by including OTT services in their own portfolio and providing access to added-value services. As OTT services require the connectivity provided by operators, offering added-value OTT services increases the use of messaging, voice and data activities on the operators' networks, enlarging telcos footprint. The growing array of apps and services ensures that telecom operators enjoy continuous growth in demand for bandwidth and mobile data services.

New revenue streams can be secured on top of the stabilization in core revenue. By including innovative technologies and services developed by startups and OTT players, operators do not need to invest time and resources in developing their own services. But operators are not the only ones to benefit from these partnerships. OTTs can start building a global user base by accessing the telcos' clientele. Startup OTTs in particular can learn about customer acquisition, mass communications, and customer touchpoints from telcos, benefit from brand exposure and financial resources, and gain market insights (See Athow 2015).

However, OTT/telco partnerships entail risks as well. These partnerships accelerate the entry of potential new competitors in the telecommunications market. Telcos might lose customer relations, which may result in a shrinking market share. For OTTS on the other hand, the risk consists of their high dependency on the connectivity provided by the telcos. A telco's potential exit from a cooperative venture would immediately shut down operations—in the short term at the very least—and significantly harm the OTT's revenue flow.

2 Telcos' Options for Competition Strategy

But how should telcos act in the face of OTTs and their growing power on the market? As described by Dörflinger and Heuermann (2015), there are three possibilities besides the zero option of "do nothing": defense, offense or cooperation.

Differentiation in analysis of the specific circumstances is required in the choice of the competitive actions. One determining factor is undoubtedly the competitive impact of the services offered by the OTT on the telco's own portfolio—is the effect one of cannibalization or reinforcement? The other is the OTT's market power. The greater its market power, the more substantial the impact of the services it offers. Market power and impact of the OTT's services can be depicted in a matrix of the OTT company types and the competitive actions for the telcos as shown in Fig. 2.

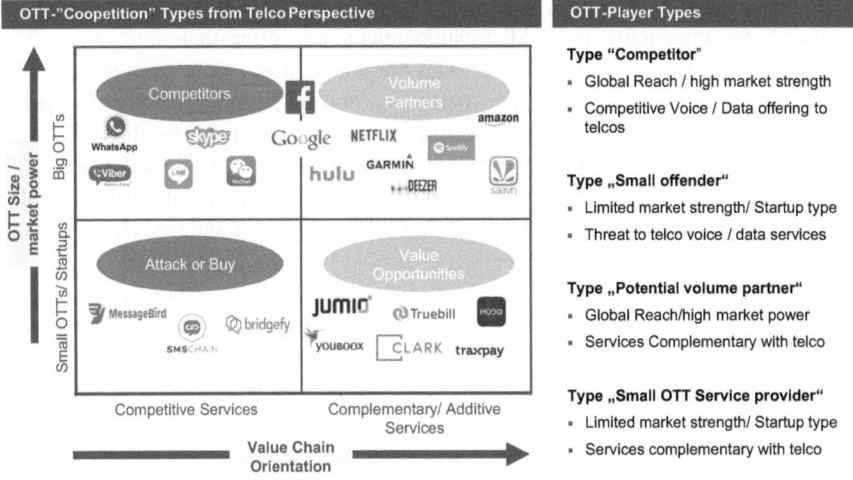

Fig. 2 OTT types and telco competition

Below are some theoretical competitive strategies that telcos might consider as a response to the different OTT types.

1. The greater the extent to which the OTT serves are competitive ("cannibalizing")—one example being WhatsApp versus the telcos' text messaging business—the more appropriate it is for the telcos to pursue a strategy of defense or offense.
2. A strategy of cooperation, on the other hand, is all the more appealing the more complementary the offered services are with the telco's portfolio, i.e., when there is little immediate overlapping. In this case, the OTT's product (app/service) has more of an additive character and may, in certain cases, even provide access to a new target group. One example of this type is entertainment services in the music sector such as Spotify or Deezer, where telcos previously had only a limited footprint.
3. Spotify or Netflix are among the firmly established OTTs that have already developed substantial market strength on the music or TV markets. This has an impact on the cooperation strategy. Telcos must treat them as big players and volume partners with the objective of limiting customer churn or of opening new market segments. OTTs in this position often expect a significant revenue share in recognition of the market strength of their products.
4. Smaller OTTs such as Truebill may serve as value partners. Truebill e.g. offers their users to track their subscriptions and enable them to pay their services using mobile accounts. They represent a value enlargement for telcos. Engaging into a partnership in an early development stage of The OTT before hype is advantageous for telcos in commercial terms, but entails a high flop risk as well.

In this article, we will be illuminating primarily the right-hand side of the matrix (in other words, the "cooperation" side). We will be looking at the cooperative strategies towards OTT types which can be used as "volume partners" or "value partners" for enhancing the added value telcos can offer their customers. The segment of OTT volume partners is well-known, as they represent the big OTT players. The segment, called value partners, represent the smaller OTTs and is mostly composed of startups. An example of a new segment of smaller OTTs are Fintechs (see Arner et al. 2015; Dapp et al. 2014). Fintechs drive digital transformation in e-Commerce, as they typically offer services in the fields of artificial intelligence, big data, digital payment systems, or contract management (see Brummer and Dorfine 2014).

Why are Fintechs of interest for telcos? Because the latter can benefit (if they act intelligently) by complementing their services and adding value to existing portfolios in vertical industries. Fintechs are therefore potential candidates for innovation partnering. Fintechs such as Jumio, traxpay, or Truebill offer opportunities for telcos to provide B2B/BSC-centric financial services such as mobile/digital payment (e.g. www.46degrees.net).

In consequence, operators such as Telefonica have set up their own startup accelerators like Wayra to get in touch with these startups. In the final stage these startup partners can also be used as value partners to refine digital products.

3 Strategies in Cooperation Management

Telcos must be extraordinarily sensitive in pursuing cooperation strategies with OTTs because, even though partners, they have different demands; moreover, in the worst case they can also morph into competitors.

The international partnership study (see Krämer—Success in Partnering): International Partnering Study (Detecon, 2017) revealed general success factors in cooperation management:

- An inspiring, coordinated, specific, and holistic partner strategy with a sustained degree of resoluteness drives success potential.
- Top performers set themselves apart through their professional partner selection. They stand out for their systematic partner and competitor analysis, and they make use of a model-based selection process with a standard of review that moves in the direction of due diligence.
- Another clear success factor is partner integration. Along with standardized and automated processes, it lays the foundation for upscaling the cooperative venture and profitable growth.
- Top performers also scored higher in systematic partner development. Above-average engagement in partner development pays for itself in higher revenues for the partner.

There must be a fine differentiation among these factors dependent on the market strength of the target partner. There is an overview of the different building blocks in these segment-oriented partner management in Fig. 3.

When **large "high-volume" OTTs** become part of the partnership, attention must be devoted to clear added-value growth stimuli in the partner strategy, the provision of clearly defined partnership models, and control as well as exit points. Telco and OTT work as peers; it is even possible that in terms of sheer company size, the OTT is larger. Accordingly, the telco must focus on persuasion during the partner acquisition and establish an appropriate "beauty" position during the initial approach. A mutually accepted business plan, secured by a partnership agreement, is an indispensable basis for management and collaboration. Professional meshing of the activities at the commercial and technical levels must be organized so that high volumes in sales and a scaling of the partnership are possible.

The partner study indicates that a short time period for getting the partner to the market is an especially important success factor. In addition, extensive technical integration to ensure smooth E2E delivery to end customers is a good way to avoid points of friction and to reach significant volume targets. Partner development here focuses primarily on partner steering. It must be supported by individual incentive

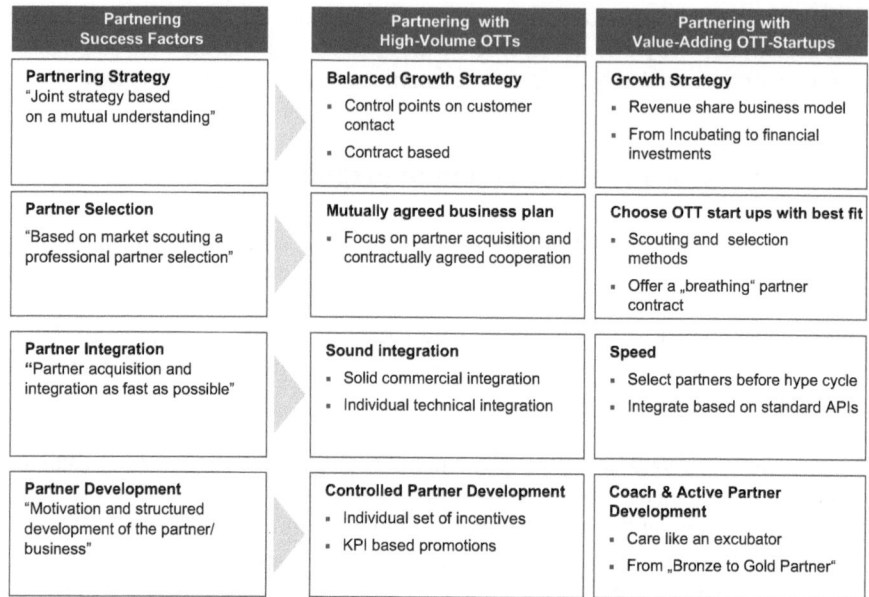

Partnering Success Factors	Partnering with High-Volume OTTs	Partnering with Value-Adding OTT-Startups
Partnering Strategy "Joint strategy based on a mutual understanding"	**Balanced Growth Strategy** • Control points on customer contact • Contract based	**Growth Strategy** • Revenue share business model • From Incubating to financial investments
Partner Selection "Based on market scouting a professional partner selection"	**Mutually agreed business plan** • Focus on partner acquisition and contractually agreed cooperation	**Choose OTT start ups with best fit** • Scouting and selection methods • Offer a „breathing" partner contract
Partner Integration "Partner acquisition and integration as fast as possible"	**Sound integration** • Solid commercial integration • Individual technical integration	**Speed** • Select partners before hype cycle • Integrate based on standard APIs
Partner Development "Motivation and structured development of the partner/ business"	**Controlled Partner Development** • Individual set of incentives • KPI based promotions	**Coach & Active Partner Development** • Care like an excubator • From „Bronze to Gold Partner"

Fig. 3 Situational cooperation strategies

structures. An in-depth analysis in the partnering study indicates that top performers guide their partners more sustainably with "management by objectives"—guiding the joint operation to success.

When **smaller OTTs or startups** are involved, the telco frequently assumes the role of coach or channel captain. The preconditions for an unrestricted growth strategy must be created here so that the OTT's business model can realize its full potential. Defined partnership models of the telco, perhaps embedded in partnership programs, serve as solid signposts for orientation.

Telcos must place especially high value on professional partner selection when initiating the partnership so that they acquire the OTTs with the best fit and high market potential as partners. They recognize partner risks at an early stage, and their choice falls on the best-fit candidate to ensure high potential for success for both parties. As they are the stronger party, the acquisition of partners is seemingly a little easier. However, telcos need to adapt to the differing corporate culture of startups and be willing to acknowledge it.

Telcos have to build up flexible structures and simple communications points to embrace the culture of these startups. Beside this they have to define assets, which are appealing to those OTTs like Fintechs, that are interested in financial support and additional resources of the telco such as selling power, access to an existing (and broad) customer base, topped by access to a business network and the interchange of new ideas.

Timing is an important success factor here because the corporate value and market position of the OTTs can change quickly. Telcos will find it to be advantageous to initiate the partnership at an early stage, before the OTT has reached hype status in its growth curve. Revenue-sharing models can be negotiated more easily when the OTT is still in the initial stage of growth. High importance must be placed on partner development within the framework of a partner program. The OTT then has motivation to grow from its own goal perspectives. Surveys in the partnering study showed that top performers also create special funds for partnering.

4   Summary and Outlook

OTTs have a pivotal role to play in the transformation of telcos from a bit transporter to orchestrator of digital added value. Telcos are respectively developing into integrated telecoms service providers, while OTTs are making a major contribution to the expansion of the added value services of telcos that are targeting key growth areas such as content and cybersecurity.

Conversely, cooperation also entails the risk that customer relationships of telcos will be undermined. A gradual improvement of the OTTs market position might occur at the expense of the telcos. The discussion of the cooperation potential leads to the following **approach for telcos**.

During the first stage, the telcos must identify the competitive power or the value-add power of the OTT offering. Does the considered OTT have more a cannibalization effect or a stimulus for growth by adding value?

Recommendation #1: **Focus strategically on partnerships in added-value areas** that can be managed and strengthened by partnering; the trend is towards music, video, TV, gaming OTT services.

If this question is answered affirmatively in the sense of a complementary business model with opportunities for advancement, telcos must manage their dealings with OTTs in a second stage holistically.

Recommendation #2: **Transport your partner strategy by means of directed use of the partnering instruments**. Key success factors of partnering (see previous chapter): a convincing partnering strategy, professional partner selection, partner development, and rapid partner integration.

In the realization phase, success will be all the greater the more the telcos give situational consideration in their cooperation management to the various requirements and unique characteristics of the relevant OTT types. Greatly simplified, it can be said that:

Recommendation #3: When OTT **volume partners** are involved, the tendency would be to view **a "control strategy"** as suitable (strict partnering governance, management by objectives). A deliberate **"care" strategy** with partner fostering and consistent partner development is more appropriate for **smaller added-value OTTs.**

Enormous sensitivity is required when telcos begin seeking out OTTs for partnering. A situational conduct of actions in partnership management by Telcos will be rewarded by a higher return on OTT partnering.

References

Arner, D. W., Barberis, J., & Buckley, R. P. (2015): The evolution of Fintech: A new post-crisis paradigm. *Geo Journal International.*

Athow, D. (2015). Why network operators must embrace innovation – and OTTs – to survive. Accessed September 09, 2017, from http://www.techradar.com/news/phone-and-communications/why-network-operators-must-embrace-innovation-and-otts-to-survive-1286103

Brummer, C., & Dorfine, D. (2014). *FinTech – Building a 21st-century regularly toolkit.* Milken Institute.

Dapp, T., Slomka, L., & Hoffmann, R. (2014). Fintech – The digital (r)evolution in the financial sector. Deutsche Bank Research.

Detecon. (2017). *International partnering study.* Köln.

Dörflinger, T., & Heuermann, A. (2015). Netzneutralität und Erosion von Geschäftsmodellen. Cologne.

ET Telecom. (2017). Vodafone offers free Netflix for a year to postpaid customers, inks carrier billing deal. Accessed October 05, 2017, from https://telecom.economictimes.indiatimes.com/news/vodafone-offers-free-netflix-for-a-year-to-select-red-customers-/59333307

Investopedia. (2017). Over the top.

Kuthari, M. (2017). Ovum Operator – OTT Partnerships Tracker: 1Q17, 2017.

Mumo, M. (2017). Facebook to hold contest for 15 Africa start-ups in Nairobi. Accessed October 09, 2017, from http://www.businessdailyafrica.com/corporate/companies/Facebook-hold--contest--15-Africa-startups/4003102-4132292-q1cgt7z/index.html

Cooperative, Connected and Automated Mobility

Overcoming the Loss of Strategic Competences by New Co-operation Models for Automotive and Telecommunication Industries

Wolfgang H. Schulz, Horst Wieker, and Bettina Arnegger

1   Introduction

The mobility megatrend is transforming entire industries and demands cross-industry and cross-product solutions. Cooperative, connected and automated mobility (CCAM) are decisive key technologies during the changes on the mobility market. Integrated networks are an alternative solution. To implement CCAM, the automotive industry has to overcome white spots in its strategy specific competences by inter-industrial co-operation. Function-specific co-operations will create new maneuvering room for the automotive and telecommunications industries as well as time advantages when adapting previous business models to the new megatrends. However, as diverse barriers and isolation mechanisms exist, and new co-operation models for automotive and telecommunication industries are required. Inter-industrial co-operation is simplified with the aid of the Institutional Role Model (IRM).

The automotive industry is one of the key drivers behind Germany's economic power. However, it is under substantial social pressure to ensure that its actions are ecological, economic, and social. Society's demand for a fundamental transformation in the direction of sustainable mobility continues to grow in intensity

W. H. Schulz (✉)
Zeppelin University, Friedrichshafen, Germany
e-mail: wolfgang.schulz@zu.de

H. Wieker
Hochschule für Technik und Wirtschaft des Saarlandes, Saarbrücken, Germany
e-mail: wieker@htwsaar.de

B. Arnegger
Zeppelin University, Friedrichshafen, Germany

NVIDIA, Santa Clara, CA, USA
e-mail: b.arnegger@zeppelin-university.de

© Springer International Publishing AG, part of Springer Nature 2019
P. Krüssel (ed.), *Future Telco*, Management for Professionals,
https://doi.org/10.1007/978-3-319-77724-5_19

(Kemp et al. 2012). As a result, the automotive industry is under enormous pressure to adapt and remain competitive because the path in which traditional business models will develop is not clear. In this transformation phase, cooperative, connected and automated mobility systems are decisive key technologies. Facing the powerful dynamics of the technological developments, the automotive industry has to overcome the loss of strategic competencies by adapting new co-operation models with the appropriately selected competence holders such as the telecommunications industry (Köhler et al. 2009).

2   The Significance of CCAM

Cooperative, connected and automated mobility system, CCAM for short—are key technologies based on C2X- and/or 5G-communication. The automotive industry prefers s short-range technology, which uses a defined band of the spectrum or radio frequencies for C2X-communication. The telecom industry, contrary, favorite an open long-range cellular system, which needs as basis technology 5G. Both technology approaches have economic advantages and disadvantages. However, at the end, it is neither the one nor the other because the business use case and the ease of use will lead to mixed technology approaches. There is no need for a battle between automotive industries and telecommunication industries about a dominant technology. Both industries struggle with their strategic competencies. Even one of both industries wins the technology battle; it does not mean that this will be the dominant technology of the future because other competitors outside of both industries will implement alternative technologies that are based solely on the customer needs (e.g., StreetScooter). The times are over that monopolistic advantages can be reached by pushing through a technology standard. Therefore both industries have to concentrate on the question, which technology fits to which business case, and which is the fastest way to reach the customer satisfaction.

The socio-political acceptance of CCAM is high because of the substantial potential for improvement of road safety and transport efficiency in the direction of sustainable mobility. There is a correspondingly high readiness of transport policymakers to support the implementation of CCAM- by making complementary investments in road infrastructure. The co-operative technologies have been defined and tested over the course of many research and development projects. The systems have been tested under actual road traffic conditions in major field trials at both the national level (Safe and Intelligent Mobility—Testfield Germany: www.simtd.de/index.dhtml/enEN/index.html) and at the European level (for example DRIVE C2X: www.drive-c2x.eu/). Moreover, results from the simTD (Safe and Intelligent Mobility—Testfield Germany: www.simtd.de/index.dhtml/enEN/index.html) project in Germany have verified that general economic benefits can be achieved. The benefits obtained from significantly enhanced road safety alone would lead to economic savings in resources in the amount of eight billion euros a year if broad-area availability were achieved by complete outfitting of all cars and trucks (Schulz et al. 2013a). Along with the road safety effects, however, improvements in transport efficiency are also possible. Transport efficiency effects arise above all from the avoidance of congestion and detours achieved by the optimization of traffic flows. These results are decisive for productivity-oriented transport, e.g., business trips in cars, and road freight traffic. The primary economic effects would be achieved by the time savings that could be realized and reductions in the amount of travel required. It was possible to demonstrate within the scope of simTD that such savings in Germany could reach a magnitude of four billion euros a year (ebenda). It must be noted, however, that these efficiency effects from co-operative systems in productivity-oriented road transport would make it possible for companies to realize further savings in transaction costs because the logistics processes can be optimized. Such savings in transaction costs result from improvements in the spatial division of labor. Furthermore, both labor and capital productivity can be increased. From an overall economic perspective, these positive downstream effects of co-operative systems are much more decisive because they would strengthen the competitive position of Germany as a location. At this time, however, there are no estimates concerning the magnitude of these overall economic effects of co-operative systems.

3   The Automotive Industry's Loss of Strategic Competences as an Obstacle to Successful Implementation of CCAM

Technological product and process innovations have in the past played a decisive role in facing the powerful dynamics of the technological development. In the meantime, however, it has become apparent that transformation must encompass the mobility sector as a whole. This transformation goes beyond the previous advances in technology and requires cross-industry and cross-product solutions in order to find an adequate response to the changes in mobility behavior and to successfully implement CCAM (Köhler et al. 2009).

Intermodal mobility, a turning away from individual car ownership, and new business models such as carsharing are only some of the buzzwords that describe the new aspects of the megatrend mobility, shaping the industry's competitive dynamics.

The decisive stimulus for competitive pressure comes from the heterogeneity of resources. Competition in the automotive industry has always been characterized by the way carmakers differentiate their positions by means of company-specific expertise such as rear-wheel versus front-wheel drive. This type of intra-industrial competition has dominated. Now new developments such as connectivity, cloud computing, big data analytics, and open data platforms are giving rise to new strategy-specific expertise, as also described with the help of the methodological analysis framework presented in Fig. 1.

However, for the adaption of previous business models to the new megatrends, the automotive industry requires new strategy specific expertise, which could be provided by various competence holders. Relevant competence holders, such as digital companies, telecommunications companies, and startups have cross-financing potential, with changes in competitive conditions already becoming noticeable in this area. Competition within the area of strategy-specific expertise

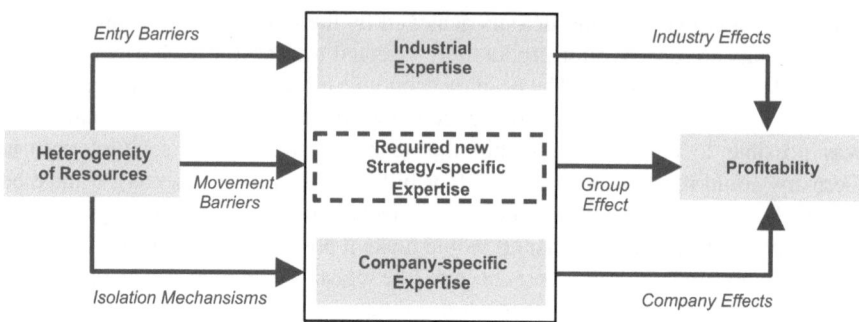

Fig. 1 Competitive dynamics in the automotive industry. Source: Own presentation based on Schmalensee (1985)

was previously found within the supplier industry; examples are Bosch with ESP and ZF with 8-speed transmissions. As a rule, carmakers were able to exploit this competition among suppliers to increase their own profitability. Now that other stakeholders are becoming relevant for this area, it remains to be seen how profitability in the automotive industry will change. The area of industrial expertise is of the highest relevance for current developments in competition. This is where it will be determined whether the "old industry"—the automotive industry in the sense of all manufacturers as a group—will be able to ward off the market entry of "new industries"—digital companies as well as the telecommunications industry. The successful market entry of "new industries" will, as a rule, have substantial impact on the locations because the agglomeration centers of the "new industries" will win out over the agglomeration centers of the losing industry. Since the management behavior patterns of the new competitors do not have the traditional contextual relationship to the automobile, the impact will be felt on the competitive action parameters of price, quality, capacity, advertising, innovation activities, and R&D investments as well. Besides this general competitive situation, however, ongoing developments in the following areas must also be considered:

- Competition will intensify because of the market entry of new providers from the BRICS countries. The market entry of German automotive companies on emerging markets in the form of local production facilities will heighten the competition between established firms and newcomers even further.
- Internal competitive pressure from suppliers striving to achieve extensive vertical integration will grow (strategic competition).
- Technological innovations will enable new players from other sectors of industry such as digital companies, telecommunications companies, energy utilities, and financial service providers to enter the market (company-specific competition).

Until now, carmakers have always found ways to block the market entry of potential pursuers from the supplier industry to end production. One example is Magna's failed acquisition of Opel. But the classic strategies for blocking market entry are useless in the case of current and future newcomers such as Tesla and Google because these providers have such power on the market (thanks to their financial strength and the opportunities for refinancing that are available to them) that battle strategies have no effect and cutthroat competition can be avoided only through co-operation.

4   The Automotive Industry's Need for Closer Collaborations with the Appropriately Selected Competence Holders Such as the Telecommunications Industry

The ability to internalize strategy specific competence from other industries by co-operations will be crucial facing the competitive dynamics of the automotive industry, as also outlined in Fig. 2.

Within the framework of external growth it can mainly be distinguished between forms of co-operations (function specific co-operations or licensing) and acquisitions (asset deals including joint ventures or share deals). Main distinguishing feature between these two main forms of external growth is the level of involvement measured by equity capital transfer and the influence on the legal and economic autonomy of the involved companies (Küting 1989). The Central benefit of entering into a co-operation is gaining access to targeted competences at decreased costs due to shared capital contributions and ultimately shared risk among the collaborating companies. Targeted competencies can hereby specifically be selected and shared comparatively within a defined timeframe. Also economies of scale by reaching a critical mass (e.g., R&D mass) can be achieved (Seubert 2010). In addition, also positive image improvement effects as a reason or side effect of cooperations shall be mentioned (Bosshart and Gassmann 1996) In contrast to acquisitions activities, partners in cooperations maintain their legal and economic autonomy and the co-operation decision remains reversible. In highly dynamic market environments like the automotive industry, this offers the opportunity for an incremental decision making process during the venture's temporal evolvement. The term "strategic alliances", which has become prominent for function specific co-operations hereby

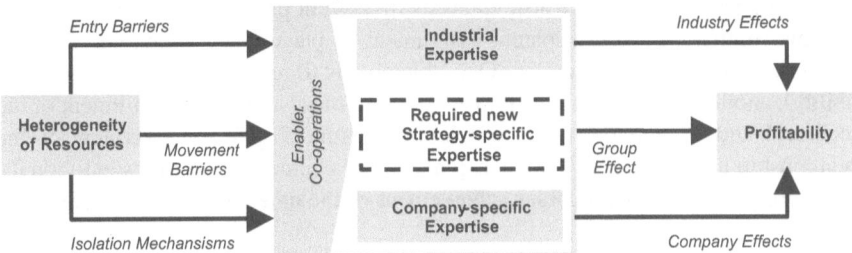

Fig. 2 Co-operations as an enabler for internalization of strategy-specific expertise for the automotive industry. Source: Own presentation based on Schmalensee (1985)

emphasizes the qualitative and strategic value-add for the co-operating companies (Seubert 2010).

One possible co-operation strategy for the automotive industry could be to make use of the communications architecture of the New Economy for CCAM instead of developing their own communications infrastructure (Wieker et al. 2014a, b).

This means, however, that the automotive industry must intensify function specific co-operations with the appropriately selected competence holders from "new" sectors of industry such as telecommunications companies, digital companies, energy utilities, and financial service providers. However as empirical studies have shown, co-operations tend to be rather unstable and require high steering and coordination efforts, which are often linked to lengthy negotiation cycles (Müller-Stewens and Lechner 2005). As the level of complexity is very high in CCAM ventures due to the demanding technical requirements and the necessary networking of a large number of public and private institutions, hence traditional models of collaboration did not offer a suitable starting point for reducing the complexity to such a degree that sustainable and functional business models, i.e. in the field of CCAM, could be established.

5 New Collaboration Models for Automotive and Telecommunication Industries put into Practice

The CONVERGE project (http://www.converge-online.de/?id=000000&spid=en) has been able to demonstrate how non-discriminatory, inter-industrial co-operation for the market launch of a communications platform in conformity with competition law must be prepared. As a result, the project was able to show that non-proprietary solutions among various industries for the implementation of CCAM are both technologically and economically possible. The telecommunications industry was represented by VODAFONE and Ericsson in CONVERGE while Volkswagen, Opel, and BMW took part on behalf of automobile manufacturers. Also, Bosch attended as a representative of the supplier industry. In CONVERGE, the theory of the institutional role models is used instead of the business modeling of the past. CONVERGE applies the "Institutional Role Model Matrix" (IRMM) for coordination of the various players, whose interests diverge in part. IRMM is an aid for the non-discriminatory and transparent assignment of players and their roles within a networked and complex situation. One objective is to reduce complexity (manageability), another is to identify the players who are most suitable for fulfillment of the technical and economical roles (efficiency). Figure 3 gives an overview of the relationship that exists between roles, players, stakeholders, institutions/individuals, and third parties and that enables achievement of the objectives using actions.

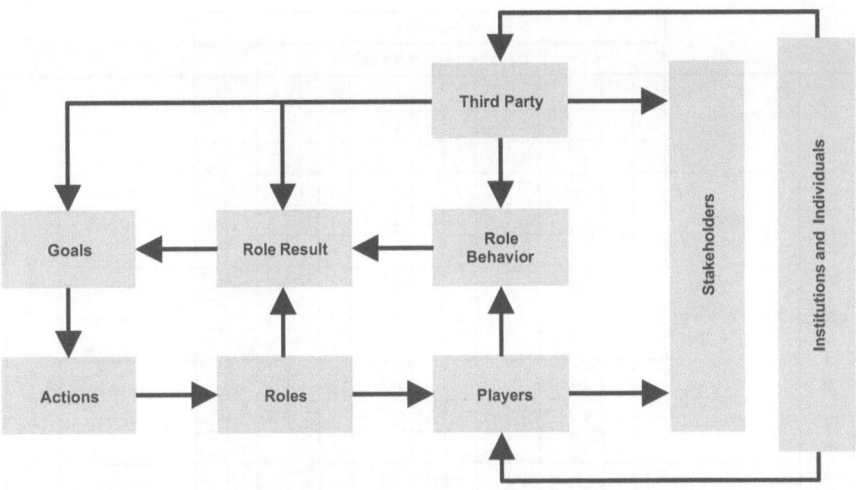

Fig. 3 Approach of the institutional role model. Source: Own illustration

6 Inter-industrial Co-operation Is Simplified with the Aid of the Institutional Role Model

The theory of the institutional role model is essentially based on the principles of institution economics (Schneider 1995), system theory (Luhmann 2002) and the theory of system dynamics based on industrial economics (Schulz 2005). The IRM approach was first taken and its concepts evolved within the framework of the research project on the subject of operator models and implementation scenarios for co-operative systems of the Federal Highway Research Institute (Schulz et al. 2013b). The matrix of solution variants for intelligent transport systems in road traffic developed by the Federal Highway Research Institute (BASt) is a part of this conceptual expansion (Lotz et al. 2014). The CONVERGE project has provided a closed use case for a cooperative integrated network (Vogt et al. 2013; Schulz et al. 2014). A broader application to fundamental issues of the critical infrastructures is carried out (Geis and Schulz 2015). In a general sense, institutions are understood to mean regulatory systems (rules) and action systems (organizations) (Schneider 1995). As a whole, regulatory systems encompass codes of conduct and customs on markets and in other organizations as components of market and company constitutions along with legal frameworks. Action systems cannot exist without acting persons. This characteristic separates the action system from the regulatory system (Schneider 1995). A regulatory system, in other words, is an abstract construct of sentences, standards, and conditions. The smallest unit of an action system is an individual; a company is a larger unit.

Figure 4 shows the final IRM matrix for the economic roles. The 15 partners in the CONVERGE project are shown in anonymous form as institutions identified by the numbers from 1 to 15. Each of the partners has defined its preferences regarding

Meta-roles		Market Phase — Development & Research																Assessment
		1	2	4	1	3	2	1	1	2	2	4	4	2	5	4	own view	
	Business Management	1	2	3	2	3	3	2	2	3	2	2	3	3	3	3	partner view	
		2	2	3	3	3	2	2	1	3	3	2	4	3	4	4	neutral view	
		1	1	5	2	3	3	1	1	4	2	1	3	3	5	4	own view	
	Data Gathering	1	2	4	1	3	3	2	3	4	2	1	4	3	3	4	partner view	
		2	2	3	2	4	4	2	1	3	3	2	3	3	4	4	neutral view	360° Assessment
		1	1	5	2	4	2	1	1	5	2	1	3	4	5	4	own view	
	Data preparation	2	3	4	1	3	3	2	3	4	2	1	4	4	3	4	partner view	
		2	2	3	2	3	3	2	1	3	3	2	3	3	5	4	neutral view	
		1	1	4	1	3	2	2	1	5	2	4	3	4	4	4	own view	
	CONVERGE-Services	2	2	4	1	4	3	2	3	4	2	1	4	4	3	4	partner view	
		2	2	4	2	3	3	2	1	3	2	3	4	3	4	4	neutral view	
		1	1	4	1	2	1	1	1	2	2	3	4	3	2	3	own view	
	Human resources	2	2	3	2	3	3	2	2	2	2	2	3	3	3	3	partner view	
		2	2	3	2	3	3	2	1	3	2	3	4	3	4	4	neutral view	
		1	1	3	1	3	2	1	1	1	2	3	4	3	3	4	own view	
	Financial Management	2	2	3	2	3	3	2	2	2	2	2	3	3	3	3	partner view	
		3	3	4	2	3	2	2	1	3	3	3	3	5	3	4	neutral view	
		1	1	3	2	2	1	1	1	0	2	3	4	2	5	4	own view	
	Controlling	2	2	3	2	3	3	2	2	3	1	2	3	3	3	3	partner view	
		2	2	3	3	3	3	2	1	3	3	3	4	3	4	4	neutral view	
		1	2	3	4	5	6	7	8	9	10	11	12	13	14	15	Institutions	

Fig. 4 IIRM matrix for the economic roles in the market phase development and research of the project CONVERGE. Source: Own diagram

role performance. The other partners have expressed their wish with regard to role performance of the other partners. The role performance of each partner is appraised objectively with the aid of an algorithm. This method enables a 360° assessment of all of the partners in terms of their role performance. The intensity of the role performance is shown on a scale of 1–5:

1. Role should not be assumed.
2. Partner could/would assume the role, but it does not have any previous experience in performing this role.
3. Partner could/would assume the role, but it has little previous experience in performing this role.
4. Partner wants to/should assume the role because it has experience in performing the role.
5. Partner wants to/should assume the role because the partner has unique selling propositions in this area.

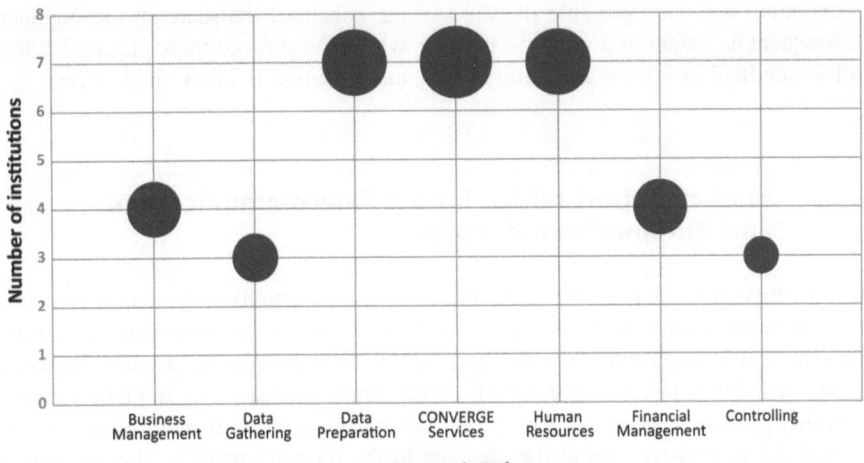

Fig. 5 Results of the IRM matrix. Source: Own illustration

The process of the 360° assessment can be illustrated clearly by an example. Institution "14" ranks its preferences regarding the performance of the role "business management" with the value 5. Institution "14" claims that it has the unique selling propositions that make it especially well qualified to perform the role "business management". The other institutions come to an average result, however, indicating the institution "14" has had only limited experience. As a consequence, the assumption of the role is assessed on average with a 3. If the average assessment is reached by broad agreement, the variance in total is zero. However, the average assessment can also be the result of an extremely heterogeneous appraisal by the other institutions. The variance in the assessment is a value that is taken into account when the "neutral view", the unbiased assessment of whether a partner is suitable for the assumption of the role, is followed. The neutral view includes, along with the variance in the partner assessment, the variables company size, declared willingness to collaborate in the implementation of CONVERGE, and the technical capabilities of the institution. In our example, the final result is that institution "14" appraises its assumption of the role with a 5, the partners assess the assumption of the role by institution "14" with no more than a 3, and the neutral view comes to the conclusion that the required role performance corresponds to an intensity of 4. The results of the IRM matrix can be transferred to Fig. 5.

The results based on the diagram are positive for CONVERGE because all of the necessary economic roles can be performed by more than one partner. The size of the group reflects here, taking into account the "neutral view", how trustworthy the readiness of an institution to perform the role is and, at the same time, how accepting the other partners will be of its performance of the role. Figure 4 makes it clear that all of the economic roles can be assumed by partners. The roles "data gathering" and "controlling", however, have an inherent risk that the partners identified as suitable for these roles might, in reality, refuse to perform the roles. Taking these results as a foundation, it is now possible (because of the enhanced transparency) to conduct subsequent negotiations among the partners within the consortium to ensure that the roles identified as critical are performed by an institution in every single case.

7   Market Opportunities for the Telecommunications and Automotive Industries

Competitive pressure is the force driving co-operation actions of the automotive and telecommunications industries. The automotive industry is under enormous pressure to adapt and remain competitive because the direction in which its previous business model will develop is not clear. OEMs could themselves turn into suppliers for new mobility providers. However, they could also transform into new mobility providers.

CCAM is decisive during the changes in the mobility market. The successful implementation of CCAM applications and the powerful dynamics of the techno-logical developments mean that proprietary solutions involve a high risk of loss, possibly even the risk of bankruptcy. In such a setting, integrated networks are an alternative solution. To implement CCAM or to establish integrated networks, the automotive industry has to overcome the loss of strategy specific competencies by inter-industrial co-operation with the relevant competence holders such as the telecommunications industry.

However, as diverse co-operation barriers exist, new collaboration models for automotive and telecommunication industries are required.

Inter-industrial co-operation is simplified with the aid of the institutional role Model. Moreover, the creation of a transparent institutional role model can establish a fair distribution mechanism among the various players so that sustainable co-operation projects and service products can appear. Co-operation with the aid of the institutional role model gives rise to greater space for action as well as time advantages for the realignment of previous business models to match with new megatrends for both the automotive and the telecommunications industries.

References

Bosshart, O., & Gassmann, O. (1996). Management Strategischer Technologieallianzen. In O. Gassmann & M. von Zedtwitz (Eds.), *Internationales Innovationsmanagement. Gestaltung von Innovationsprozessen im globalen Wettbewerb* (pp. 187–211). Vahlen: München.

Geis, I., & Schulz, W. H. (2015). Critical infrastructure: Making it private or public – An institutional economic discussion on the example of transport infrastructure.

Kemp, R., et al. (2012). Introduction: Sustainability transitions in the automobility regime and the need for a new perspective. In F. W. Geels, R. Kemp, G. Dudley, & G. Lyons (Eds.) *Automobility in transition? A socio-technical analysis of sustainable transport* (pp. 3–28). London: Routledge.

Köhler, J., Whitmarsh, L., Nykvist, B., Schilperoord, M., Bergman, N., & Haxeltine, A. (2009). A transitions model for sustainable mobility. *Ecological Economics, 68*(12), 2985–2995.

Küting, K. (1989). *Unternehmerische Wachstumspolitik: Eine Analyse unternehmerischer Wachstumsentscheidungen und die Wachstumsstrategien deutscher Unternehmungen* (Vol. 38). Berlin: Erich Schmidt Verlag GmbH & Co KG.

Lotz, C., Herb, T., Schindhelm, R., & Vierkötter M. (2014). Matrix von Loesungsvarianten Intelligenter Verkehrssysteme (IVS) im Strassenverkehr; Berichte der Bundesanstalt fuer Strassenwesen. Unterreihe Fahrzeugtechnik (F 97) 978-3-95606-111-0.

Luhmann, N. (2002). *Einführung in die Systemtheorie.* Heidelberg: Carl-Auer-System-Verlag

Müller-Stewens, G., & Lechner, C. (2005). *Strategisches management. Wie strategische Initiativen zum Wandel führen.* Stuttgart: Schäffer-Poeschel.

Schmalensee, R. (1985). Do markets differ much? *The American Economic Review, 75*(3), 341–351.

Schneider, D. (1995). *Betriebswirtschaftslehre,* Bd. 1: Grundlagen, 2. Aufl. München.

Schulz, W. H. (2005). Application of system dynamics to empirical industrial organization – The effects of the new toll system. *Jahrbuch für Wirtschaftswissenschaften [Review of Economics], 56,* 205–227.

Schulz, W. H., Joisten, N., & Mainka, M. (2013a). *Entwicklung eines Konzeptes für institutionelle Rollenmodelle als Beitrag zur Einführung kooperativer Systeme im Straßenverkehr.* Bergisch Gladbach: Bundesanstalt für Straßenwesen.

Schulz, W. H., Joisten, N., & Mainka, M. (2013b). Volkswirtschaftliche Bewertung: Wirkungen von simTD auf die Verkehrssicherheit und die Verkehrseffizienz. Accessed XXX http://www.simtd.de/index.dhtml/object.media/deDE/8136/CS/-/backup_publications/Projektergebnisse/simTD-TP5-Abschlussbericht_Teil_B1-B.pdf

Schulz, W. H., Wieker, H. & Kichniawy, J. (2014, April 6). Research joint ventures as a European policy instrument beneath directives and action plans: Transitions, interlocking and permeability of political, technological and economical requirements. *Interlocking and Permeability of Political, Technological and Economical Requirements.*

Seubert, C. M. F. (2010). Build, Ally or Acquire: die strategische Entscheidung über den Entwicklungsweg (Vol. 125). BoD–Books on Demand.

Vogt, J., Fünfrocken, M., & Wieker, H. (2013). Converge-ITS communication architecture for future mobility. In 20th ITS World Congress.

Wieker, H., Fünfrocken, M., Scholtes, Vogt, J., & Wolniak, N. (2014a). CONVERGE – Future IRS-Infrastructure as open service networks. Presentation, ITS World Congress 2014, Detroit, 07–11.09.2014.

Wieker, H., Fünfrocken, M., Vogt, J., & Eckert D. (2014b). CONVERGE – ITS design like Internet. Presentation, 10th ITS European Congress, Finland, Helsinki, 14–19.06.2014.

Part V

Regulation, Wholesale and Wholebuy

Regulation of Telecom Markets and OTTs in Europe is Uneven with Impact on Network Operators

Martin Lundborg

1 Uneven Regulation of Telecommunication Providers and OTTs

In the EU, telecommunication providers are regulated by the sector specific regulation, i.e. regulation which is implemented mainly for telecommunications providers only. As there is no comparable regulation for the OTTs (Facebook/WhatsApp, Microsoft/Skype, Apple/Facetime), this regulation has led to an unleveled playing field for telecommunication providers and OTTs (over-the-top) although both types of providers are offering communications services.

There are many regulatory provisions which lead to the unleveled playing field (Lundborg 2016):

- Specific market regulation of dominant providers to remedy anti-competitive behavior is imposed on dominant telecommunications providers (also referred to as SMP, Significant Market Power regulation). In addition to the competition law regulation, which is applied on any provider, the telecommunications market regulation defines regulatory obligations ex ante (i.e. in advance), instead of competition law regulation which remedies anti-competitive behavior after it has been conducted (ex post). According to the telecommunications market regulation (SMP regulation) the dominant players are forced to offer certain (wholesale) products/services at regulated prices and on a non-discriminatory basis. The SMP regulation has a regulatory toolbox of six standard obligations which are normally imposed on providers in four main telecommunication markets (access markets and interconnection). Details are included in the section "Sector specific regulation of dominant telecommunication providers" below.

M. Lundborg (✉)
WIK GmbH, Bad Honnef, Germany
e-mail: martin@lundborgs.de

© Springer International Publishing AG, part of Springer Nature 2019
P. Krüssel (ed.), *Future Telco*, Management for Professionals,
https://doi.org/10.1007/978-3-319-77724-5_20

- In 2015, EU decided to additionally introduce the net neutrality regulation, limiting all telecommunications networks providers in offering internet services. This regulation has the aim to reduce the abilities for telecommunications providers to offer services with premium customer experiences compared to services offered over the "open internet" (e.g. "OTT services"). The section "Net neutrality within the EU" outlines the details of this regulation within the EU.
- The telecommunications regulations included consumer protection and universal service obligations targeting the telecommunications sector only and are parts of the European regulatory framework. This regulation does typically not apply to OTT players. This means that the telecommunications providers e.g. have to inform their end-users about consumer rights and obligations and might be forced to offer their services in unprofitable areas (universal service obligations).
- While the OTTs are acting globally, the telecommunications providers are national in scope and have to obey to national laws and taxation. As OTT players offer their services internationally, they have the possibility to optimize the burden of taxes and fees by choosing where the revenues and corporate profits are subject to tax payments.
- Data protection regulation within the EU imposes obligations on telecommunications providers which limits their ability to "sell data" as a part of the business model. Instead of enabling data as a revenue source, the data retention regulation imposes additional costs on telecommunications providers.

This situation leads to an uneven regulation of telecommunication providers as e.g. the telecommunications market regulation as well as the net neutrality regulation is aiming at the telecommunication operators only and the OTTs has the chance to avoid national laws, regulations and taxes.

2   Regulation of Telecommunication Providers

A couple of decades ago, the Telecommunication markets were state owned monopolies in most countries around the world. This changed with the WTO (World Trade Organization) agreements, according to which telecommunications markets should be opened up to competition. With the WTO rules, every WTO country is obliged to establish a regulatory authority, open its telecommunications markets and implement obligations imposed on telecommunication providers. The first WTO commitments were made with the Uruguay round 1986–1994 and further extended in the post-Uruguay round until 1997. The regulation of telecommunication networks of dominant operators has since been a cornerstone of the market regulation implemented to fulfill the WTO obligations.

The aim of the telecommunications market regulation (SMP regulation) is to facilitate competition in the markets by preventing network operators from anti-competitive

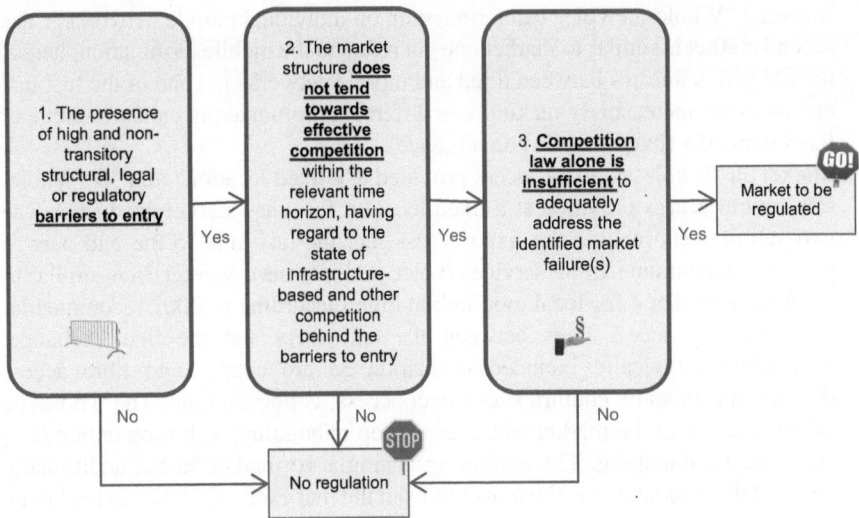

Fig. 1 The 3-criteria test

behavior (ex ante[1]), which would harm other telecommunication providers and consumers. This regulation, also referred to as sector specific regulation of operators with SMP (significant market power), includes e.g. tariff/price regulation and access regulation (obligations to offer access to networks as wholesale services).

The specific markets (i.e. products and services) which are subject to the telecommunications market regulation in the EU are defined in market reviews conducted by the national regulators based on recommendations from the EU Commission. For a product or service to qualify for regulation, the three criteria test[2] must be fulfilled (Fig. 1).

The EU Commission has conducted its own market review and recommends four markets to be regulated in the EU (EU Commission 2014). The national regulators must pay specific attention to this recommendation and hence, these markets are regulated in most EU countries, including e.g. Germany:

- Market 1 "Wholesale call termination on individual public telephone networks provided at a fixed location": The market 1 refers to voice call interconnection in the form of call termination on fixed networks. This wholesale service is bought by telecommunication providers in order to provide calls from own customers to customers on other fixed networks.

---

[1]In difference to competition law regulation, the sector specific telecommunications regulation imposes regulatory obligations before to anti-competitive conduct to prevent this to happen.

[2]The three criteria are (a) the presence of high and non-transitory barriers to entry; (b) a market structure which does not tend towards effective competition within the relevant time horizon; (c) the insufficiency of competition law alone to adequately address the market failure(s) concerned.

- Market 2 "Wholesale voice call termination on individual mobile networks": The second market is similar to market one but refers to the mobile termination, hence, the EU still delineates between fixed and mobile networks in spite of the fact that end users are increasingly making less difference from making calls to mobile or fixed networks (fixed mobile convergence).
- Market 3a "Wholesale local access provided at a fixed location" and 3b "Wholesale central access provided at a fixed location for mass-market products": The two sub-markets include the fixed access, i.e. the last mile to the end user to provide telecommunication services (voice and internet). Market 3a is similar to the former market 4 for local loop unbundling (according to 2007 recommendation), i.e. the access lines between the end users and the first exchange. Technologies typically included in market 3a are copper and fibre access (FTTx), and in some countries also duct access, collocation etc. The wholesale offers included in the market 3a are local loop unbundling, sub-loop unbundling and virtual unbundling. The market 3b is similar to market 3a but additionally includes the backhaul, e.g. the transport from the first exchange towards backbone networks. While the services in market 3a are typically for passive access to the network, the services included in market 3b are providing active access on layer 2 or 3 (Ethernet or IP).
- Market 4: "Wholesale high-quality access provided at a fixed location": The market 4 refers to the market for wholesale point-to-point (or point-to-multipoint) connections, also referred to as leased lines or managed services. The market mainly addresses wholesale inputs to carriers and ISP for their services to business/corporate customers. Relevant technologies are SDH/PDH and Ethernet.

Several national regulators have decided to regulate additional markets which are not or no longer included in the EU market recommendation. These include e.g. the SMS termination market in France, the retail markets in Ireland and access to the inhouse/drop cable for FTTH in France.

For the main provider in these (and other) markets, the regulators impose certain regulatory obligations (EU Directive 2002):

- Access obligations, i.e. the requirement to provide (wholesale) services giving competing providers and/or resellers access to the regulated operators networks.
- Price regulation, e.g. obligation to offer prices which shall be cost oriented or reasonable
- Transparency obligations, including obligations to publish regulated (wholesale) standard offers, publication of prices and contractual conditions as well as network information
- Non-discrimination obligations, implying that all customers of the SMP operator shall be treated on equal terms
- Cost accounting obligations/accounting separation obligations requiring the SMP operator to provide financial transparency, e.g. in order to identify prices which are below costs (cross-subsidization)

- Structural/functional separation obligations, i.e. that the (wholesale) business units are separated from other (retail) business units. In case of functional separation, the business units are separated in independently operating business entities. Thereby the wholesale/network business entity shall supply access products and services to all undertakings, including to other business entities within the parent company, on the same timescales, terms and conditions, including those relating to price and service levels, and by means of the same systems and processes. Structural separation goes one step further. In this case, the firm that operates a network and provides services over it, is split into (a) a company owning the local access network and providing wholesale access to service providers; and (b) another company providing retail services. In this case the firm is literally divided in two entities with different ownership, management, etc.

With the regulatory obligations imposed on telecommunication operators in the (wholesale) markets recommended by the EU, the regulatory intensity is high for basic communication services including the provision of telephony and internet access as well as voice calls. The telecommunication operators must basically offer wholesale services enabling other providers to provide the same retail services. For these wholesale offers, only cost oriented prices (enabling a sufficient rate on return) are allowed. Further, the wholesale offers must be provided with reference offers, which require lengthy regulatory approval procedures in front of the national regulatory authorities. This limits the flexibility of the telecommunication providers and increases the time to market. With the agile business models of the OTTs, this is a major drawback for the telecommunication providers.

3 Net Neutrality Regulation

In addition to the market regulation described above, the EU has additionally implemented net neutrality regulation facilitating an "open internet", which regulates all telecommunication providers offering internet services and not only the dominant players. This net neutrality regulation introduces a regulation of the relationship between the telecommunication providers and the OTT players such as Google, Facebook/WhatsApp, Apple and Microsoft/Skype, Spotify etc.

The legislative changes decided by the EU include amendments of Art. 8(4) of the framework directive, Art. 20–21 and 22(3) of the Universal service directive as well as Art. 5 of the ePrivacy Directive. Based on the changes the national regulatory authorities (e.g. BNetzA in Germany and OFCOM in UK) shall promote the ability of end-users to access and distribute information or run applications and services of their choice. Further, the national regulatory authorities shall ensure that the degradation of service and the **hindering** or **slowing down** of traffic over networks are prevented by setting minimum **quality of service requirements** on telecommunication providers.

The implications of the changes are that every end user must be able to have access to the "open internet" and all content providers including the OTTs shall be able to provide their services via a high-quality "open internet". Further, all traffic on the "open internet" must be **treated equally** and the EU has communicated that there will be no such thing as paid prioritization, e.g. requiring end users to pay more to access a specific internet service, application or service provider such as Youtube or Spotify at a higher bandwidth (EU 2015).

The net neutrality regulation does not though prevent the offering of **special services** such as IPTV, high-definition videoconferencing or healthcare services. A precondition to offer the special services is that these are offered in addition to the "open internet", with a minimum quality of the "open internet" to be guaranteed.

Under the network neutrality regulation, reasonable **traffic management** will be allowed for as long as this is justified by technical criteria and it must be independent of the origin, destination and type of traffic. This can inter alia be measurements to preserve security and integrity of networks, e.g. due to denial of service attacks, viruses and malware. Temporary traffic management due to congestions is also allowed, but only if the **network congestion** is temporary or exceptional.

Zero rating, i.e. where the internet providers do not count the internet traffic related to particular applications or services is also not prohibited, though it must be in line with the net neutrality regulation, e.g. the zero rating shall not have a negative impact on the "open internet" traffic.

In Germany, Deutsche Telekom has introduced the service "StreamOn" which is an offer which actually treats content on the internet differently and hence, could be seen as a violation of the net neutrality concept. StreamOn is an add-on for certain mobile tariff plans. Customer signing up can stream e.g. voice or voice and video without using the traffic volumes (GB) included in the tariff plan (i.e. this is a zero rate service). The video quality of the stream is limited to DVD quality (i.e. no HD). The StreamOn is only valid for content for Telekom partnering providers such as Apple Music, Deezer, Netflix, Youtube, Zattoo, ZDF Mediathek. In order to avoid being regulated and fall under the net neutrality regulation of the EU, Deutsche Telekom is offering the partnership to all content providers on a non-discriminatory basis. Further the quality of the streams is lower than in the open internet, making it a different service compared to the open internet prescribed by the net neutrality regulation (Deutsche Telekom 2017).

Die BNetzA decided through in October 2017, that StreamOn violates net neutrality, because it treats video and music streaming different and the service is not available in case of roaming. It is expected that Deutsche Telekom will appeal to court against this decision (Heise 2017).

With regard to the level playing field for telecommunications operators and OTTs, net neutrality will rather intensify the differences. For the telecommunication providers, net neutrality will have a twofold impact:

- The telecommunication providers will have less negotiation power when these are dealing with OTTs (partly), because the end users are guaranteed to access all applications and services of the OTTs by regulation.
- A limitation of the possibilities to prioritize traffic or discriminate over the "open internet" means that telecommunication providers first have to invest in the "open internet" before any special services can be offered. This increases the required amount of investments and limits the market potential of services and applications with prioritized traffic.
- When telecommunication operators enter into partnerships with OTTs or other market players, they must offer such partnerships to basically everyone on equal, non-discriminatory conditions.

4   Impact of the Unlevelled Playing Field

The regulation of the telecommunication sector was introduced to open up the markets to competition before internet was a common retail infrastructure and before OTTs were offering communication services. With the market entry of OTTs, the market situation has changed and now, OTTs are competing with the telecommunications providers. This development has led to an unlevelled playing field created by the telecommunications sector regulation.

The outcome of the unlevelled playing field is must relevant for the voice calls markets (market 1 and 2 of the EU market recommendation). In these markets, the telecommunications operators are obliged to offer interconnect at prices and reference offers approved by the regulators. This means that the telecommunications operators have problems adapting to the rapid change in the markets for voice call.

With regard to the provisioning of the broadband access services, which is at the core of the telecommunications providers operating models, these are intensively regulated (referred to as market 3 and 4 according to the EU market recommendation). If the OTTs would replicate these services, they would also be regulated.

One main problem for the telecommunication providers is that they need to invest in the upgrade of their networks to new technologies (FTTx in fixed networks and 5G in mobile networks), which requires the significant amount of CAPEX. With regulated wholesale offers and prices, there is a cap on revenues generated by the telecommunications providers, which means that these have problems to finance new investments.

One way out of this dilemma for telecommunications providers would be to offer the broadband/mobile access as bundles with other services. This has e.g. been done in Germany (StreamOn). Thereby, the telecommunications operators can generate revenues from the additional value created to end users of having higher bandwidths. The net neutrality regulation further limits the telecommunications provider though, designing such offers. The regulation also means that the telecommunications providers must in effect offer any such solution to the OTTs as well, and this on a non-discriminatory basis.

The net neutrality also requires that the telecommunication providers transport all OTT services over their networks at a minimum quality. Hence, the regulatory framework favors the OTT over the telecommunication providers by creating an unleveled playing field. It increases the likelihood that the telecommunications providers are turned into so called bitpipe providers, i.e. providers who only offer the basic internet access but no additional services on top.

The net neutrality does preclude the special services from the regulation. This means that the markets for special services such as IPTV are not included in the EU recommendation on markets susceptible to sector specific regulation and the telecommunication operators are allowed to offer these services the way they like, as long as it does not have a negative impact the "open internet". As the telecommunication providers are obliged to offer an "open internet" though, carrying the OTT services (including streaming services such as Amazon Prime or Netflix), the possibilities to provide differentiated special services is limited, but still remains a window of opportunities for the telecommunication providers.

When it comes to the markets for messaging services, these are less regulated and hence, the unlevelled playing field is less of an issue. Problematic for the telecommunications providers is though, that the OTTs with WhatsApp and Skype have already succeeded to capture a significant fraction of this market. Interesting to see is that in France, where SMS has been regulated, the success of OTT messaging services has been limited. Hence, sector specific regulation in the messaging markets has had little negative effect on the telecommunications operators. The reasons for lost market shares in the messaging markets are more likely to be found elsewhere than in the field of regulation.

References

Deutsche Telekom (2017). Retrieved August 18, 2017, from https://www.telekom.de/unterwegs/tarife-und-optionen/streamon

EU (2015). Memo updated 27th October 2015. http://europa.eu/rapid/press-release_MEMO-15-5275_en.htm

EU Commission (2014). Commission recommendation of 9 October 2014 on relevant product and service markets within the electronic communications sector susceptible to ex ante regulation in accordance with Directive 2002/21/EC, 2014/710/EU.

EU, Directive 2002/19/EC of the European Parliament and of the Council of 7 March 2002 on access to, and interconnection of, electronic communications networks and associated facilities (Access Directive).

Heise (2017). Retrieved October 17, 2017, from https://www.heise.de/newsticker/meldung/Bundesnetzagentur-untersagt-Details-des-StreamOn-Tarifs-der-Telekom-3852918.html

Lundborg, M. (2016). Die unausgewogene Regulierung des Telekommunikationsmarktes, OTTs in Europa und die Auswirkungen auf Netzbetreiber. In P. Krüssel (Ed.), *Future Telco III – Powerplay für Telekommunikationsunternehmen*. Köln: Detecon.

The Rise of OTT Players: What is the Appropriate Regulatory Response?

Markus Steingröver, Edgar B. Cardozo Larrea, and Nikolay Zhelev

1 Competition and OTT Business Models

One of the bigger challenges of traditional telco operators is the issue how to compete with OTTs, who face low barriers to entry and are able to leverage global economies of scale. A global OTT Media player faces lower costs per MB (storage & hosting) and is able to negotiate better content deals compared to local players. However, the OTT's business model requires regulated access from telco operators. Thus, the revenue loss should be part of any equation concerning wholesale price levels.

Meanwhile the impact of OTT Media is that a sizeable portion of internet traffic is now solely concerned with streaming media. Infrastructure providers need to invest massively in higher capacity infrastructure (both access and core) to meet this demand. The impact is aggravated by the fact that many Telcos' business models include flat rate data plans.

Figure 1 below summarizes the business models behind OTT use cases. Many OTTs now have successfully solved the monetization problem. They are however still using infrastructure paid for by Telco consumers within the framework of flat-rate data plans. This model is the kiss of death for traditional Telco services for which the consumer has to pay on a per-use basis, but which offer no apparent advantages over their free alternative.

As a consequence Telcos are losing revenues, but no one is gaining them. This is destructive competition resulting from flat rate business models.

Content producers suffered in a similar way. OTT media services provide flat rate music or video streaming—content previously supplied on a unit price basis. Following major legal battles new services have emerged where the media industry

M. Steingröver (✉) · E. B. Cardozo Larrea · N. Zhelev
Detecon International GmbH, Cologne, Germany
e-mail: markus.steingroever@detecon.com; EdgarBruno.CardozoLarrea@detecon.com; Nikolay.zhelev@detecon.com

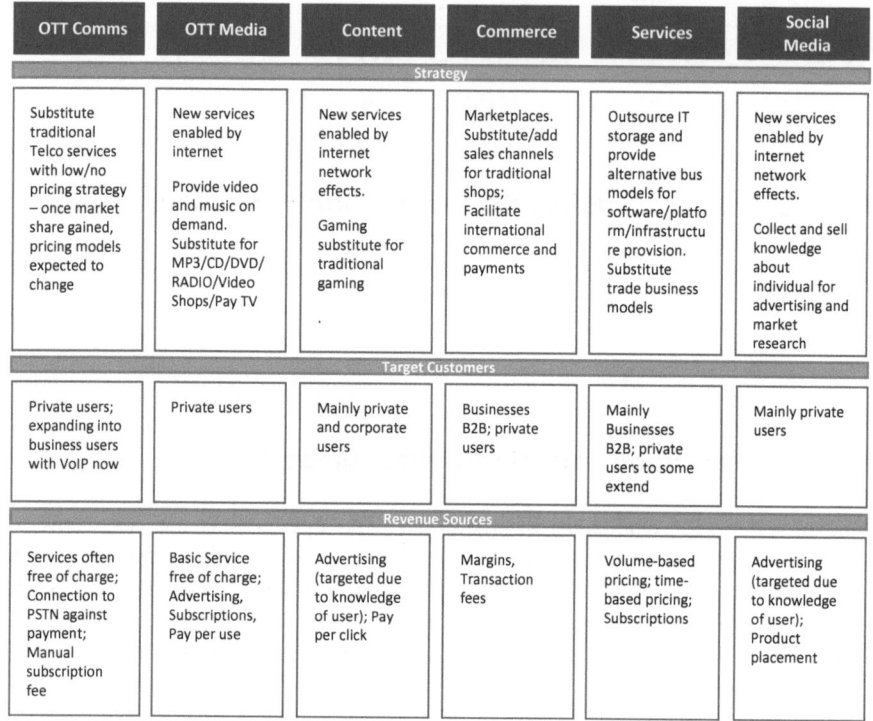

OTT Comms	OTT Media	Content	Commerce	Services	Social Media
Strategy					
Substitute traditional Telco services with low/no pricing strategy – once market share gained, pricing models expected to change	New services enabled by internet Provide video and music on demand. Substitute for MP3/CD/DVD/RADIO/Video Shops/Pay TV	New services enabled by internet network effects. Gaming substitute for traditional gaming .	Marketplaces. Substitute/add sales channels for traditional shops; Facilitate international commerce and payments	Outsource IT storage and provide alternative bus models for software/platfo rm/infrastructu re provision. Substitute trade business models	New services enabled by internet network effects. Collect and sell knowledge about individual for advertising and market research
Target Customers					
Private users; expanding into business users with VoIP now	Private users	Mainly private and corporate users	Businesses B2B; private users	Mainly Businesses B2B; private users to some extend	Mainly private users
Revenue Sources					
Services often free of charge; Connection to PSTN against payment; Manual subscription fee	Basic Service free of charge; Advertising, Subscriptions, Pay per use	Advertising (targeted due to knowledge of user); Pay per click	Margins, Transaction fees	Volume-based pricing; time-based pricing; Subscriptions	Advertising (targeted due to knowledge of user); Product placement

Fig. 1 Summary of the OTT business models

is partnering with OTTs in a way which is less destructive for the industry—offering streaming options and premium pricing for advertisement-free services.

Traditionally network operators invested in networks because they generated revenues with the provision of content. Assuming that excess capacity would always be available, flat rate tariff plans emerged. This is no longer true. Telcos are paid flat rates for the use of their capacity while their role is reduced to that of a wholesaler. At the same time prices for wholesale services are often regulated at a restrictive cost-oriented level. On top of this net neutrality rules made it impossible to offer differentiated QoS and prices. Profits and thus the incentive to invest came to a certain extent from peering/transit agreements—and from income earned with the network operator's own service and content provision. This is no longer true.

In the specific case of cloud services this problem is particularly acute. Cloud services need a highly resilient network and the synchronization of data to numerous devices demands significant capacity. This in turn requires investments which are paid back after years. But the increasingly wholesale nature of the Telco's business, the international nature of many cloud service providers (giving them access to numerous alternative network operators) and, last but not least, the upcoming demand for and obligation to provide data portability combine to ensure that Telcos have no planning certainty concerning the income to be earned if investments are made.

2   Regulatory Imbalances and Options

OTT services—particularly OTT media and content provision—are increasing demand on capacity. In a competitive market—which the access market in many countries is—this problem should be solved using the principles of supply and demand. Access providers should rebalance their prices to reflect volume usage. If the customer wants additional volumes, then they should be willing to pay for it. If this is not the case regulators should carefully analyze why the market mechanism fails. Unless the business models of the market players (mainly the network operators) are adapted to suit the new market structures, investments in the network will ebb.

To enable the network operators to function in a competitive manner it may be necessary to change the regulatory guidelines which they face. Existing regulatory requirements must be reassessed within the framework of the new market situation to re-establish a level playing field and incentives to invest.

- Rebalancing of tariff structures—away from flat rates and strict network neutrality and towards traffic and/or quality of service-oriented schemes for data. If competition would be working, Telcos would change their tariff plans in line with market demand. As this is not happening, competition is apparently dysfunctional. The recent move of mobile operators to add unlimited streaming to their tariff plans underlines this assessment. Research shows that this is a situation found throughout the world and for which there is not yet a patent solution—although the root cause apparently lies in the combination of flat rate tariffs based on business plan assumptions that no longer hold true in combination with non-sustainable competition from OTT players and regulatory obligations which make it impossible for the Telcos to react freely to the commercial changes demanded of them. It must be a central regulatory task to analyze this dysfunctionality thoroughly and to introduce measures to eradicate the problem. In many cases network operators are subject to tariff regulation for their wholesale rates. The regulator will need to take action to rebalance or even redefine these regulated rates to enable SMP operators to behave conform to the demands of the market.
- The question of if and how to license new internet market players has occupied regulators for more than a decade. The term licensing includes in this case the EU notification procedure that comes with some telco-specific obligations attached. Definitions are the name of the game when determining who should be regulated and how. Is, for example, an OTT voice provider a provider of voice services? Skype has a very clear position here: "Skype does not have any operations in Singapore [author's note: or anywhere except Luxembourg]. Users of Skype simply download the Skype software from our Luxembourg operated website...". Recently a slight withdrawal from the "light" regulation trend in

Europe can be seen, as the subject of security is becoming more prevalent. The recent draft of the European Communications Code (ECC) which intends to reform the 2009 Regulatory framework includes a new definition of electronic communications services. It distinguishes between (i) Internet Access, (ii) Interpersonal Communications Service and (iii) Other Services. Other Services refer to M2M and Broadcasting signal transmission. Interpersonal Communication Services are subdivided into number-related services and services that do not use numbers such as many OTT services. The service definition leads to the following regulatory implications:

– Regulation is only foreseen in case services are provided free of charge. The important question whether free of charge includes add- or data-financed services remains to be answered.
– Regulatory obligations remain in place for number based services
– Services that do not use Numbers are not subject to authorization, dispute resolution and consumer protection regulation such as Quality of Service requirements. Moreover, there are no interoperability requirements foreseen for these services

• In summary Telco operators cannot expect a tighter regulation in the EU. However OTT providers will increasingly become targets of requests from local law enforcement authorities, as was the case for Skype in Belgium.
• With regard to licensing of OTT services other parts of the world—especially emerging markets and developing countries still take a different much more interventionistic approach. Some recent approaches to regulate OTTs are:

– The Tanzanian government recently drafted regulations for online content producers and users on social media. The Tanzania Communications Regulatory Authority (TCRA) will licence all content providers, including bloggers and deregister any provider and block users deemed to have broken the rules.
– Thailand's regulatory watchdog NBTC made bold moves in mid-2017 to regulate OTT media services trying to force them into registration followed by a U-turn on the categorization of OTT video-on-demand services as a broadcast business.
– Indonesia just announced in October 2017 to publish a ministerial regulation on OTT services this year to set a level playing field for local and international online messaging service providers. Target date for enacting the regulation is end of 2017.

• While in the past disputes between some EU member countries (e.g. France, Spain) have blocked OTT providers when offering voice services that connect to the PSTN. Justification is that the OTT is then behaving like a Telco and should fulfill the obligations of a Telco too (offer emergency services, LI, pay USO etc.). The very nature of IP communications means that the VoIP connections are often location-independent and reliance on a functional electricity supply makes them inferior in some disaster scenarios. In an attempt to encourage OTT voice providers to participate in legal intercept and emergency call access the UK offers them for example geographic numbering if they agree to comply with Telco obligations and provide such services. Otherwise they are assigned numbers from a specific range which is clearly identifiable as not being "ordinary" telephone numbers. It is questionable whether the OTT providers regard the different numbering as a real problem, or reason enough to take on the costs of the obligations.

- A further use case benefitting from regulatory imbalance is OTT Media. Here it is the traditional broadcasting companies which are subject to strict content and copyright restrictions while the OTT media providers enjoy comparative freedom. The situation here is further complicated by the convergence of ICT and broadcasting issues—which is leading to the logical convergence of the different regulatory instances e.g. in the UK—where Ofcom has been established as the result of the convergence of the ICT regulator (Oftel), spectrum management (RA), the regulator for private television (ITC), the standards commission (BSC), the regulator of independent radio services (Radio Authority) and the overseers of the BBC. In this way consistency in the treatment of different cases can be guaranteed and competence discussions avoided from the outset.

- The strict Net Neutrality definition which requires equal treatment for all bits acts in fact as an obstacle for Telcos to adopt new business models. When Skype first took off in the UK, several network operators blocked its use. Ofcom intervened and used Net Neutrality as argument to make them provide access again. In the last years regulators distanced themselves from absolute net neutrality. The BEREC guidelines on Net neutrality ban on blocking or slowing some apps but acknowledge special services such as live broadcasting and are open to zero rating (subscriptions in which use of specific services not affects data caps). The potential for differentiated quality based business models is shown by the rise of content delivery networks (CDNs). This promising business model potentially available to network operators has mainly been snapped up by third parties.[1] OTT providers pay for these services, so the traditional network operators can monetize the relationship by providing this service—and the skills required are already available as CDNs are a natural extension of the transport business.

- Finally, data portability is a regulatory aspect which may aggravate present imbalances if not correctly handled. This concept is akin to that of number portability and is intended to protect consumers from lock-in effects, especially in the case of cloud data services. Although a positive concept in terms of consumer protection, its implementation must be planned carefully to ensure that all market players, whether network operators, ISPs or application service providers, are subject to data portability requirements and thus to prevent a distortion of competition. In the EU the portability regulation will become applicable in all EU Member States from 1 April 2018. The Regulation grants consumers the right to access their portable online content services when travelling in the EU.

---

[1]These providers offer local data storage to the OTT providers so that latency is reduced and the consumers' joy of service increased. Technology developments such as EDGE computing and IoT/5G push this further.

- With regard to tax avoidance EU initiatives are under way to limit tax avoidance such as loan/interest-tax reductions via low tax high tax country transactions. Moreover led by France and formally backed by 10 EU countries the EU Union intends to align its tax policy with more modern and technologically focused businesses. A tax agreement on the digital industry should be reached by mid-2018.

3 Other Regulatory Options

Regulators and policy makers can also facilitate the roll-out of broadband networks by several other options that determine network operator's business models instead of easing the establishment of new ones.

One option is the structural separation of the markets for network and service provision. Australia and Singapore have both chosen this option and set up broadband companies specifically to provide nationwide broadband infrastructure.[2] Italy is just discussing this option and merge TIM access network with the national broadband network. Network operators are then required to sell capacity on a wholesale basis at regulated prices. The rationale here is that the provision of the broadband infrastructure is not a competitive market and will not develop to the benefit of the economy if not supported by regulatory intervention. Such a major intervention into the market can therefore only be recommended if a detailed cost study reveals that a natural monopoly situation exists, which is resistant as well, and such intervention is thus justified.

If reduced to the role of wholesale provider, Telcos lose their customer contact. If broadband capacity is seen as a commodity by the customers, brand loyalty will with time fall to zero. Competition may be purely price-based and customers will cherry pick from each service offering. Such behavior which will be further facilitated by consumer protection activities such as data portability regulations and market transparency provided by the internet.

On the other hand, if regulatory practice softens network neutrality, this will enable some differentiation again—particularly in combination with requirements from the regulators that the users are well-informed of the differences in the quality of broadband connections (as in Singapore). As the role played by internet services continues to grow in society, the willingness to pay for high quality will develop in a greater range of customer segments.

Network operators can accept its role as a commodity supplier and adjust its business model accordingly to maximize production efficiency to provide broadband connections to the mass market at minimum price, or it can search for options for

---

[2]The National Broadband Network in Australia is to provide 93% of homes, schools and businesses with a fiber-to-the-premises broadband connection of up to 100 Mbit/s. The other 7% are to be served with wireless and satellite connections. The Next Generation NBN in Singapore has separate companies providing dark fiber network and ducts ("Opennet") and active infrastructure ("Nucleus Connect"), while the services are provided to the users by retail service providers (RSPs).

differentiation in broadband provision and lobby for regulatory freedom to act as a commercial entity. In addition Network Operators have the alternative of entering into commercial agreements with internet application or content providers to offer, for example, value added packages to the customers with enhanced quality. Many operators have adopted a symbiotic approach with partnerships with the OTT players (e.g. Mobily in Saudi Arabia). In this case, the applications are natively installed on the device, and traffic from these applications is zero-rated when specific bundles are purchased. Although this does not fully compensate for lost revenues, it offers the customers an attractive alternative which may increase loyalty.

Further alternatives could be to enter into service agreements with OTT providers to provide QoS at a price—a possibility now that the concept of net neutrality has been negated—or to use Apps as a distribution channel for Telco services.

A further option for regulators and policy makers to ensure roll-out and funding of broadband networks would be the introduction or extension of existing universal service obligations to cover broadband access—potentially including a funding mechanism for the provision of broadband to be carried by all market players.

4   Coordination and Combination of Regulation

With Internet applications facilitating convergence and digitalization of business models progressing, establishing coordination procedures between different regulatory authorities such as financial service regulation, data privacy and protection regulation, broadcasting/publishing regulation and communications regulation will be necessary. Only such a process can ensure that measures taken are consistent and coherent. The internet is already playing a central role in both business and private life, and this can be expected to become even more important in the future. In some countries (e.g. Germany), there are even calls for an "Internet Ministry" to be established. Even if not take so far, coordination of existing bodies is vital. Figure 2 summarizes minimum coordination requirements between different verticals.

Convergence of the broadcasting, communications (and publishing) markets turns the convergence of their regulation into a logical next step. As the barriers between the markets are blurring this would help ensure consistent treatment of market players. However policy making is far behind digital transformation. The OECD is for example diagnosing a large gap between technology 4.0 and policy 1.0.

Digital governance and regulation should follow three targets:

1. Enabling agile regulation and governance: agility and flexibility as central principle for governance and regulatory action; for example: create "regulatory sandboxes" to stimulate innovation in Fintech sector. Examples for this approach can be found in Bahrain or Switzerland.
2. Organization of governance and regulation: Reorganization of governance structures to reflect the central and elementary importance of digital issues in every aspect of life. This can be achieved by creating a "Digital Transformation" body with a horizontal role across sectors. The body should have easy access to

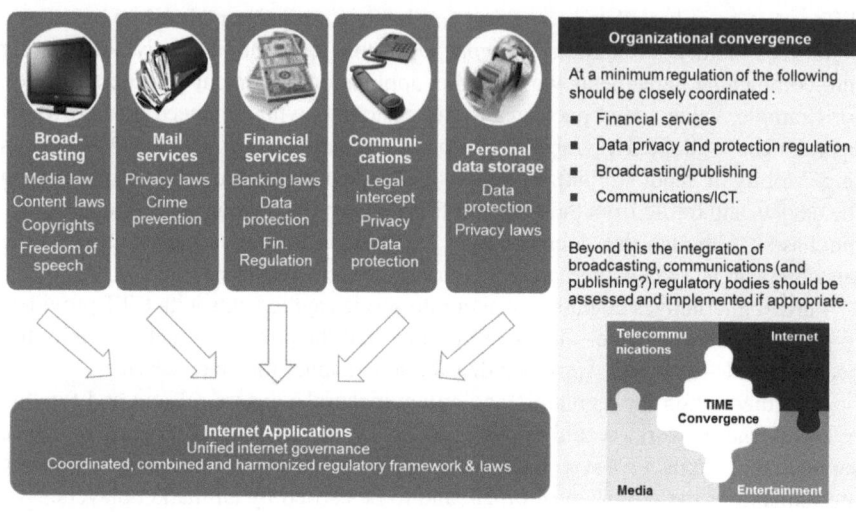

Fig. 2 Internet and convergence of governance

appropriate departments responsible for sector specific initiatives and a clear mandate to consult with all relevant stakeholders to create an inclusive strategy. This setup promotes looking beyond individual projects to embark on overall systemic changes and to bring policy-makers closer to the ongoing transformation.

3. Redesign of regulatory and legal frameworks: Review of existing acts, legislations and frameworks to reflect rapid technological, economic and societal change in the light of digital transformation:

The latter point comprises two dimensions: (a) a horizontal dimension which refers to policy fields that are relevant for the entire economy or society, such as privacy and security and a vertical dimension which is referring to sector-specific policy fields that are relevant to particular industries such as telecommunications, finance industry or transportation. The vertical digital transformation policy should follow three targets:

1. Efficiency: Leveraging technology for improved operational efficiency. This is accomplished through the digital transformation of processes and services;
2. Usability and accessibility: Increasing and easing access to governmental services and requirements for citizens and businesses. This might occur for example by creating a single digital access point for relevant services;
3. Impact and promotion: Enabling and promoting digital transformation in production, education and society to increase economic growth and safeguard direct structural and societal digital change. This includes industrial policy as well as economic regulation to safeguard competition.

5   Take Aways

OTT players substantially changed the competitive landscape—not just in the Telco market but in many other sectors. The answer this can only be a substantial change in regulation and digital governance. This concerns the sector specific regulations as well the entire governmental structures in which regulation operates. With regard to the telco market policymakers or regulators should:

- Re-assess the broadband market to determine whether political rollout targets are commercially viable, develop/adapt policy as appropriate.
- Implement a review of regulations faced by network operators to ensure that these reflect the changed market situation and rebalance obligations as found to be necessary.
- Determine whether the provision of specific (free) OTT services represents unfair competition and is detrimental to the development of the market as well as take action as required.
- Determine whether a lack of competitive pressure on the fixed broadband providers is preventing the market from functioning. If so, take steps to open up the market to more competition.
- Define a framework for net neutrality regulations to enable commercial service offers, restructuring of telco business models and cost-oriented market pricing while protecting the consumers' interests.
- Update the license/operating conditions of existing operators and service providers to create a level playing field and modernize the net neutrality obligations. Recent EU policy initiatives point into this direction. More efforts will be required.

With regard to the regulation of Digital Transformation policymakers should at a minimum, establish coordination procedures between financial service regulation, data privacy and protection regulation, broadcasting/publishing regulation and communications regulation to reflect the convergence resulting from the internet and to ensure that all regulatory measures are consistent and coherent. At a maximum, consider the integration of broadcasting, communications (and publishing) regulation. Moreover governance of the internet and digital transformation should follow three targets:

1. Enabling agile regulation and agile governance.
2. Reorganize governance structures to reflect the central and elementary importance of digital issues in every aspect of life.
3. Review of existing acts, legislations and frameworks to reflect rapid technological, economic and societal change in the light of digital transformation.

OTTs Versus Telcos: Network Neutrality and Operator Strategies

Arnulf Heuermann

1 Convergence of OTT and Telco Markets

1.1 Erosion of Traditional Telco Business Models

The business models pursued in telecommunications companies today must take a substantially more complicated market environment into account than in the past. Previous business models are under attack from the retail and wholesale side. Traditionally Telcos offered network access often internally subsidized by retail service usage and incoming traffic from other Telcos was creating wholesale revenue in particular via termination fees. Internet access is part of their offering but there is little Telco produced content over the public Internet and consequently little retail revenue from services. Services offered by Telcos are usually "reserved services" with a defined bandwidth and other specific network parameters, even if they use the IP protocol.

Internet content is largely produced by global "Over-the-Top" players (OTTs) as exclusively web-based services. The term "Over-the-Top" means any transmission of Apps or services via the (public) internet where the Telcos or ISPs are just transmitting the IP-packages over their infrastructure, but have no control over content distribution, rights and duties.

End customers pay Telcos for their Internet access, and broadband access revenues are booming in line with the rising demand for OTT services. But a significant portion of the retail revenues goes straight to the Internet content providers. OTTs are offering their services on a global scale and some have gained more customers than the biggest Telco Groups. Telcos are still producing their retail services on a national level which is limiting economies of scale and global adoption.

A. Heuermann (✉)
Detecon International GmbH, Cologne, Germany
e-mail: Arnulf.Heuermann@detecon.com

© Springer International Publishing AG, part of Springer Nature 2019
P. Krüssel (ed.), *Future Telco*, Management for Professionals,
https://doi.org/10.1007/978-3-319-77724-5_22

Also on the wholesale side Telcos are under pressure. Today voice services are causing a small share of total transmission traffic, only. Costs for dimensioning access networks are determined by IP-based video services. Voice based interconnection charges (often regulated on a cost based level) are shrinking and are not really off-set by data related charges.

OTTs obtain access to the Internet via hosting providers with content delivery networks (CDN). In addition, IP transit providers and the providers of Internet exchange points (IXP) transport the IP traffic. OTTs are typically paying wholesale prices for hosting providers and CDNs as well as IP transit providers. Internet traffic among telecommunications companies (last-mile ISPs), on the other hand, is handled usually free of charge on a "peering" basis. Telcos are usually in the role of a last-mile-ISP and therefore generate no or little wholesale revenues from OTTs.

As a rule OTTs follow other business models for revenue creation. In contrast to traditional Telcos, OTTs pursue different monetization strategies than rental and usage prices. A substantial source of revenue for Facebook, Google, and YouTube is targeted and customer-specific advertising, while Spotify or Netflix charge in App rental fees for online movies and music streaming. "Freemium" business models give users the opportunity to purchase content in addition to the free "basic" service for a fee. One example of this model is the Japanese chat platform "Line"; the service offers a free messaging service, but a major part of its revenues comes from users' purchases of "stickers" that they use to enhance the appearance of their messages.

Consequently, the relationship between telecommunications companies and OTTs is ambivalent. On the one hand, the strong worldwide growth for fixed and mobile broadband connections of the Telcos would not be imaginable without the many attractive web applications offered by the OTTs. On the other hand, the telecommunications companies find themselves confronted with increasingly stiff competition from OTTs. The latter are moving in on the "traditional" sovereign domains of telecommunications providers and cannibalizing the achievable revenues in certain select service categories.

From a Telco point of view OTT services can basically be subdivided into two categories: Complementary offerings and substitutive offerings (Fig. 1).

Complementary services include online social media and search engines, gaming, and music services because they help to enhance the service portfolio of a telecommunications provider without overloading the networks. Depending on the Telco portfolio, cloud services can have a complementary or competitive nature. "Hosted storage" services, for instance, might be in competition to Telco offerings for business customers.

Video services are in essence also complementary, however they are extremely data intensive. The share of data traffic attributable to services such as YouTube has been rising to over 70% in 2016. In particular, mobile networks where global data traffic is doubling every 2 years, have to be dimensioned to IP-video traffic and CAPEX and OPEX are rising sharply. Due to the usual peering arrangements the OTTs who are generating the retail revenues do not contribute to the network expansion costs. Unless Telcos have managed to finance their networks basically

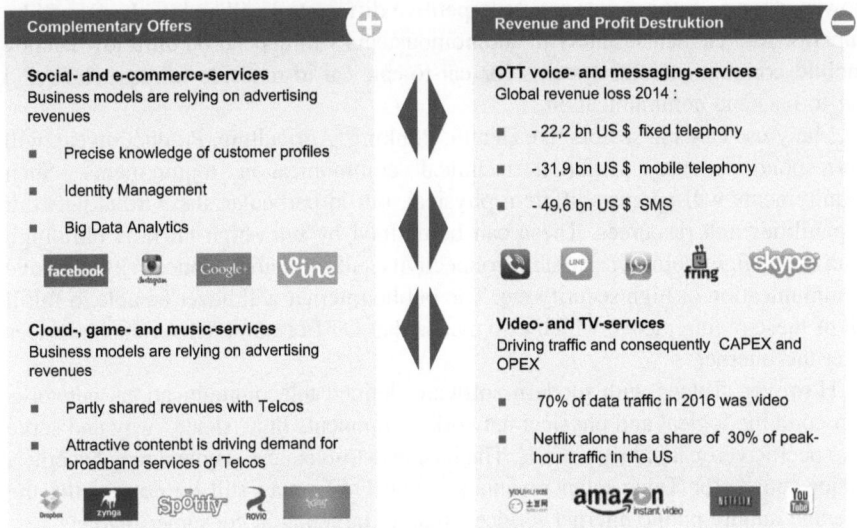

Fig. 1 Ambivalent Telco–OTT relationship from the Telco perspective

out of retail access charges or offer their own "reserved" video services profitably, OTT-video services are destructive to the Telco profitability.

Messaging and voice-based services such as Skype, Viber, WhatsApp, Line, or Twitter clearly fall under the heading of cannibalizing or substituting services. A prominent example is the intensive use of messaging services such as WhatsApp, Facebook Messenger or "broadcasting services" like Twitter. Telcos suffer from a direct loss of SMS revenues. OTTs finance their messaging service free of charge or any fees for utilization for users and thus are the preferred option. The erosion of traditional messaging a voice services of Telcos has significant volume.

- In 2015, the revenues generated by the sending of text messages declined worldwide by €41.3 billion. Since the communications behavior of users does not fundamentally change within a single year, we can assume that a large part of this decline is a consequence of the use of alternative (mobile) messaging services offered by OTTs.
- A decline of about €18.5 billion was recorded in the fixed voice segment in 2014, caused by users taking advantage of alternative telephony services such as Skype.

1.2 Information Society and Vertical Industry Offerings

On the way to an Information Society the digitization of all industries is progressing quickly. Telecommunications is becoming an integrated element of the service and product offering of all major industries. While a classical car some years ago was mainly differentiating himself from competition by mechanical quality and engine power, a future autonomous car's competitive differential will be largely defined by superior ICT elements. Safety in autonomous cars will depend on ultra-low latency mobile communications systems for car-to-car, car-to-roadside infrastructure and car-to-backend communication.

Many use cases in sectors like Health, Banking, Agriculture, Production etc. will correspond to very different technical communication requirements. Such requirements will address different physical, but in particular also virtual network capabilities and resources. These can be defined by ultra-high latency, ultra-high data rates, high mobility, seamless connectivity, ultra-reliable connectivity, massive communication or high security etc. The public internet will never be able to fulfill all of these requirements, which also means that OTTs cannot deliver such services over the Internet.

However, Telcos with modern software defined telecommunications networks can combine logical and physical network components in a "sliced" way and serve the specific vertical industry needs. The business-to-business segment may become a major source for Telco retail revenues, while OTTs may still be dominating the average quality public internet services, mainly targeting at consumer markets.

2 Strategies Under Net-neutrality Restrictions

On a world-wide scale a policy for "Net-Neutrality" has been developed, similar but different in detail in the EU, the US and elsewhere. Net-neutrality is kind of a guard rail for strategic options of Telcos and OTTs. It has basically two major components,—a user protection concept and a network design concept.

In terms of user protection net neutrality can be seen as the right of all end-users to access and distribute legal content, applications and services of their choice without any blocking of content by OTTs or Telcos. This right has been fixed for example in the EU "Telecom Single Market Directive".

The network design aspect of net neutrality was basically formulated by Media Professor Tim Wu of the Columbia University in 2003: "Network neutrality is understood to mean the principle of service providers and governments treating all data on the Internet equally, excluding any price or technical discrimination among users, content, applications, platforms, locations, type of equipment used, or type of communication".

Both aspects are now part of the regulatory framework in many developed parts of the world, however the implementation usually follows some compromises. Basically net neutrality rules everywhere prevents ISPs from discriminatory blocking, throttling and filtering of traffic and discriminatory OTT-Telco partnerships with zero-rating.

2.1   Network Neutrality and Freedom of Information

Every ISP must ensure that its customers, without exception, have access to all of the content on the Internet. This concerns both the obligation of the OTTs' search engines to process information "neutrally" and the ISPs' obligation not to block or favor specific sites technologically.

Search engines are not search neutral. Besides the frequency of users' clicks, it is above all the payments made by providers to search engine operators which decide about the chance to appear at the top of the list for the information being sought. In many cases, the search engines "customize" the hit information for users by prioritizing their individual areas of interest on the basis of their past user behavior. Google's large market share (85%) has made them a target for search neutrality litigation via antitrust law in the US. The Federal Trade commission (FTC) opened an investigation, but finally concluded that Google's "practice of favoring its own content in the presentation of search results" did not violate U.S. antitrust laws. They determined that even though competitors might be negatively impacted by Google's changing algorithms, Google did not change its algorithms to hurt competitors, but as a product improvement to benefit consumers.

Therefore, it can be concluded that OTTs are rather free in their strategies to prioritize content along commercial rules.

Apart from prioritization a general obligation of last-mile ISPs and OTTs to provide unrestricted and complete access to Internet content does not exist in reality. On the contrary, ISPs in almost every country are required to ensure that "criminal" or "undesirable" content as defined by local media, press, or Internet laws is not accessible. An exception to this is found in only a few of the poorest countries in the world, where primarily the lack of authority and budgetary resources for security authorities guarantees free access to all Internet content.

The interpretation of what exactly constitutes access-limited content in the areas of pornography, cybercrime, terrorism, anti-Islam content, gambling, file-sharing in violation of copyright laws, child abuse, glorification of drugs, excessive violence, insults to persons or institutions, undesirable religions, or political incitement differs completely in many countries. That is why the content is typically blocked for the specific country by the use of content filter systems, or the governments conclude agreements with the ISPs for the blocking of content.

Recently the discovery of massive disinformation campaigns and possible interference into general elections supported the proponents of a stricter regulation of Internet content in many countries. On June 30, 2017, Germany approved a bill criminalizing hate speech on social media sites. Among criminalizing hate speech, the law states that social networking sites may be fined up to 50 million Euros if they persistently fail to remove illegal content within a week, including defamatory "fake news."

Such type of legislation is causing considerable costs to OTTs. Facebook alone is planning to increase the number of employees dealing with deleting content from 4500 to 7500 world-wide.

2.2   Network Neutrality and Telco-OTT Partnering

In 2016 the number of Telco-OTT partnerships in terms of bundled retail service offerings rose to more than 800. More than 50% of the partnerships were closed with Google, Facebook and WhatsApp. In general net-neutrality is limiting last-mile ISPs to include OTT offerings in their tariff packages with a zero-rating component on the maximum data volume.

It is a widespread concern that companies with a dominant position in the market should not be allowed to block access to OTTs or specific content, and they should be subject to monitoring by the competition and media authorities. This is especially true for exclusive agreements or vertical integration with large OTT media corporations.

Zero rating in some countries is regarded as discriminatory in general, e.g. in Chile. The European Commission finally did not forbid zero-rating in general but obliges NRAs to investigate case-by-case whether the zero rating is distorting competition. A typical way out taken by Telcos is to offer zero-rated service bundles on an open platform basis, where all OTTs have the right to enter at non-discriminatory conditions.

2.3   Network Neutrality and Service Differentiation

If the principle of network neutrality and equal treatment of all IP packets were applied to all applications, network operators would not be allowed to distinguish between the transport of IPTV, telephony, and the downloads of texts or emails. This would prevent any sensible management of tools for monitoring, load management, or spam protection. This makes little sense from a consumer perspective. "streaming" services such as telephony or television require quality parameters for latency or bandwidth different from what is needed for emails. While a delay of a few seconds in the delivery of emails during peak times or when capacities are in short supply in mobile and fixed access networks is insignificant for customers, interruptions in video transmission or phone calls present a substantial impairment of utilization quality.

This is why services such as telephony or IPTV are typically offered as managed services by telecommunications companies. Managed services are not governed by the "best effort" principle like other applications on the public Internet; instead, they have guaranteed minimum quality parameters which transport the data packets as virtual circuit-switched service. In this case, the IP packets are not forwarded in the router according to the "first in, first out" principle; priorities are assigned to certain traffic classes.

Traffic management is therefore generally not regarded as violating net neutrality. Traffic can be prioritized by reserving fixed bandwidth or flexibly during peak hours for bandwidth hungry applications, only.

2.4   Network Neutrality and Traffic Differentiation

Managed services and other Internet services share available network resources. At this time, 27% of the total IP traffic is managed IP, 67% fixed and 6% mobile Internet, although the latter displays a sharply rising tendency. Despite the relatively low share of managed services, network bottlenecks may result in conflicts and competition among them. In technologically highly developed countries, this happens especially in mobile access networks and the old copper cable access networks originally designed for PSTN, while in other regions such as Africa the backbone networks or international connections can also be affected.

The extent to which OTTs require network capacities varies to a very high degree. In 2007, YouTube was already transmitting more data volume in 3 months than all television, radio, and cable television stations in the world in a year. Google, on the other hand, despite its billions of users, utilizes only a relatively low data volume for its search engine. For instance, the OTT provider Netflix generates more than 30% of the peak load traffic in the USA while Amazon, Facebook, and Hulu together make up only 5%. This is evidence that the network dimensioning by telecommunications companies (last-mile ISPs) must be oriented essentially to these providers. In addition, while Netflix is responsible for 32% of the downstream traffic, only 4% of the upstream traffic is attributable to this service. Nevertheless, Netflix insisted on a peering agreement providing for exchange of data at no charge between the two providers during its famous dispute with the telecommunications provider Comcast. Ultimately, Comcast was able to ward off this demand and secure co-financing of the infrastructure expansion in the form of an IP transit agreement—not, however, until after customers had protested about the sharp decline in service quality.

When telecommunications companies conclude appropriate wholesale rate plans that cover their costs with certain OTTs that verifiably cause substantial investment costs in the networks, this is simply the application of the principles of cost causation and free market. This is another case in which network neutrality means an agnostic and not an arbitrary distinction between OTTs.

3   Telco Operator Strategy Options

The impact of OTTs on traditional telecommunications business has become considerable. Telcos have three fundamental strategic options (aside from the "zero option"—the option just to do nothing) to deal with competition with OTTs.

These fields of action can be broken down into sub-categories based on specific "tactical" approaches (see Fig. 2).

The zero option: No specific "countermeasures" are initiated, based on the premise that OTT services always contribute to rising data utilization. This is a disadvantage for flat-rate contracts, but is beneficial for "pay as you use" contract models.

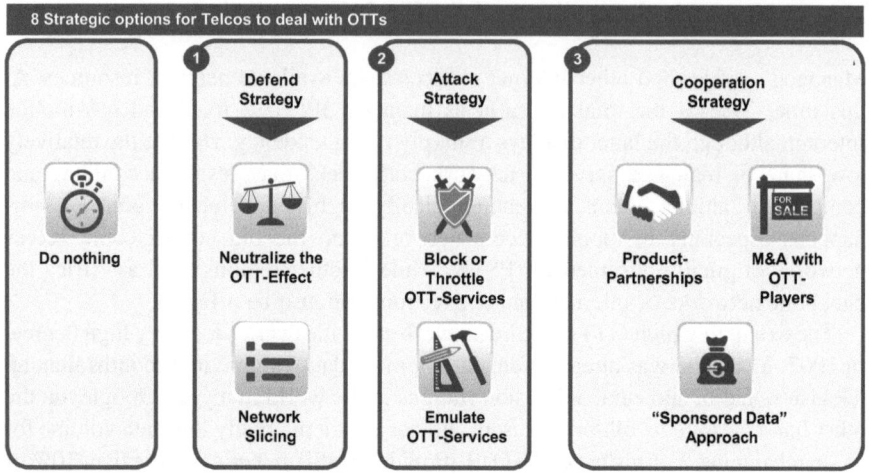

Fig. 2 Strategic options for action for telecommunications providers

3.1 The Defense Strategy

3.1.1 Data Capping

This approach emphasizes the implementation of attractive integrated rate plans (telephony, text messages, and data use) that make the utilization of OTT services ineffective from the customer perspective. Alternatively, the introduction of data caps is an option for reducing the traffic generated by OTTs to a certain level. In particular in mobile networks the growth of public WiFi networks is limiting the opportunities for data capping.

International examples are Telkom South Africa that introduced soft caps for a fixed broadband data volume (20GB). AT&T, Vodacom, DT have soft caps on mobile broadband plans.

Typically, regulatory authorities accept this strategy, if additional costs can be proven for rising data traffic, as it is the case in mobile access and partly for legacy copper networks. However, costs will not increase in fiber networks.

3.1.2 Network Slicing

The possibility to offer specific business customers network slices with specific resources and network related services features in terms of speed, reliability, security latency etc. are a characteristic of software defined networks. In particular future 5G network architectures will incorporate this opportunity.

OTTs by definition are using the public internet which has a "best effort" quality of service and therefore cannot fulfill specific requirements vertical industry solutions may have. Network slicing therefore is a powerful instrument of Telcos to serve future needs of the Industry 4.0 with or without partnerships with OTTs. Of cource, OTTs may also be paying clients of Telcos for network slices specifically designed for the OTT applications.

OTTs are also seeking for possibilities to open their own last-mile network elements but are currently focusing on global coverage in under-served areas mainly in developing countries. But this business is asset heavy with long time-to-market and low returns,—the contrary of the business models OTTs are used to.

In the contrary, network slicing is building on the clear strength of Telcos, their know how to invest in long-term network assets. It may become the major strategic tool for Telcos to create revenues from industrial partners in the B2B2C segment on a global scale.

3.2 The Attack Strategy

3.2.1 Blocking and Throttling

The blocking or throttling of OTT services is certainly one option. But this variant can prove to be relatively risky for an operator because customers could take a highly negative view if some of their beloved services are blocked or throttled. If other market players do not pursue the same strategy, customers may well head for greener pastures.

Market examples are AT&T blocking Skype calls over 3G/4G networks and Deutsche Telekom offering VoIP over 3G usage only with a special tariff. In some MENA countries VoIP services are generally blocked in fixed and mobile networks.

Modern net neutrality regulation would typically not allow this tactical approach.

A softer attack with lower legal boundaries could be the blocking of advertising in OTT services with the consent of Telco subscribers.

3.2.2 Emulate OTT Services

Another option for operators is to strengthen their position by adding their own services and platforms that are competitive with OTT services and can be integrated right into rate plans. However, this presumes that operators develop their own proprietary services at great costs of time and money, and success is uncertain; the circumstances should be analyzed in detail before this step is taken. In terms of time to market OTTs still outperform larger Telcos.

There are several market examples with limited success, like the GSMA lead "joyn" messaging service initiative and the China Telecom proprietary messaging App "YiChat". Regulatory restrictions for this tactical approach are small.

3.3 The Cooperation Strategy

3.3.1 Product Partnerships

The one cooperation variant relies on strategic partnerships with one or more OTTs aimed at securing a useful supplement to the provider's own service portfolio or to determine directly the price structure for the utilization of the various OTT services by implementing innovative rate plans. An additional advantage for the telecommunications provider would be that it could benefit from the OTT's brand name and positioning.

A market example is the product partnership between Deutsche Telekom (DG) and music and video streaming OTTs. Some years ago DT opened it's zero rating offer with Spotify which was zero-rated and very successful. After the publication of the EC net neutrality rules this individual partnership was closed and DT opened a "StreamOn" platform for all Video and Audio OTTs with zero rating. The zero rating product was accepted by the national regulator, however some features were not approved, in particular that TV signals are reduced to DVD quality, but audio not. Also the restriction of the tariff to German territory has not been accepted, the service has to be opened to all EU countries.

The advantage of this partnership from the Telekom perspective is in the enhancement of the service portfolio in conjunction with a rate plan designed specifically for this purpose. For Spotify and the other OTTs, the partnership represents an additional distribution channel. Both profit from this partnership in terms of brand awareness, but the opening of the DT platform to all OTTs limits severely the benefits.

3.3.2 Sponsored Data Approach

Another tactic is the "sponsored data" approach, seeking to offer preferred access to selected OTT services to consumers. In this case, the costs are borne by the OTT provider or are "sponsored" by the usage.

A good example is the Vodafone "Netzclub" sponsored mobile data approach. Telefónica Germany has joined forces with the "Netzclub" brand to position itself as the provider of web-financed smartphone rate plans "free of charge". Currently available as a prepaid model, customers receive a free SIM card and unlimited mobile use of data on the Telefónica network· (100 MB/month high-speed data volume, subsequent use throttled to a speed of 32 Kbit/s). Additional costs are incurred from telephony (9 eurocents/min) and the sending of text messages (9 eurocents/text message) to all German networks. The costs for the data transmission traffic are borne by sponsors. The German regulator did not interfere into this case.

3.3.3 Acquisition of OTTs

The complete or partial acquisition of OTTs is another cooperation option for increasing market shares, differentiating further the service portfolio, and integrating competitors into a company's own corporate structure. However, this approach presumes considerable financial willingness to drive mergers and acquisitions (M&A). The fragmented nature of the OTT landscape makes the success of specific acquisitions questionable, and any such actions must be assessed with great care.

A market example is the acquisition of Intel Media by Verizon, a business division developing Cloud TV products. Antitrust authorities did not intervene.

4   Summary

Telco management must quickly make decisions about their strategic orientation based on some of the options for action described. Major revenue streams from their legacy retail and wholesale business is eroding quickly and is shifting to OTTs.

The path which telecommunications providers should take—especially in consideration of the principles of network neutrality—is highly dependent on what part of their own business model they want to protect and develop. In our view a "Do nothing" strategy will not work and result in Telcos becoming a "dump-pipe" with strong effects on employment and profitability. Many "attack strategies" are blocked by modern regulation. "Cooperation strategies" are a good short term tool, but may not differentiate much in competition if many Telcos follow such an approach. Therefore "defense strategies" with data capping in the short run and network slicing in the medium term may be the most recommendable way of Telcos to further growth and prosperity.

Open Access via Mobile Wholesale Network: A New Approach to Broadband Deployment: The Case of Mexico

Ulrike Eberhard and Arnulf Heuermann

1 Background: Market Environment in Mexico

Mexico is the second largest country in Latin America with a population of about 127 million people, of which 21% are living in rural areas. According to the World Bank classification, Mexico is a "upper middle income" country, with a nominal GDP per capita of more than 15,000 USD, an unemployment rate below 5% and low but still stable economic growth (World Bank 2017; OECD 2017a, p. 16).

The structure of the mobile telecommunications market shows three major players: AT&T, Telefónica and Telcel. Telcel continues to dominate the mobile market with around 70% market share in terms of subscribers and belongs to America Móvil, which also owns the incumbent fixed line operator Telmex. Since 2007 fourteen Mobile Virtual Network Operators (MVNO) have entered into the market. However, their market shares are negligible and do not reach 1% in total (IFT 2017b). In the fixed market there are six major players besides Telmex, who are offering fixed voice, broadband, enterprise services, cable TV and WiFi Hotspots. Four of them do not offer any mobile services, while two of them have relatively unsuccessfully entered the MVNO business. Telmex held around 63% of the fixed voice line and about 57% in the fixed broadband market by end of 2016 (IFT 2017a).

Competition between the mobile network operators (MNO), is intense, with little infrastructure sharing and co-operation. Before the reforms in 2013 and 2014, regulatory decisions have been blocked through long lasting court cases. A major concern for the Mexican government was the ICT market performance, showing a comparatively poor development. Availability of services was limited. In 2014, fixed telephone penetration of households was at 59%, fixed broadband household penetration of households below 40%. In 2014 the mobile SIM penetration was at approximately 85%, very low when compared to similar countries like Argentina,

U. Eberhard (✉) · A. Heuermann
Detecon International GmbH, Cologne, Germany
e-mail: Ulrike.Eberhard@detecon.com; Arnulf.Heuermann@detecon.com

© Springer International Publishing AG, part of Springer Nature 2019
P. Krüssel (ed.), *Future Telco*, Management for Professionals,
https://doi.org/10.1007/978-3-319-77724-5_23

Brazil or Chile where penetration rates between 135% and 155% had been reached. Quality of service of mobile networks were low while prices ranged well above most OECD countries. There was a poor performance in internet usage: In 2014, while the Mexican economy was ranked second in Latin America, the country was ranked 20th in terms of internet users (IFT 2017a; OECD 2017b).

2   Mexican Government's Objective and Regulatory Initiatives

In 2013, the Mexican Government and main political parties agreed to a constitutional reform which, amongst other provisions, also targeted at changing the legal framework of the telecom sector. The major objectives related to the telecom sector were to give Mexicans access to better and cheaper telecommunications services and to raise the competitiveness of the Mexican economy.

The reform introduced an individual right for the Mexican citizen to have access to timely information from multiple sources and to seek, receive and impart information and all types of ideas by any means of expression. This right includes, that the State shall guarantee the right of access to communication and information technologies, as well as broadcasting and telecommunication services, including broadband and internet access.

Furthermore, the reform led to a change of the institutional framework in telecoms and broadcasting. The old regulatory body was to be dissolved, and a new "Federal Telecommunications Institute (IFT)" has been created to improve the predictability and enforceability of regulatory decisions. In addition, an antitrust commission (COFECE) and specialized courts for ICT cases were to be established. Also the status of the IFT as an independent and autonomous body was made explicit by the new constitution. Now, any IFT decision can only be appealed through juridical review and cannot be suspended for the duration of the court case.

The revised regulatory environment foresees a strict ex-ante regulation of dominant players and ex-post regulation of all other players. Following the constitutional reform, Telcel and Telmex have been declared by the IFT as "preponderant" and are obliged to offer interconnection free of charge until its market share falls below 50% in the future. In addition, Telcel and Telmex have to offer numerous wholesale services to its competitors on conditions to be fixed in reference offers, including access to passive infrastructure (Ovum 2017).

The cornerstone of the reform is the project Red Compartida (RC), a constitutional mandate to create a shared network through a wholesale-only operator to providing services that are unbundled and non-discriminatory. The details of the Red Compartida project were specified through the revised law for telecommunications and the specification of contract conditions and wholesale license conditions during the subsequent public tendering process led by the Mexican Ministry of Telecommunications (SCT) in collaboration with the new IFT from end of 2014 until end of 2016.

The vision of the project is to provide broadband access for all. The mission is to deploy a shared wholesale network that enables the provision of telecom services through existing and new service providers. The goals are to increase coverage of mobile broadband services, to promote competitive prices, and to raise quality to international standards (SCT 2017).

3   The Concept of a Public-Private- Partnership Based Mobile Wholesale Only Network

Red Compartida (RC) is implemented as a Public Private Partnership (PPP), where the State is represented by Telecomm and the newly created entity OPRITEL. Nonetheless, RC is regarded in its essence as a private venture, where the Mexican Government neither is part of the shareholder base nor is it involved in network design, deployment or its commercialization. These are the responsibilities of the "Developer", i.e. the private entity.

OPRITEL is a special organization within the Ministry of Communications which was created to hold the 700 MHz spectrum license provided by the regulator (IFT) and to pay the respective spectrum fees to the Ministry of Finance. Through another lease scheme OPRITEL leases the spectrum to the Developer of RC and takes receipt of the leasing fees from the same. Furthermore, OPRITEL controls the wholesale and coverage obligations of the Developer.

The rights to use the premium, unencumbered contiguous spectrum of 2×45 MHz of the 700 MHz band through OPRITEL at 0.002 USD/MHz/inhabitant, which is well below international benchmarks, is the major input from the State to the PPP. In addition, through Telecomm, the Developer of RC receives the right to use one pair of fiber optics of the Federal Electricity Commission's (CFE) fiber optic network mounted to the poles of the national power grid as a backbone service. The total length of CFE's fiber network is about 30,000 km.

It is the mandate of RC to share its entire infrastructure and to provide the unbundled services and capacities, exclusively to marketing firms and telecommunications network operators under conditions of non-discrimination and competitive prices. Consequently, the business plan of RC is mainly based on the following services:

- National Roaming services for established MNOs to cover gaps in their network capacity and filling rural and indoor coverage.
- MVNO services. These can be differentiated into two groups: "Full MVNOs" like fixed operators with a portfolio gap in mobile will build a complete network infrastructure except radio access and mainly buy from RC radio access services. "Light MVNOs" are like retail chains who sell SIM cards with their own logo and will require a number of additional network services from RC, including internet gateway access, transport, billing etc.
- Sharing of transport network and radio access network with other operators may also be a part of RCs portfolio, but certainly with a minor revenue potential.

RC is not allowed to sell directly on the retail market. Sub-leasing or sharing of spectrum by OPRITEL or the Developer of RC is also not permitted to ensure the implementation of the network according the constitutional mandate. Prior to the final public tender, which was published end of January 2016, the Government was considering to allow sub-leasing of RC's spectrum to other MNOs. This provision was much debated, as it was principally against the idea of RC. Sub-leasing would lead to a slicing of the spectrum between many operators and a parallel roll-out of infrastructure—which contradicts the idea of having one efficient network with lower costs and higher data rates. It is also unlikely that spectrum could be sub-leased at prices which would compensate the Developer for services with a higher value-add like national roaming, thus putting the viability of the project into doubt.

The Government, as specified in the tender conditions and PPP-contract, expects that the Developer commits to the following population coverage targets until 2023:

- 30% of the Aggregate Population as of March 31, 2018
- 50% of the Aggregate Population as end of year 2019;
- 70% of the Aggregate Population as end of year 2020;
- 85% of the Aggregate Population as end of year 2021;
- 0.5* (85%+Offered %) of the Aggregate Population as end of year 2022, and
- Offered % of the Aggregate Population as end of year 2023.

Since the sole winning criterion of the public tender was the population coverage for the end of year 2023 the Government expected bidders to propose more than the minimum of 85%.

Measuring population coverage is not trivial. Apart from problems of outdated census data and differences between digital maps a clear definition of population coverage was the basis of the winning criterion. The Government has defined a minimum cell-edge-data rate of 4 Mbps download and 1 Mbps upload measured at peak hour as the criterion for acceptable coverage. These data rates do not seem very high for high-density areas, but compared to a cell-edge speed of 512 Kbps the number of sites has to be increased by 30% just to meet the coverage criterion.

4   Economic and Technical Evaluation of the Concept

The pure commercial roll-out of broadband networks in Mexico started in the major cities and then slowly penetrated smaller cities and rural areas. This typical development also left some unprofitable regions as "white spots" or underserved areas.

The rationale behind this was two-fold. The wealthiest customers, as well as those who are most innovation-oriented, live in the largest cities and can afford high-speed ICT services. The CAPEX per user in telecommunications networks is highly dependent on the customer density—which means metropolitan centers like Mexico City have low costs per user while sparsely populated areas can only be connected at exponentially rising costs.

With the rising importance of ICT services and the underlying telecommunications infrastructure, governments around the world cannot easily accept the existence of large white spots in their countries. There are both supply side and demand side arguments that underline their attempts to reduce white spots to a minimum.

ICT service production and ICT network operations have different cost characteristics.

- The millions of apps in various App-Stores show that service creation is a typical market with few market entry barriers and low fixed costs, where customer segmentation, specialization and innovation are king. Thousands of competitors may offer such services at very affordable prices—a free competitive market structure delivers the best performance.
- ICT network operations, and in particular building access networks with nation-wide coverage, have very high fixed costs. Once a street has been opened for a fiber cable, or a region is covered by a mobile site, adding an additional subscriber invokes very low marginal costs. If such infrastructure is duplicated by a second or third network operator the revenue from a proportionally reduced market share will typically not cover the costs, in particular in low density areas. Theoretically a monopoly could actually serve the market at lowest cost per user, the inefficiencies of monopolistic conduct are however well-known and may over-compensate for technical efficiency gains.

From the demand side the political participation of all citizens of a state in access to information, as well as the capability to participate in interactive ICT services for eGovernment, eHealth, and eLearning services is often regarded as a basic right that should be offered to everyone at affordable prices. In addition, a high penetration and usage of ICT services has positive "external effects" on the economy as a whole, as can be measured by incremental growth of GDP for example. This GDP growth may not benefit the network operators directly, but may result in higher tax income for the government. Even from a purely economic point of view it is rational for a state to support the ICT sector over and above the purely commercial level which would result from free market forces.

Governments have typically chosen the following options to foster the development of their national ICT sectors:

1. Profit gap financing,
2. Mandatory infrastructure sharing,
3. National public broadband network expansion, and
4. National wholesale network operator.

A national mobile wholesale network operator like in the Mexican case is a new approach, which has the following technological and investment advantages driven by physical characteristics:

- A single operator who gets an un-sliced portion of the whole 700 MHz spectrum should have the lowest cost for covering the country. The cell size in this band is substantially larger than in bands with higher frequencies and therefore for customers on the move (along roads etc.) and in low density rural areas fewer cells have to be constructed to provide coverage. This saves substantial CAPEX.
- With a larger amount of bandwidth compared to sliced smaller spectrum lots a single wholesale operator can offer higher data rates per cell than several operators could. Broadband targets can be achieved much easier.
- The 700 MHz spectrum penetrates walls much better than spectrum in higher frequency bands, therefore in-house coverage is less costly to achieve.

Next to these advantages of a "natural monopoly" on the physical network side, an intensification of service competition in the retail markets could be motivated. The smaller MNOs will get relatively affordable and fast access to currently non-covered areas at much lower cost than they have to pay to the incumbent. RC is a specialized network to host all types of MVNOs. This will lead most likely to a boost of the MVNO market in Mexico with increased price and quality competition for specific customer segments.

5 Current Status of the Red Compartida Project

The project has been being planned for a long time in Mexico, and even after having achieved political agreement and a related constitutional reform, the concrete implementation has taken several years. The complexity of implementing a totally new concept should not be underestimated by the many observing countries, which also intend to implement similar solutions.

5.1 Major Steps of the Public Tender

- In 2013 the constitutional reform came into effect which amongst others defined the project "Red Compartida".
- In July 2014 the Federal Telecommunications and Broadcasting Law was changed to accommodate Wholesale Open Access, with much discussion about the pure wholesale mandate of the Open Access Provider and a possible participation of the existing retail operators.
- In Summer 2014 the Ministry (SCT) and the independent regulator (IFT) signed a contract to jointly implement the project Red Compartida. In December 2014 a USD 7 m contract to design the process was signed by the SCT with an advisor.
- In January 2015 a consultancy agreement for the design of the regulatory framework was signed between the IFT and an external advisor.
- In March 2015 a request for Expression of Interests (EOI) was published and answered by several interested parties.

- In July 2015 a first request for information (RFI 1), was released and commented by the interested stakeholders, followed by a modified request for information (RFI 2) in September 2015. The RFI included the drafts of all relevant legal documents, in particular the
  - Invitation to the International Tender,
  - Preliminary bidding rules,
  - Templates of the PPP agreement,
  - Rules of IFT to prevent consortia with anti-competitive influence, and
  - Spectrum concession title and wholesale concession title.

During the consultations the conditions regarding roll-out obligations, the selection criteria, the minimum technical quality criteria etc. have been modified several times.

The final Request for Proposals (RfP), was published end of January 2016, followed by public consultations in February and March 2016. The bidders were requested to apply for the Anti-Trust opinion at the IFT latest by early August 2016 and the final bids were due by early September. Even though the SCT promoted the project at different events and during a dedicated roadshow in the US, Europe and Asia only two consortia, Altán Redes and Rivada Networks, submitted their bids. The selection of the winning bidder, Altán Redes, was proclaimed in November 2016. The PPP contract was signed in Q1 2017. Finally, in March 2018 Altán Redes launched operations of Red Compartida with an initial 32% of population coverage.

5.2   Regulatory Considerations

The creation of a mobile wholesale-only Open Access Provider is a new phenomenon. It is not easy to find international examples of a successful regulation. However, if the European Commission's framework for market analysis would be applied the following considerations should be taken into account.

Although RC will have a monopoly in one spectrum area, all other MNOs are able to compete in the LTE wholesale markets using other spectrum. In the Mexican market for mobile service termination the existing MNOs even have a position of economic strength and joint SMP, because amongst others:

- They are vertically integrated and thus control access to retail markets
- They control the existing 2G, 3G and LTE access networks and thus an essential facility for RC
- Market shares, overall size and access to capital markets is extremely asymmetric
- There is low countervailing demand side market power for RC

It therefore might be expected that oligopolistic tacit collusion could occur amongst the MNOs with them leveraging their vertical market power through

- refusal to deal, in particular in national roaming and thus trying to economically "dry out" RC,
- exclusivity contracts, in particular for international roaming partners or existing MVNOs, bundling and tying in particular retail and wholesale services,
- delaying tactics in negotiations with RC about infrastructure sharing, interconnection etc.,
- price or quality discrimination, and
- cross-subsidization etc.

The national regulator therefore has a massive task to get RC running as a start-up company. A particular concern may be exclusivity contracts. MNOs with SMP should therefore not be allowed to set exclusivity clauses in commercial contracts for

- International roaming services
- National roaming services
- Host MNO services for MVNOs

Due to a lack of Significant Market Power (SMP), Red Compartida should be free in the start-up phase to set commercial wholesale prices. Ex-ante regulation might be limited to transparency and non-discrimination.

MNOs should be obliged to offer non-discriminatory open access products including national roaming to Red Compartida for a certain period, and this has to be enforced in a reasonable time-frame.

It will also be very important to limit the influence of the existing retail MNOs on RC's operations, or at least to balance this influence to avoid one of them abusing the new market constellation.

6   Summary and Outlook

There are few examples of functioning wholesale-only operators. The carrier "Yota" in Russia could be compared. Yota operated a wholesale LTE network with 30 MHz in the 2.5 and 2.7 GHz band, covering 180 cities with about 70 millio people. However, in 2013 Megafone acquired 100% of the shares of Yota and integrated operations into its existing mobile operations Interestingly Megafone decided to use the Yota brand for its own MVNO. In Rwanda a mobile wholesale only network was launched under a PPP scheme together with Korea Telecom in 2014. The Kenyan government started with a similar approach but with more management involvement by the public sector slowing down the process. The Kenyan mobile operator Safaricom already announced not to wait any longer for the public network but to roll out its own LTE network eventually.

Another example is the Czech Republic. In June 2015 PPF, the new majority owner of O2 in the Czech Republic split the company into a NetCo called Cetin and a ServCo still called O2. The minority shareholders were offered shares for O2 worth 31% of the combined share price and 69% for Cetin. Cetin offers fixed and mobile wholesale products to all operators in the country. It is interesting to note, that the share price of Cetin stayed constant until end of 2015, while the virtual mobile operator value more than tripled over that same period.

The final version of the public tender for Red Compartida was published end of January 2016. The Government undertook considerable communication to attract national and international investors. There were three groups active setting up their consortia, securing the necessary funding and preparing their bids. None of them was supported by an international telecom operator. Two of the groups were specialized in funding infrastructure projects, pursuing a state of the art wholesale network approach. The third group, led by Rivada Networks, promoted its bid with an innovative business model centered around their dynamic spectrum and capacity arbitrage technology. It allegedly enables the dynamic selling of excess network spectrum or capacity to commercial tenants, i.e. an approach similar to the today known exchanges for energy. Rivada Networks also participated in the bid for FirstNet, a nationwide public safety LTE network in the 700 MHz band of the US Government.

Finally, the consortium "Altán Redes" won the bid end of 2016 by offering 92.2% population coverage to be achieved by end of 2023. The consortium was built by the following members (% shareholding):

- Mexican investors: Hansam, the original promoter of the project (9.35%), Axtel, a fixed network operator (4.01%), Megacable, (4.01%), and invexBanco (6.54%).
- International investors: Morgan Stanley (33.38%); IFC/Worldbank (26.7%), and the Canadian Pension Fund CDPQ (12.68%).
- Spanish Investor Eugenio Galdón through Isla Guadalupe Investment (IGI), who previously developed the Spanish fiber optical based broadband network operator ONO, which was acquired by Vodafone Spain in 2014.

Several other countries are observing Mexico with the potential intention of copying the approach. In particular South Africa is also planning to have a "national wholesale operator" in mobile. If Red Compartida becomes commercially viable the whole idea of national wholesale operators as a new instrument to boost national broadband roll-out may spread to many other countries.

References

IFT. (2017a). *Cuarto Informe Trimestral Estadístico, Mayo 2017*. Mexico City: IFT Publication.

IFT. (2017b). *Análisis sobre el Mercado de Operadores Móviles Virtuales (OMV)*. Mexico City: IFT Publication.

OECD. (2017a). *Economic surveys Mexico 2017*. Paris: OECD Publication.

OECD. (2017b). *Telecommunication and broadcasting review of Mexico 2017*. Paris: OECD Publishing.

Ovum. (2017). *Regulatory scorecard 2017: Central and South America*.

Secretaría de Comunicaciones y Transporte (SCT). (2017). Accessed October, 25, 2017, from http://www.sct.gob.mx/red-compartida

World Bank. (2017). The World Bank Data, World Development Indicators Data Base. Accessed October 25, 2017, from https://data.wordbank.org

Part VI

Customer Centricity

Making the Difference for Customers: Three Levers to Achieve Superior Experience in Telco Business!

Andreas Penkert, Julia Steffens, and Sascha Krpanic

1 Understanding the Importance of Customers Experience

Innovation has always maintained an unshakable position on telcos' agenda. The objectives of initiatives related to innovation are also congruent with one another as a rule: the development of unique and outstanding products or services, ideally a comparative advantage over the competition that is supposed to be followed by commercial success. But, is this strategic focus still sharp enough today? Cost pressure from falling prices, stagnating sales because of the substitution of important cash cows in core business, the growing competition from OTT providers with an eye on future business models, and the diminishing loyalty of customers are factors confronting telcos with many and varied challenges.

Nowadays innovative, proprietary products and services are no longer sufficient alone to achieve differentiation on the market. Even aggressive price strategies have reached the end of their effectiveness in creating leverage against the backdrop of rising cost pressure. Nothing less than the strict orientation of all added-value activities and processes to outstanding, distinctive customer experience will suffice. The authentic satisfaction of rational and (above all) emotional expectations and needs of customers, at all times and at every touch point of the company, is today and will be in the future a decisive driver in competitive, saturated environments.

How can telcos achieve excellent customer experience at every touchpoint now and in future? Furthermore, which methods, technologies and organizational changes can telcos apply to ensure best customer activities and offerings to differentiate from competitors?

A. Penkert (✉) · J. Steffens · S. Krpanic
Detecon International GmbH, Cologne, Germany
e-mail: Andreas.penkert@detecon.com; Julia.steffens@detecon.com; Sascha.krpanic@detecon.com

© Springer International Publishing AG, part of Springer Nature 2019
P. Krüssel (ed.), *Future Telco*, Management for Professionals,
https://doi.org/10.1007/978-3-319-77724-5_24

2   Creating Experience by Understanding Your Customer

Facing this environment, *customer centricity* is one of the most challenging part for telco companies nowadays. This implies a deep understanding of customers and their individual needs as well as the implementation of this mindset into development and adaptation of products, services and touchpoints. Thereby, *customer centricity* is shifting the value chain from product to customer perspective through starting with the customer himself.

Personas is one important step towards *customer centricity*. The method identifies, understands and penetrates the typologies of customers. Typically, telcos view their customers by target groups with mainly socio-demographic dimensions such as age, income, occupation, or family status. As part of the digital revolution, new clusters such as digital natives, best agers and silver surfers have emerged. However, this form of segmentation has largely worn out, lacking of deep insights. Customers are now looking for telcos especially based on how well their own needs are tailored individually or even overfilled.

In order to understand the needs of a customer in detail, it is useful to develop the various relevant customer typologies as *personas*. The *persona concept* does not understand a customer or a customer group as a single mass. It is possible to analyze target groups much more specifically and more detailed than conventional segmentation models do. Thereby, the concept presents typical users of products or services as *real persons*—in all relevant facets, such as individual CVs, character traits, needs, interests, problems/fears, etc (Christoph 2014). To model and develop a typical customer in this way, it is easier to discuss the meaning and nonsense of a product/service, the effectiveness of customer approach or a marketing campaign. Moving around in the context of a *persona* forces you to think authentically target-group-focused instead of product-focused. The application of the concept is comparable to the work of so-called profilers, as they are known from criminal films: Essential relevant personality traits and characteristics are identified. They help to get to know the "unknown essence of the customer" better, and to draw conclusions about his behavior. The customer is presented here as a fictional person (persona) and represents a whole group of consumers with similar characteristics, living environment as well as matching needs. A *persona* should cover all customer groups being developed ideally based on real empirical data and findings generated, for example, from partly structured in-depth interviews. Such interviews are the most promising methodology to identify genuine experiences and to understand customers' emotions. Alternatively, customer shadowing, analytics and feedbacks can provide important data and information as well. The characteristics and the depth of detail in a *persona* are dependent on the industry and product context. There is, however, a basic framework, which might be followed. Besides name, age, place of residence, profession and material status, it also comprehends character traits, individual needs and interests of a customer. And—last but not least—it should reveal his digital affinity and channel preferences to meet new innovative distribution channel expectations.

The *persona concept*, in principle, can be used wherever one is concerned with target groups, e.g. product and service development, software development, marketing and commercialization, human resources. Particularly in the telecommunication industry, *personas* offer themselves as a method for target group determination. On the one hand, because the pace of innovation and product diversity are steadily increasing, and on the other hand, because the needs and uses of communication technologies and services differ significantly between different types of customers.

Having fully understand your customers' needs is essential to reach customer centricity. As a next step, it is mandatory to also understand how customers move inside a company's ecosystem. With a *customer journey* telcos can understand all interactions, recognitions and feelings of a customer while using their products and services at any touchpoint.

For deriving a complete *customer journey* the following dimensions have to be comprised: (Maechler et al. 2016)

- *Episodes*: A journey is a specific, discrete experience in the customer life cycle, consisting of different parts. The act of simply purchasing a product in a store is a touchpoint within a *customer's journey*. Researching and then buying a new product and getting it up and running at home would constitute the full journey as the customer sees it from end-to-end.
- *End-to-end experience:* It is not enough to measure customer satisfaction on any single touchpoint; what matters is the customer's experience across the entire journey.
- *Language:* A journey event has to be described in the way a customer would articulate (e.g. "upgrading my product or service"), but not in a company language (e.g. "shipping new equipment").
- *Channels:* Multi-touch and multi-channel journeys need to be considered. In nature, a "new-product onboarding" journey might begin with a website visit, then a sales call, then a second website visit, followed by a store visit, then a technical-help call during the activation or installation stage.
- *Duration:* Journeys are often longer than expected. For example, the onboarding journey can extend through two or three billing cycles.
- *Repetition:* Journeys are repeatable—and can be repeated for a meaningful percentage of customers.

The *customer journey* demonstrates at which touchpoint or at which stage excellent customer experiences are realized or whether there are still pain points. Identified issues can then be customized and consistently optimized to the desired end-to-end customer experience using the *customer journey mapping*. Then, products, services and touchpoints can be better created and developed consequently from the customer's perspective, as his needs, user preferences as well as his behavior in the environment of the company's ecosystem are transparent based on the *customer journey* method.

3   Technology to Boost Your Customers Experience

By achieving customer understanding with all their needs and desires, a company has to use this insight knowledge to provide customers with their expected experience. The development of unique and need-oriented products, services and customer touchpoints are essential to stay competitive in market for telcos. With present technological inventions, companies got enabled to boost their productivity and to increase their customer experience by involving more and more digital innovations.

In this context, well-discussed technological innovations are augmented and virtual reality (AR/VR). AR got first boosts of excitement by consumers in the past, latest with airing the mobile game Pokémon Go[1] with 164 mio. users in 2016 (Lynch 2017). AR is applied to create users experiences by extending the real world with some virtual elements that supplement an environment with computer-generated sensory inputs (Javornik 2016). By using VR companies have the possibility to engage customers like never before, drawing them into the complete immersion of a head mounted display and peripherals, most commonly driven via computer, smartphone or specific VR glasses (Backhaus et al. 2014). In general, business applications based on VR/AR are nothing new, but now for the first time smartphone and tablets developed the performance levels to support and applicate an AR/VR ecosystem.

By using AR/VR technology telcos have the possibility to reinvent or improve their product and service offering at any touchpoint in a modern and unique way. It offers for example the possibility to develop 3D digital showrooms and sales presentations for their products and services. Consumers can virtually use the product and therefore they can trust its performance and shown features. They do not have to rely on salespersons' descriptions or even leave their house to get real insights of the product. A presentation based on VR/AR is visual, technically proficient and can be personalized for each customer. Telco companies are often distributing virtual and intangible products, like mobile and phone tariffs, cloud products, smart home applications etc. They can use a 3D digital showroom to show customers all benefits of a product and services with a real look and feel. A company's service activities, as an essential customer touchpoint, can strongly benefit from VR/AR usage. Swisscom provides a telco specific example about the usage of AR technology to improve service. The Swiss telco provider started to use an AR based Smart Assistance solution to support their field service agents. Some Agents got equipped with special glasses. The glasses provided the agents with information to make their work more efficient and helped them solve complex system problems faster while being on customer's premises (Dufft 2017). This

---

[1]Pokémon Go is an interactive, mobile game that uses your phone's GPS data and clock to show Pokémon hidden near your current, physical location via AR Technology. The Pokémon which appear on-screen in the app can be captured. As you go to different locations and explore different cities and towns, the types of Pokémon you encounter will change. The game was developed by Niantic

scenario provides a good example how AR/VR can help telco companies to reduce deployed resources.

Consistent with introducing customer centricity, as mentioned before, there is great potential to integrate VR and AR technologies into product and service development. With virtual product design, consumers have the ability to test-drive products and service in various stages of development. Virtual product design provides a more simplified and probably less-expensive way of a *minimum viable product* (MVP)[2] like a *virtual prototype* (Westhoff et al. 2014, p. 64). With 360° perspectives and full-features product shows customers can provide companies with detailed feedbacks and impressions without witnessing it in real. For example, Swisscom established an innovative development processes to cause best customer experience. It involves customers and uses repeatedly interactions with them during the entire product and service development process. Every phase is using customer feedbacks from previous stages as input to generate ideas or improvements for (new) services or products (Detecon Project Experience 2017). By using MVP's Swisscom has the opportunity to gain feedback and insights for product, services and touchpoints to continue or discontinue their development.

Gaining knowledge from customer analysis and observation and using it to improve products, services and touchpoints are among other things results from previous mentioned methods and technologies. In that context, AI technology is able to shift or rather assist companies in their customer experience creation. It describes a technology that emphasizes the creation of intelligent machines that can work and react like humans. It mostly includes image and speech recognition, learning, planning and problem solving elements. In our customer context, it is now possible to effectively, and accurately, automate customer interactions and predict further customer needs through deep learning and predictive analytics of customer data. At the same time companies can provide their customers with a more immediate and personalized response to their request (Conick 2017). Nowadays many people are using partial AI through Apple's Siri or Amazon's Alexa.[3] Moreover, chatbots, a natural language procession technology of AI, are becoming the new norm for online and in stores service because they offer the immediate and convenient experience that shoppers crave.

By implementing AI companies can leverage many customers related activities like visual search, natural language processing, usage and shopper behavioral analytics and online-shop analytics. Furthermore, AI algorithms can be used for cross-platform advertising, intelligent e-mail and content-aware video advertising, among others things. With endorsing AI technology telcos can address their customers individually, demand-focused and at the right moment without the need to bring more human capital into it. Especially, telcos need to deal with many service requests, even if they are often minor request like lost PIN or PUK numbers. With

---

[2] A minimum viable product, conducted from lean startup methodology, is a first product version with minimal functions to test-drive it with costumers (Ries 2011).

[3] AI Technology based on natural language understanding and functioning as a personal assistant.

the help of AI technology human involvement can be reduced without losing service quality while the availability of service offerings extends. In doing so AI does not need to replace every human interaction in service. Many AI technologies help to improve service quality for customers while working tangent with human agents. *Machine learning* for example can help streamline processes, which helps humans to focus on tasks that actually require human input. Right now AI augmented messaging, AI organized email inquiries and AI enhanced customer phone calls are taking over simple serving tasks, so humans can focus on more complex service requests (Walker 2017).

As seen in the examples above businesses and furthermore telco companies are implementing more and more technological trends and innovative development methods, which are all aiming one goal: To provide their customers or prospects with the best, most personal experience and most fitting offerings. Technological inventions are providing aids to fulfill and possibly exceed customer expectations by supporting products, services and touchpoints.

4   Enabling Customer Experience by Organization and Governance

As already indicated in prior parts of this article: concepts such as *persona* and *customer journeys* have evolved to fundamental elements when identifying customer needs and the attended development of products, services or touchpoints. Moreover, the integration of the mentioned megatrends AI as well as VR and AR are becoming tremendously important levers to boost customer experience. Despite the fact that the application of similar concepts and technologies has often become common practice, telcos are still not acting customer-centric.

In our experience, this clash can often be attributed to telcos' organizations and underlying governances that are defining the actual behavior of a company. In these days, company segments decide business activities with customer impact in own organizational silos and in the sense of their business interest without bearing in mind the entire organization with all its cross connections. While doing so, customer targets are mostly not in focus but rather steering elements consisting of pure efficiency targets, short-oriented revenue improvements or pressure resulting from competition to bring a new product to the market (Detecon Project Experience 2017).

The quintessence? The pure application of customer-centric marketing concepts is only the first step in the direction of real customer orientation. If the organization and the underlying governance do not support and facilitate resulting implications due to their steering framework, telcos are not be able to deepen the relationship to their customers.

In order to encounter this issue, companies need a central unit assuming responsibility for a customer-oriented behavior within their organization. Consequently, such an unit would be in charge for analyzing customer needs, for decisions which measure might be executed based on customer needs as well as for consequent tracking of an effective implementation—or, in other words: if the voice of the customer yields to short-term oriented (and non-customer-centric) objectives of single business segments.

From our point of view, a possibility to achieve consequent customer-oriented behavior would be the implementation of a *customer board*. It orchestrates as well as comprehensively steers all company activities affecting the customer experience as a neutral decision-making committee from customers' perspectives. As a result, all company activities that influence customer experience need to pass this board.

The target of this committee is to create transparency regarding companies performance in terms of main customer KPIs (e.g. satisfaction indices or net promoter scores) on a company, product, service and touchpoint level as well as regarding already existing and planned measures throughout the customer journey. Moreover, the customer board should be equipped with a decision and recommendation mandate in order to design diversified, customer-orientated and up-to-date product portfolios. This implies the decision-making authority to release new or adapted offerings for market entry, to eliminate existing offerings from the product portfolio, to prioritize necessary changes and new product requests and thus to define central customer segments for marketing and sales.

However, telcos need to consider some basic principles for a successful implementation of such a customer board. One major success factor is the consideration of customer KPIs inside companies overall steering logic as well as in individual targets of all relevant units and decision-makers. Moreover, the customer board should be positioned on a comprehensive level in the organization of telcos. In this context, it is common to allocate the customer board directly below the executive board level and to integrate executives into the group of participants in order to strengthen the board's position and acceptance. Only with organizational autonomy, the customer board is able to function independently from *traditional* business unit objectives and to take decision over products, services and touchpoints beyond business silos by considering customers perspectives.

In addition, the customer board should assume responsibility for and decide over all existing product units. This allows on the one hand the already mentioned creation of transparency regarding planned measures. On the other hand, it enables to break up silo thinking in the organization. Thus, the customer board possess a comprehensive view on the product portfolio and is able to define as well as to prioritize implementation decisions based on the overall business strategy. Against this backdrop, it might also make sense to equip the customer board with a decision mandate regarding the corresponding allocation of financial and human capital in order to invest company resources throughout the entire organization reasonably along customer needs.

For the personal composition, this implies that the customer board should consist of all representatives with voting rights from every product unit. In addition, important representatives from units with customer contact (for example sales, call center, administration of customer-relevant processes) and from a central customer unit should be considered.

In case of the task allocation, our experience shows two different implementation options for a customer board. Consequently, it is one option to implement one customer board solely for decision-making over development and adaption of product, services and touchpoints. A second option is to implement three customer boards based on the described decisions. Although especially option two results in additional complexity of an already challenging idea, it is a valuable concept to share and divide decision-making power within the organization. Moreover, it aims to reduce the perception that too much power and control is allocated to an already newly introduced committee (Detecon Project Experience 2017).

Briefly, the customer board completes the persona and customer journey concept to create as well as the technologies AR/VR and AI to boost customer experience and consequently builds the organizational fundament.

5   Aiming a Customer Experience Ecosystem

All three dimension exemplified before are providing a structure to enable a customer experience ecosystem into a company's organization. By focusing on approaching customer centricity into development and the organization itself companies are enabled to better understand their customer demands and expectation. Companies have the opportunity to provide customer a unique and digital experience as nowadays expected, by combining customer centricity with innovative technologies at any customer touchpoint. Moreover, new innovative trends present opportunities to provide customers with unique experiences, therefore to differentiate from competitors. The establishment of a customer board builds a fundament for telcos to overcome organizational barriers and to fulfill every activity to deliver excellent customer experience. After all, companies have to adapt all three levers together to build an effective customer oriented value chain in their own ecosystem. By managing the customer experience from end-to-end telcos are able to make the difference in market and stay competitive in the long run.

References

Backhaus, K., Japser, J., Westhoff, K., et al. (2014). Virtual reality based conjoint analysis for early customer integration in industrial product development. *Procedia CIRP, 14*(25), 61–68.

Conick, H. (2017). The past, present and future of AI in marketing. *Marketing News, 17*(51), 26–35.

Christoph. (2014). Die Zielgruppe erkennen mit dem *Persona-Konzept*. Accessed October 8, 2017, from https://www.marktding.de/marketing/die-zielgruppe-erkennen-mit-dem-persona-konzept/

Dufft, N. (2017). Die digitalen trends 2017: Von der Vision zur Realität!. Accessed October 7, 2017, from https://www.swisscom.ch/de/business/enterprise/downloads/digitalisierung/digitale-trends-2017.html

Javornik, A. (2016). What marketers need to understand about augmented reality. *Harvard Business Review Digital Articles, 16*, 2–5.

Lynch, M. (2017). *Global equity research: Number of active users of Pokémon go worldwide from 2016 to 2020, by region (in millions)*. Bank of America. Available via Statistica. Accessed October 9, 2017, from https://www.statista.com/statistics/665640/pokemon-go-global-android-apple-users/

Maechler, N., Neher, K., & Park, R. (2016). *From touchpoints to journeys: The competitive edge in seeing the world through the customer's eyes*. McKinsey Customer Experience Compendium 2016.

Ries, E. (2011). *Creating the lean startup*. Accessed October 10, 2017, from https://www.inc.com/magazine/201110/eric-ries-usability-testing-product-development.html

Walker, J. (2017). *Use cases of AI for customer service – What's working now*. Accessed October 7, 2017, from https://www.techemergence.com/ai-for-customer-service-use-cases/

How Carriers Can Use Protection Services to Safeguard the Digital Lives of Their Customers

Carolin Obernolte, Joachim Hauk, and Clemens Aumann

1 The Daily Dilemma: Convenience Versus Disclosure of Data

The Internet connects all of us. It offers an almost infinite variety of different services of which many of them are for free. Consumers in particular profit from the variety of services that make their lives easier or more interesting. The diversity of services ranges from complimentary worldwide telephony using for instance Skype to the exchange of news on Facebook, the sharing of photos with friends on Instagram and communication via video messages on Snapchat. Online shopping services allow purchases to be made flexibly from home, and smart home solutions connect the entire home in a single network, so that a smartphone can be used to operate multiple devices remotely, be it lamps, radiators or televisions. Isn't it convenient to stay at home all the time not having to leave the house to take care of shopping or standing up to turn devices on and off?

But that is not all. Wearables such as smartwatches allow people to collect and analyze their vital signs; the results can point the way to a healthier life style. Not knowing your way around a city does not mean you have to go to the tourist information office or ask passer-by for directions; just turn on your smartphone's location function and let the navigation app direct you to the desired destination. These use cases are only some of the many examples from our everyday lives. The benefits of all of the use cases can be summarized under key words such as digitalization, sharing economy, and convenience. The usefulness of these services, however, is offset by certain risks that consumers face. Everything has a price—and in this case it is not only a question of money. Consumers pay the highest price by disclosing their data.

C. Obernolte · J. Hauk · C. Aumann (✉)
Detecon International GmbH, Cologne, Germany
e-mail: Carolin.Obernolte@detecon.com; Joachim.Hauk@detecon.com; Clemens.Aumann@detecon.com

© Springer International Publishing AG, part of Springer Nature 2019
P. Krüssel (ed.), *Future Telco*, Management for Professionals,
https://doi.org/10.1007/978-3-319-77724-5_25

Providers of Internet services generate enormous quantities of personal data, then analyze them for the purpose of preparing detailed customer profiles. These profiles are used by providers to tailor advertising and, in general, all types of information to the interests of specific consumers. The problem is that these activities on the part of providers are not transparently revealed to consumers. All the customers can see—based on the advertisements and messages tailored for them specifically—is that inferences are constantly being drawn about their personalities from the generation and analysis of their data. But the collection and analysis of the data are not the only risks for consumers. Many providers of Internet services store their customers' data and may even disclose them to third parties. Consumers have no idea of what may later happen with these data; they are also at risk because the providers of Internet services cannot fully protect their customers from hacker attacks and espionage as these companies are themselves often victims of massive cyber-attacks.

In short, consumers are no longer the masters of their personal data. They can neither control what data are recorded and analyzed nor determine when and how long their data are stored. If they want to avoid these risks, consumers have to do without Internet services. Conversely, they do not receive any economic compensation for the use of their data or any share in the added value generated for the companies using the data, although a study by the Ponemon Institute (2015) reveals that consumers are definitely aware of the value of their data.[1]

So what can people do who want to enjoy the benefits of Internet services, but nevertheless want to protect their personal data? Can protection services offer the appropriate protection model to consumers according to their individual need for protection? Would such a protection model be at all helpful and desirable from the customers' perspective?

1.1   Key Theses About Protection of Personal Data from the Customer Perspective

1. **Lack of transparency for consumers regarding the use of their data:**
   There is a lack of transparency, ranging from extensive to complete, for consumers concerning the recording, processing, analysis, and storage of their data. According to the Eurobarometer study on the subject of data protection, 50% of the respondents replied that they had only partial control over their data (Schiavoni 2015). 35% of the respondents stated that they did not have any control whatsoever over the data they share (Schiavoni 2015).

---

[1]Study of consumers from Europe, the USA, and Japan according to various categories. Example: health measures, $35; buying history, $17.80; employer/training status: $8.50; current geographical location: $5.10 (Ponemon Institute 2015).

2. **Risk of complete surveillance and spying:**

Consumers are increasingly aware of the risks of surveillance and spying when using Internet services (Bonneau 2016; Bartley 2017). The Eurobarometer study reveals that 50% of the respondents are concerned about becoming victims of the misuse of their data (Schiavoni 2015). 32% of the respondents fear that their information is used or even stolen (29%) without their knowledge (Schiavoni 2015).

3. **Lack of trust of consumers in the companies processing their data:**

Consumers must entrust their data to the Internet service providers if they want to utilize the offered services. Yet there is a heightened mistrust on the part of consumers regarding these companies when the issue is the handling of their personal data. The Eurobarometer study showed that 78% of consumers find it hard to trust the companies that process their personal data (Schiavoni 2015). Even though almost 80% of the surveyed consumers mistrust the data processing companies, the majority of these consumers continue to use the services of these providers because they do not want to do without the benefits or conveniences they obtain from these services (Schiavoni 2015). Facebook's acquisition of WhatsApp at the beginning of 2014 is a good example illustrating this situation. When it was announced that Facebook would be acquiring WhatsApp, there was a wave of revolt because Facebook's data protection provisions are not the friendliest for consumers (Radke 2016). There were concerns that WhatsApp would be forced to align itself with Facebook's data protection provisions, and this was seen as a risk to the security of the user data (Radke 2016). Many consumers, looking for a more secure chat alternative, decided to go to Threema. Even though Threema actually does offer significantly user-friendlier handling of its customers' data (Stiftung Warentest 2014), this service was in the middle term not able to compete with WhatsApp and has not achieved user figures (4.0 million) that are anywhere nearly comparable to those of WhatsApp (per Jan 2017) (Statista 2017a, b) (Fig. 1). This could be explained by the unwillingness of the majority of users to make the effort to look for and use an alternative secure solution.

4. **Growing unease when using Internet services:**

There have been a number of incidents in the past in which companies have lost customer data because of security gaps and hacker attacks. In October 2013, for instance, three million credit card records of Adobe customers were stolen (Little 2014). These and other such incidents have not only raised the level of fear among consumers regarding malware and cyber-attacks, but have also increased their awareness of the value of their data. Their fears of becoming potential victims of data misuse incidents or identity theft are on the rise. The information policies of many companies in the event of data misuse or security breaches are often way too defensive and reinforce these reservations (Rybak 2015). But despite these fears, the majority of consumers do not handle their data carefully. A study conducted by GSMA on the subject of awareness of data protection

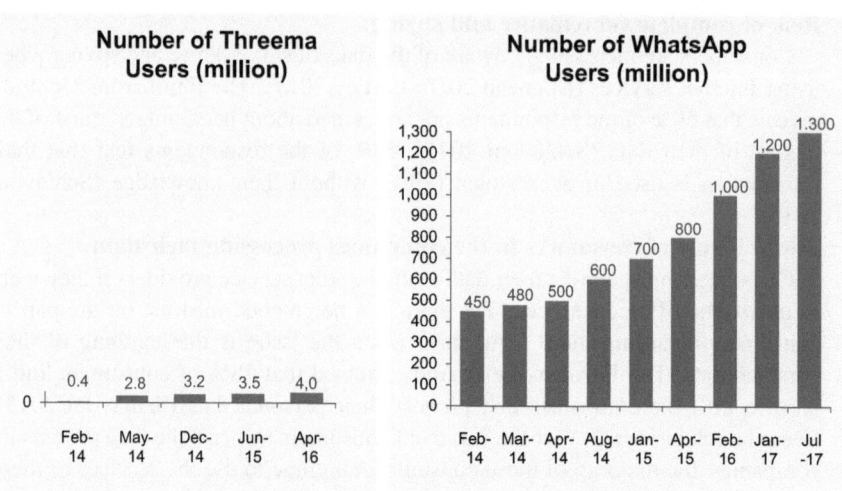

Fig. 1 Comparison of user figures for Threema and WhatsApp (Statista 2017a, b)

provisions of mobile Internet users determined that 80% of the users of Internet services or apps accepted the data protection provisions without reading them because they are too long or contain too much legalese (Schiavoni 2015). This implies that consumers are often overwhelmed by the current data protection provisions.

5. **Growing demand for protection services:**

Consumers are not only demanding more and more security during the use of Internet services, but are also insisting more persistently that they want to be able to decide where and to what extent they disclose what data. According to the GSMA study, 60% of the surveyed consumers want a standard body of rules for the protection of their data and want all providers to comply uniformly with these rules (Schiavoni 2015). Even though most customer solutions are still not understandable and are too technical for the average user, more and more consumers are concerning themselves with the subject of security because the significance and necessity of security solutions are rising in the eyes of the consumers (Mohr-McClune 2015). Until now, however, consumers had to become active on their own initiative if they wanted to protect their data because they did not receive any active or extensive help or the appropriate services for this purpose. A study conducted by Orange on the subject of behavior change among consumers in relation to data protection revealed that 37% of the respondents had the feeling that companies or organizations did not give them any instructions for personal data management (Schiavoni 2015). Consumers are more likely to have the feeling that they are dependent on the good intentions of the providers despite the previous agreement of data protection provisions.

"[Customers] are concerned that companies are using their data for more than was initially agreed." (Schiavoni 2015).

6. **Less trust in OTTs than in telecommunications providers:**

Consumers regard OTTs as companies that want to profit from the data of their customers (Schiavoni 2015). Telecommunications companies, on the other hand, are perceived as the "clearing agents" of data and are therefore seen to be more trustworthy where the handling of personal data are concerned (Schiavoni 2015): "[. . .] [Operators] are often seen as more trustworthy than Internet companies or other service providers and can position themselves more strongly in terms of protecting their customer's privacy." According to the GSMA study (Schiavoni 2015), telecommunications providers are even regarded as the consumers' contacts when there are problems related to the subject of protection of data or privacy because 58% of the respondents ask telecommunications providers for help whenever they have these kinds of problems (Schiavoni 2015). A recent study by Syniverse (2016), however, shows that the trust consumers have in mobile providers has declined. Respondents were asked to state whether their trust in mobile providers with regard to the protection of their personal data had changed.[2] Half of the respondents indicated that over the last 3 years they have had "less" trust, 35% have had "just as much" trust, and 15% have had "more" trust in mobile providers (Syniverse 2016). This implies that consumers have become increasingly skeptical about mobile providers when it comes to data security over the last 3 years. A basic finding, however, is that consumers of providers who give them more control and transparency in the management of their data are regarded more positively than other providers who resist or even refuse to give transparency: "Consumers appear increasingly to trust and to use companies that are willing to offer them greater control through tools that are easy to use" (Schiavoni 2015).

This customer perspective clearly shows that consumers always want services that collect, analyze, and store as little of their personal data as possible. Since in reality there are almost no offers of any such services at this time, customers must accept the loss of control over their data. Nevertheless, they are becoming increasingly sensitive to the topics of security and data protection so we can deduce there is a fundamental need for protection services. Since many consumers often have difficulties in understanding the content of data protection provisions, such services should aim to provide intuitive security services that are simple to understand and simple to use and that customers can use to manage the security of their data. Protection services are a measure that builds trust among consumers as it gives them transparency about and control over the utilization of their data (Schiavoni 2015). Conflicts of interest severely limit the credibility of

---

[2]The attitude of consumers with respect to data protection was surveyed in this study. More than 8000 consumers in eight countries were questioned.

the providers of services. Since customers use a number of providers at the same time, clarity suffers. In our view, only regulatory authorities, network operators, or completely new companies with a security focus will be able to provide a relevant protection product that encompasses all of the services and at the same time offers impartial protection. We believe that network operators are in the best position to provide this: the traffic flows all come together in their purview, they have the customer relationships, and are less sluggish than government authorities. In comparison with the startups that are appearing, they (still) have the advantage of greater reach from their clientele, brand awareness, and trust in the brand.

2   Key Arguments in Favor of Offering Protection Services from the Telecommunications Providers' Perspective

2.1   Regional Representation and High Reachability

Large telecommunications companies cover a broad geographical territory through their own subsidiaries or partners. In the countries where they provide service, they can generally be reached easily because of the full-area coverage provided by a network of shops and their excellent accessibility on digital and phone channels served by large service (Agresti et al. 2016) units operating these channels; they are consequently well prepared to respond and act, especially if there are problems or in crisis situations. In their position as local telecommunications service providers who are subject to domestic jurisdiction (and therefore directly addressable by legal action despite their multinational character) they can build a trust positioning more credible than competitors from other industries.

2.2   Data and Communication Security Is a Part of the Core Business

The security of the networks and the communications that flow through them have long been a part of the core business of telecommunications providers. They never stop thinking about these aspects and have the corresponding expertise at the technical, procedural, and regulatory levels. They are especially qualified to incorporate the growing regulatory pressure related to data protection and data security operatively and productively (Little 2014; Agresti et al. 2016). They can control data security in the utilization, processing, and transportation layers of their ecosystem, and this contributes additional credibility to their portfolio. Moreover, they usually have proven and sensitive anti-fraud processes in place. This combined with the capability to limit or (in case of major security breaches) even totally stop data traffic with a device in general or a specific app, gives telcos a special advantage in safeguarding their customers' data and protecting their interests (Copigneaux 2016).

Nevertheless, this presumes the appropriate sensitization and emphasis on the subject as well as the adaptation of the appropriate plans for actions and emergencies in the event of a security incident (Rybak 2015).

2.3   IT Competence and Trend to Cloud Products

Most telecommunications providers already have a pronounced IT competence, in some cases including even their own divisions or branch businesses whose core business includes IT development and operation for customers. As cloud products become increasingly important and large telcos acknowledge this by expanding their portfolios accordingly, these business offers will gain additional impetus, but they must also meet heightened security demands (Newman 2017).

2.4   Simple Processability of Security Services in the Business Model

The character of communications services as a continuing obligation is consistent with the business model of protection services as they also represent a commitment that is ongoing or related to a period of time. Protection services of this type can be offered simply as a supplementary option to existing contracts or as a stand-alone product. The matching billing opportunities are existing as well as capacities for third-party providers or partners already established (Clark-Dickson 2014). The necessary distribution and service competencies are already in place as well or can be added with little effort.

Overall, telecommunications service providers have an excellent position, from the customer perspective as well as on the basis of their branding and the required competencies, to assume the role of guardian of customers' data security (Copigneaux 2016). The next question concerns the possible form of such a service, i.e., what design elements and levels are desired on the customers' side and would represent sensible components of an attractive service.

3   Seize the Opportunity: Build Up a Protection Portfolio Step by Step (Protection as a Service)

Monetarization can be realized as explicit protection services as well as in the form of a general premium price model based on perceived brand dimensions. In our opinion, a logical approach is a step-by-step build-up that combines both concepts.

The fundamental axis is the impact depth of the protection. Only shallow impact depth will be realized if the information customers receive from their providers:

- Is provided only on rather rare occasions, e.g., when the contract is concluded or if significant risks appear;
- Is related primarily to the services and data offered or used by the provider itself;
- Is mostly very generalized;
- Concerns more general risks.

Starting from such a basic foundation, a protection portfolio can be developed gradually along three key dimensions:

1. Timeliness, nature, and scope of the risk assessments, i.e., the question whether these are only services that the carrier itself offers or services that go (slightly/ extensively) beyond this scope;
2. Degree of personalization of the security information, i.e., assessment or information only if the installed and utilized services are affected;
3. Action intensity, i.e., information only or a concrete request to take action up to an automatic action triggered by predefined action points agreed with or defined by the customer.

We assume that greater depth of the protection service will go hand in hand with increased willingness to pay. The portfolio evolution depicted in Fig. 2 begins with the creation of transparency as a "measure to build trust" and develops step by step into an avatar that can assume a broad range of virtual identities for customers in their relationships to other virtual transaction partners (Deuker et al. 2011). We see only the positioning and the brand goodwill with parallel premium pricing as a monetarization approach for the portfolio cluster "Transparency and Self-determination". In our opinion, the opportunity to offer this approach as an independently priced service appears with the cluster "Passive Protection".

Carriers have not taken general possession of this protection function yet. But there are already OTT providers who are actively positioning themselves in the direction of protection function and data security.

One example is Digi.me, an app that has developed from the function of a private storage facility for personal social media content into a security function and a central profile with controlled release of personal profile data (Bonneau 2016). This app is already being used by around 400,000 customers in various countries and will now be further strengthened through a merger with another startup called Personal which offers a company security platform with the target of building a personal data ecosystem (O'Hear 2017). There is also the competitor Datacoup, which sells profile information released by customers to interested data buyers and compensates the customers with payouts. Datacoup is at the moment active only in the USA, however (Datacoup 2017). These are indications that pioneers are already staking out positions in this gap. Telcos must act quickly and seize the "window of opportunity" before it closes. Some telcos have set up programs with more or less

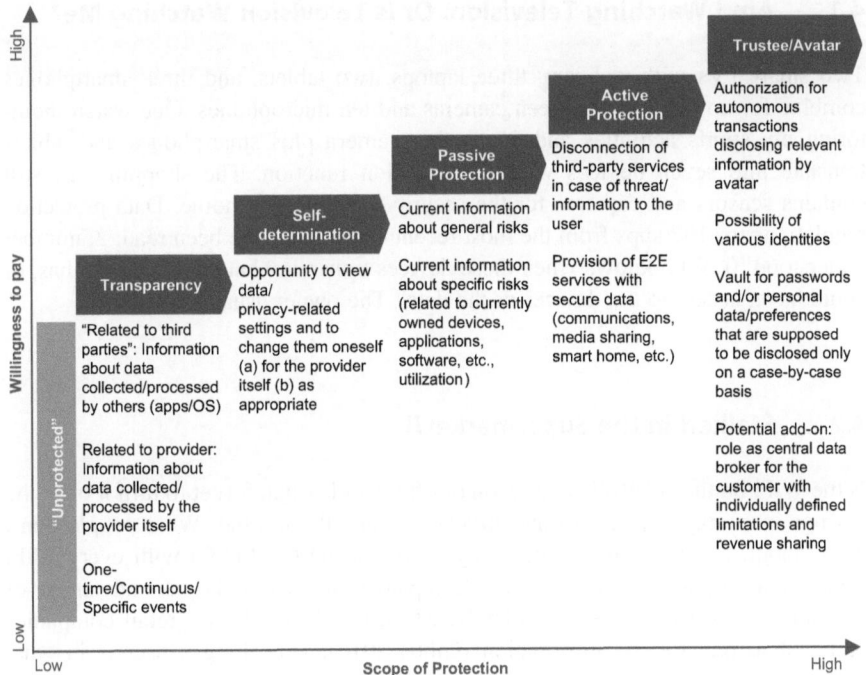

Fig. 2 Potential protection services for carriers

ambitious protection focus. Telefonica for example has set up a high level program called "Aura" which aims to improve customer interaction through AI combined with a customer side agreement to which extend data can be used and possibly shared with third party service providers. First use cases are planned to go live in the first quarter of 2018 (Alvares de Souza Soares 2017; Kompany 2016). Deutsche Telekom has been working on Open ID solutions for some time (Copigneaux 2016) and has merged all its Security and privacy products and competences in a new business unit T-Sec in 2017 which already has a substantial security portfolio in place and is now rapidly enhancing and connecting its offerings (T-SEC 2017). Other carriers have also started respective programs of various scopes and sizes. Which ones will be fast and sustainable enough to establish themselves on this now forming market remains to be seen.

4 Demand Today and Tomorrow: A Quick Check

At first glance, the avatar described briefly above may seem like a vision from a far-distant future. But if we look at the developments that have already taken place or the ones now in the pilot phase (and consequently within reach), we quickly have the impression that we already have one foot "in the matrix".

4.1   Am I Watching Television: Or Is Television Watching Me?

Two smart TVs with webcam, three laptops, two tablets, and three smartphones combine for no fewer than fifteen cameras and ten microphones. One watch monitoring my sports activities and one action camera plus smartphones and tablets translate into seven devices with GPS/location function. The shopping list still contains sensors and cameras for the security of the smart home. Data protection regulations for 107 apps from the most recent count that have been read: 2; number understood: 0. Who knows when these devices transmit what data and who has, or could have, access to the devices or the data? The owner usually does not.

4.2   Stalked in the Supermarket?!

Is the WiFi on the smartphone now on or off? People often forget to turn it off—the savings in energy appear too insignificant to take the trouble. What people don't think about: the smartphone cheerfully shares its ID (its MAC) with every WiFi access point it passes during the day. Companies such as Euclid take advantage of this and generate movement profiles based on the device ID for retail companies prepared to pay for the information (Gibbs 2016). How long a user is in what supermarket, how long he or she stood in front of what shelf, and the points the user rushed passed become visible. Users should receive bonus points entitling them to discounts for the provision of these data. As a minimum.

4.3   Paying with a Beautiful Voice Instead of Your "Good Name"?

The broad acceptance of Amazon's Alexa represents a clear leap forward in the voice base man-machine interaction, leading even to a new category called "voice commerce". Voice is becoming increasingly popular as a convenient means of control and personal identification. Also, services like the "hands-free" payment by voice command tested by Google appears to be a logical and convenient evolutionary step. In terms of convenience and voice identification surely wins over a PIN entry. However, it results in devices constantly listening to you (Edwards 2017). Is it possible to determine whether the camera and microphone are really turned off—and stay off? Are the service providers always eavesdropping?

At any rate, the European Commission in its proposal for a new ePrivacy guideline as of January 2017 plans to expand the regulation scope to OTT communications services besides the traditional communications services as provided by telcos (Schiavoni 2017). This shows that politics begin to see the need for regulating a wider scope of players in the market besides the traditional telco players.

4.4   What Does the Future Look Like?

Growing connectivity of personal devices of all kinds, an increase in the connected devices for improved company processes—all of this will lead to exponential rise in the density of the sensors surrounding us. Moreover, the number of network access points for precisely these sensors will increase significantly. The probability of our communicating with others, whether we are aware of it or not, in all areas of our lives will mushroom significantly.

The situation becomes disquieting when we give serious thought to what the linking of all these data can do. The television knows what we are watching, the heartbeat monitor records our reactions to what we see—outstanding for the measurement of the effectiveness of the advertising! The car insurance company believes that it can use the data to determine the extent to which we obey the laws of physics and the highway code—and adjusts the premiums accordingly. The health insurance company believes it knows whether, when, and how much we exercise or go shopping in our local wine shop. Will we still be able to get insurance if we refuse to allow ourselves to be tracked?

All mentioned scenarios illustrate, that even within regular use cases, users are depending on the security and privacy ethics of their business partner (even if new regulation initiatives might steer providers to some extent). But: whatever is connected, can—and probably will be—hacked. What happens, if your autonomous car dashboard urges you to pay ransom to regain control?

At the moment, the risks do not appear to frighten users very much. Despite the criticism heard from many different sides, the user numbers for large platforms such as Facebook and WhatsApp have not suffered significantly. The network effect—the decisive point for the individual: that the majority of his or her contacts also change—plays right into the hands of the large players. Moreover, the risk is still considered relatively low from the user perspective. WhatsApp users, for instance, saw the risk of unsecured messages as generally negligible for a long time. As digitalization continues to advance, however, this perspective will become more disproportionate: fully digitalized everyday life cannot be anything but fully documented and analyzable everyday life.

We see the greatest risk in the curtailing of information neutrality. What happens if the seamless record of a person's behavior is used as the basis for providing only that information that others (systems) deem to be relevant? Instead of being able to see the full range of world events and shopping opportunities that is available, a person will be given only an enhanced mirror image of what are seemingly his or her interests and inclinations. The "classic" risks such as identity theft, transaction fraud, or extortion are, and will remain, more tangible and encountered in daily life. All of these risks will be multiplied by the progressively deeper penetration into personal everyday life by digital services in the future.

5   Carriers Should Stake Out a Position: And Soon

In summary, we have prepared five hypotheses that serve as guideposts for carriers for the step-by-step determination of their need for action:

1. Drivers on the demand side: The insecurities experienced by users of telecommunications services as digitalization progresses will escalate rapidly.
2. Opportunity for carriers: In the current market system, carriers are fundamentally the most qualified entities to create services that respond quickly across all sectors and to provide transparency and protection to a broad range of customers.
3. Time pressure for carriers: The longer carriers hesitate to stake out a position "a safeguarding complement" to OTTs, the more they will be viewed as the latter's supporters.
4. Assessment and development requirements for carriers: The monetarization of the possible aspects related to customer data is viewed at this time almost exclusively through the "classic" big data glasses. Our experience indicates that the aspect of protection in terms of its possible added value has not been appraised.
5. Need for carriers to act: Not every carrier is today regarded by its customers as being adequately qualified to serve as a trustworthy protective body.
6. Rebalance the rules of the game: Carriers should not only demand the extension of communications privacy rules to OTTs in equal terms as telcos but play out the regulation adherence head start they have compared with OTTs in the market.
7. Teaming up: Participating in cross-industry alliances striving to create alternative trustable choices for consumers and businesses alike should be considered as an option. Initiatives like Verimi might in return also strengthen attractiveness—as alliance partner products can be integrated to widen the operators' offerings.

It is therefore urgently necessary for carriers to determine now the positioning that is initially relevant and possible for them. The gradual emancipation from the OTT Big Brother model that is also credible in terms of branding and the empowerment of telecommunications users are possible. The process can be oriented to the evolutionary model for protection services described above. Startups are already addressing relevant elements of this model. In view of the required build-up of competencies and the product development phase that is to be expected, the time to act is now. The need for protection of the digitally illuminated customers is a great opportunity to compensate the losses at the customer interface to the OTTs. In our opinion, it is possible to compensate these losses completely—and even more.

References

Agresti, G., Faggiano, A., & Strusi, G. (2016). *Telecom retail & consumer protection*, ADL, p. 8.
Alvares de Souza Soares, P. (2017). S O^2 S Hohe Rendite, aber schlechte Qualität. *Manager Magazin, 09*, 40–43.

Bartley, P. (2017). *Privacy as a business advantage: How building controls for privacy can empower the workforce and maximize the value of information*, Ovum, p. 5.

Bonneau, V. (2016). *Privacy business: How will privacy issues affect internet business models*, IDate, p. 14, 61f.

Clark-Dickson, P. (2014). *Data Mobile operators' consumer mobile security strategies*, Informa, p. 6, 12, 18.

Copigneaux, B. (2016). *Digital identity: Opportunities for telecom operators*, IDate, p. 5, 48ff, 60f.

Datacoup Inc. (2017). Accessed October 12, 2017, from https://datacoup.com/docs#data-profile

Deuker, A., Aumann, C., Albers, A., & Duschinski, H. (2011). Bekommen statt Suchen – Warum wir unsere Interaktion zukünftig an Smart Agents übergeben. *Detecon Management Report, 3*, 8–17.

Edwards, H. S. (2017). Alexa takes the stand: Listening devices raise privacy issues. Accessed October 20, 2017, from http://time.com/4766611/alexa-takes-the-stand-listening-devices-raise-privacy-issues/Accessed

Gibbs, S (2016). Shops can track you via your smartphone, privacy watchdog warns. Accessed October 20, 2017, from https://www.theguardian.com/technology/2016/jan/21/shops-track-smartphone-uk-privacy-watchdog-warns/Accessed

Kompany, R. (2016). *Telefónica is engaged with the platform economy*, Analysys Mason Report, p. 3.

Little, M. (2014). *Personal data and the big trust opportunity*, Ovum, p. 14.

Mohr-McClune, E. (2015). *Zeitgeist communications: Speaking in confidence*, Current Analysis, p. 3.

Newman. (2017). *Cloud BSS: The migration begins*, TMF, p. 29.

O'Hear, S. (2017). Digi.me and personal merge to put you in control of the nascent 'personal data ecosystem'. Accessed October 06, 2017, from http://techcrunch.com/2017/08/17/digi-me-and-personal-merge/

Ponemon Institute. (2015). Privacy and Security in a Connected Life: A Study of US, European, and Japanese Consumers, p. 17. Accessed October 15, 2017, from https://www.trendmicro.de/cloud-content/us/pdfs/security-intelligence/reports/rt_privacy_and_security_in_a_connected_life.pdf

Radke, J. (2016). WhatsApp: Hinweise auf Zusammenführung mit Facebook. Accessed October 16, 2017, from http://www.heise.de/newsticker/meldung/WhatsApp-Hinweise-auf-Zusammenfuehrung-mit-Facebook-3082755.html

Rybak, N. (2015). *In the age of cyber smash and grabs: Safeguarding customer loyalty, ring-fencing customer data, current analysis*, p. 3f.

Schiavoni, L. (2015). *Data protection tracker 4Q15*, Ovum, p. 6, 19, 22, 25.

Schiavoni, L. (2017). *The EU's proposal on ePrivacy puts OTTs firmly on the regulator's radar*, Ovum, p. 2.

Statista. (2017a). Anzahl der Nutzer des Schweizer Messengers Threema von Februar 2014 bis April 2016 (in Millionen). Accessed October 17, 2017, from http://de.statista.com/statistik/daten/studie/445619/umfrage/nutzer-des-schweizer-messaging-dienstes-threema/

Statista. (2017b). Anzahl der monatlich aktiven Nutzer von WhatsApp weltweit in ausgewählten Monaten von April 2013 bis Juli 2017 (in Millionen). Accessed October 17, 2017, from http://de.statista.com/statistik/daten/studie/285230/umfrage/aktive-nutzer-von-whatsapp-weltweit/

Stiftung Warentest. (2014). WhatsApp und Alternativen: Datenschutz im Test. Accessed October 17, 2017, from https://www.test.de/WhatsApp-und-Alternativen-Datenschutz-im-Test-4675013-0/

Syniverse. (2016). The mobile privacy predicament, p. 12. Accessed October 17, 2017, from https://www.syniverse.com/assets/files/custom_content/Mobile-Privacy-Predicament-Report.pdf

T-SEC. (2017). Magenta security portal. Accessed October 13, 2017, from https://security.telekom.com/

Telco Focus on Business Customers

Carsten Glohr

1 Overview

Telecommunications companies must constantly adapt their business models to maintain their position in the business customer segment. The three major business segments are network services, software services, and computing services. Each of the three business segments is facing its own imminent disruptive effects that are explained in more detail in the following:

- Disruption in the network service segment by Intercloud/network alliances
- Disruption in the software services segment: IoT and digital Darwinism
- Disruption in the computing service segment: cloud brokerage and automation

2 Disruption in the Network Service Segment by the Intercloud and Global Network Alliances

The forging of international alliances has always been an important factor for the success of providers in the business customer segment. In most cases, a provider can be successful in a bidding competition only if it covers international lines with the help of low-cost price structures from its international partners. Most network tenders contain international sections and can be won only in cooperation with experienced international partners who can be subcontracted quickly enough at competitive prices. Once only the case for the large customer segment, this situation has now become true for small midsize business customers as well. Even these businesses often have a large number of offices abroad that must be equipped with low-price WAN lines as part of an offer.

C. Glohr (✉)
Detecon International GmbH, Cologne, Germany
e-mail: carsten.glohr@detecon.com

© Springer International Publishing AG, part of Springer Nature 2019
P. Krüssel (ed.), *Future Telco*, Management for Professionals,
https://doi.org/10.1007/978-3-319-77724-5_26

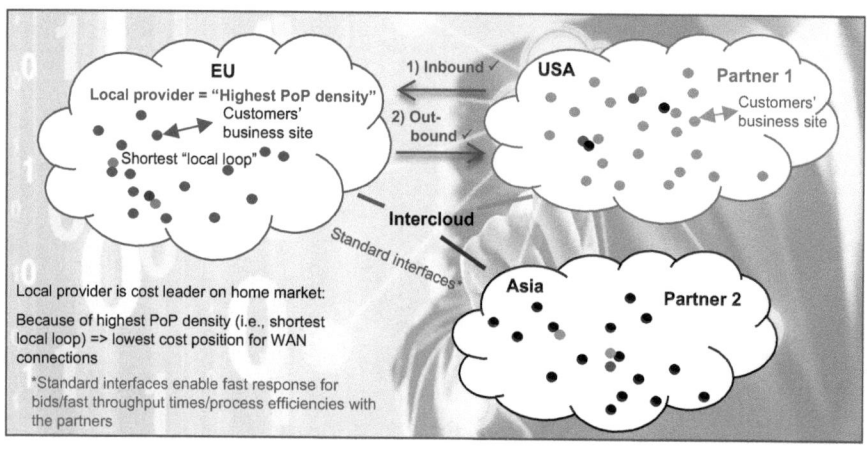

Fig. 1 Intercloud-based global alliance formation with local cost leaders

To paraphrase Porter (1980),[1] a company will not be capable of survival unless it seeks a position as either a cost leader or a service differentiator in comparison with its competitors. Both at the same time are not possible. Moreover, the chosen position must be one that cannot be imitated—or can be imitated only with great difficulty—by competitors.

The cost leaders for specific lines in the bidding competition are usually local WAN providers who are at home in the particular region and consequently have the highest PoP density. Point of presence or PoPs are access point to the internet. In most cases providers with the highest PoP density have the shortest "local loop" for the connection of the customers' business sites over the expensive "last mile" and are therefore the least expensive, as depicted in Fig. 1. The special finesse of the "Intercloud" strategy is that these local cost leaders can be interconnected forming a global alliance. The Intercloud alliance partners can then offer the lowest price at all times and in any region. This "market-clearing" characteristic is interesting above all in the business customer segment because larger business customers often require a large number of international WAN lines to connect all of their business sites with one another. Even small midsize businesses are usually active in many countries today. Globally operating telcos therefore have a decisive advantage in the business customer segment. They are the only providers who can offer competitive prices for the international aspects of tenders. Locally operating telcos, on the other hand, can easily price themselves out of the market because they cannot offer the international services except at prices that are inflated several times over or with poor response times because of complicated partner structures. International alliances in the business customer segment have always been a reliable means of winning contracts in bidding competitions, of course. But Intercloud is generating completely new

---

[1]Michael Eugene Porter is a visionary and university professor for economic science at the Institute for Strategy and Competitiveness of the Harvard Business School.

dynamics here. Since the forging of strategic alliances with international partners has in the past been individual, pragmatic, and quite frequently different from one customer to the next, the approach has led to an impenetrable jungle for larger telcos with over 100 partners and to interfaces among the various partners that are in most cases significantly less than professional.

The Intercloud has a standardizing effect in this case. Any alliance partners all use the same standards. The Cisco Intercloud, for instance, is based on "Open Stack", a cloud framework that among other features has a standardized mimic for provisioning services. Price information can also be shared quickly among the alliance partners—an important speed advantage during bidding competitions when time is often critical. Other processes as well as the preparation of quotations can be strongly accelerated. This is especially true of service management and delivery processes. Commercial and technical orders such as SLA and billing data can be exchanged automatically. If the partners have software-defined networks that have been calibrated with one another, even the provisioning of lines in just seconds is possible under certain conditions. An invaluable competitive advantage!

Intercloud alliance partners can in this way jointly improve their international competitive position. The local partner receives additional "inbound"[2] business from alliance partners whose customers require WAN lines on its home market. At the same time, the local partner is in a position to offer competitive prices in the additional "outbound"[3] business for lines that it could normally not offer competitively without the alliance. Assuming a high speed of implementation, the result is a superior business model for telecommunications services in the business customer segment.

3   Disruption in the Software and Service Segment: IoT and Digital Darwinism

Digital business models have long had a disruptive impact in many industries—including industries where IT has previously been viewed as a "commodity". There are many products that from now on will no longer be competitive unless they have been digitally enhanced.

In the future, however, not one single industry will be able to evade the far reaching impact of digitalization. IT service providers must understand the extent to which their customers' established business models are being questioned and traditional corporate structures are being challenged. If they succeed in reaching this level of understanding, it will open the door to tremensdous opportunities to provide

---

[2]Inbound: A foreign company buys a line on the provider's local home market for the connection of its business site.

[3]Outbound: A domestic company buys a line abroad (as a rule, via the foreign partner of the domestic provider) for the connection of its foreign business site.

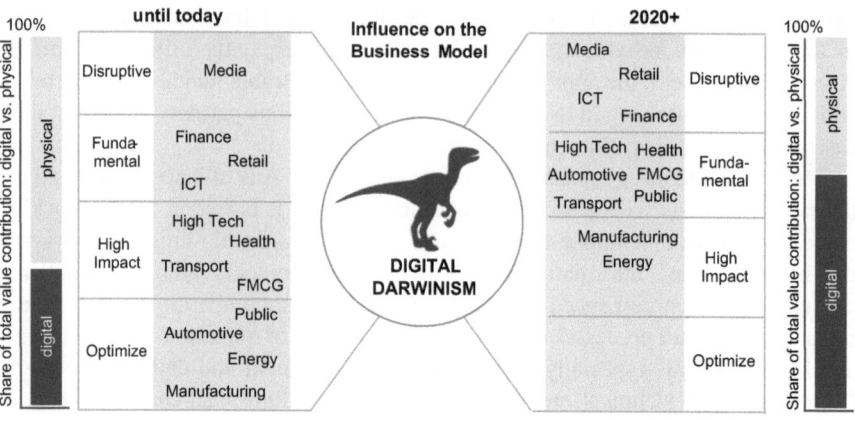

Fig. 2 Digital disruption

innovative services enabling digital transformation for their customers and assuring their own success.

The first wave of digital Darwinism affected above all those industries whose products and services could be completely digitalized. If an industry's products were not physical—MP3 from the media industry and banking services are examples of such products—they could be distributed completely in digital form. The consequences were highly disruptive effects and the unparalleled triumph of online business models. Most business processes today are handled by the customers themselves using fully automated self-services without in volving any bank personnel at all and at a fraction of the transaction costs previously charged.

Industries that produce physical products were initially less vulnerable to the substitution of digital business models as shown in Fig. 2 on the left side. Traditional industrial business models in manufacturing, automotive, or energy industries appeared to be comparatively safe because their products are physical and digital distribution is hardly possible. Moreover, many of the products in these industries are highly complex and expensive so that the availability of information on the Internet offered little differentiation in competition.

But the second wave of digital Darwinism that is on its way will have a massive impact on these industries as well. Even traditional industry is feeling the increasing competitive pressure from smart service providers such as Google and the American domination in this sector. Physical products will also be increasingly interconnected by the Internet of Things (IoT). Complex, expensive products in particular contain a high proportion of IT value generation. Car IT or machine controls are two examples. The relevance of these digitally charged products will only grow greater and play an increasingly significant role for differentiation in competition. The technology corporation Bosch, for instance, is investing about €500 million annually in projects far removed from its core business so that it will be able to compete with Google, Apple, and other companies in networking business and the Internet of Things.

Increasingly greater shares of the IT budgets are flowing now into the interconnection of customers, products, machines, and means of transport. Simultaneously, many "things" are also producing huge quantities of data. A modern elevator with IoT capability, for instance, is equipped with well over 100 sensors that generate a continuous flow of data, so big data solutions are definitely a growth field along with IoT solutions. Above all, those providers who in this sense are able to create standardized platforms and secure their intellectual property through licenses and theft protection will profit from this development. "Partnering" is the supreme event here because virtually no one acting alone is able today to cover comprehensively all of the required IoT core competencies such as industry applications, connectivity platform with identity management, interconnectivity, end device integration, or big data. Even the "big boys" of the industries cannot usually manage without the cooperation of partners.

It is true that there will be opportunities in the IoT environment for all possible business models of the ICT providers. But above all the software providers will be the ones who are able to move into a truly good competitive position. In the spirit of Porter, long-term success will be achieved only by those providers who offer functions or technical services that can be permanently protected as intellectual property rights or are difficult for competitors to copy because their source codes are secret.

These providers can earn good profit margins through licenses or SaaS models. SaaS solutions in the IoT sector are especially promising. SaaS/IP protected providers often offer system integration services as well, but solely in the form of opportunistic cross-selling products. By themselves alone, software integrator business models will hardly be able to survive unless they can offer comparative cost advantages such as low labor costs of offshore providers.

In many cases, system integrators have a chance for survival, but will be forced to seek price leadership on a low-margin market (producers). This will become all the more apparent the faster IoT standards become established. The expenditures required for system integration will decline further with every IoT component that has "plug and play" capability.

In the long run, software providers will dominate the IoT market because they have the best software development competence and already today have profound knowledge of customers' business processes. This puts them on the inside track for optimization of customer processes through the interconnection of products, means of transport, and machines.

Telecommunications companies can best maintain a permanent place on the market by creating their own connectivity platforms and providing actor-sensor data. These kinds of solutions can be sold most skillfully through sell-side partnering. The connectivity platforms are integrated into the software solutions for optimization of customer processes. Every time a software package is delivered by the sell-side partner, the corresponding connectivity platform can be sold as well. Buy-side partnering may play a role here as well to compensate for the lack of in-house competencies, but it weakens a company's own USP to the benefit of the buy-side partner.

Networks with low latency times, high security standards, and precise mobile location services can be differentiating capabilities in the area of network technology. Such "tactile networks", however, will most likely become standardized very quickly. In the spirit of Porter, cost leadership will be the sole means of securing a permanent competitive position because on a highly standardized market service differentiation will become increasingly difficult and even tactile network connections will be nothing more than a "commodity". A decisive competitive advantage, however, can come from fast and highly-automated provisioning of services with the aid of software-defined networks.

4   Disruption in the Computing Service Segment: Cloud Brokerage and Automation

A ruthless price war has long dominated the computing services segment. Traditional outsourcing solutions in the IT infrastructure segment have been suffering from a rapid decline in prices for many years. It is not at all unusual to see the market prices for bandwidth, storage, and computer services fall at a double-digit percentage rate within a single year. The continually rising offshore quotas result in falling prices even for services requiring intensive personnel resources.

Since the products such as computing performance and storage are highly standardized and virtualized, the lowest-price provider on the market (cost leader) frequently wins. Providers who can continuously underbid their competitors in the price war in the "commodity" segment because of comparative cost advantages or effects of scale are successful. Even for them, however, the market is one of low margins and intense competition.

There is a probability that so-called (cloud) broker models will appear, above all in the computing service sector. Based on customer requirements related to service levels or performance, broker software attempts to find the appropriate IaaS (Infrastructure as a Service) solution, checking as well the public cloud services offered by Eucalyptus, Amazon Web Services, Microsoft Azure, and other providers. As necessary, existing internal on-premise installations can also be integrated. Many providers have already started setting up brokerage services of this nature.

The competitive advantage is in the fast, fully automated, and therefore error-free provisioning that, as a bonus, does not require any personnel. Modern IT factories can operate virtually free of any manual system administration activities and are based on comprehensive full automation. IT employees instead realize administrator activities as source code in an application that autonomously and fully automatically provisions the software programs and other infrastructure resources. Where providing a server previously took 100 days, the provision is now possible immediately without any expenditure of manual labor and in error-free quality through applications or shop solutions. The important point for modern IT factories is rigorous standardization, modularization, and consolidation of services; this in turn has a positive effect on a large number of different cost drivers.

These highly standardized cloud provisioning paths, however, must be enriched and refined by the appropriate services because many large customers demand solutions tailored to their needs, even though a standard service is at the heart of the solution.

Full-service providers must decide on a reasonable production depth. If they can fall back on sufficient cost advantages, they can operate as cost leaders—but this will presumably be true of only a very few. Otherwise, they are forced to assume a broker role for their customers, ideally with largely automated, integrated solutions from the cost leader partner with the lowest price. This is how, for example, highly automated IaaS and PaaS (Platform as a Service) services are provisioned. The data are then migrated and tested (service refinement).

This trend is not really new and the compulsion toward perfected industrialization is a task that every provider is compelled to tackle by creating the most optimal set of internal processes possible. It is becoming evident, however, that traditional outsourcing business models are no longer viable as future alternatives because of the accelerating innovation cycles on the market. Corporate applications, for instance, are more and more frequently being offered in the form of SaaS (Software as a Service) models from the cloud by SaaS providers with the consequence that end customers no longer procure their IT operations from their traditional outsourcing partners, but acquire them instead as a SaaS package from the software provider. The traditional outsourcing providers are rapidly losing their infrastructure-related business to SaaS providers, and the outsourcing business is being replaced more and more by the cloud business. It is only a matter of time until traditional IT service providers will have lost major shares of the IT operations of their current clientele to SaaS providers. Traditional computing service providers find themselves more and more without a viable business model and have no choice but to reinvent themselves.

Since many software providers view the development of software as their actual core business and have little competence and interest in operating IT systems for their customers, new outsourcing opportunities are appearing here. Computing service providers have the opportunity to sell IaaS and PaaS services to SaaS providers or to integrate attractive production packages across the entire SaaS, PaaS, and IaaS stack into a partner model. The structure of the ICT provider's clientele will also undergo major change as a result, however. There will be fewer and fewer large outsourcing deals with end users; they will be replaced by more partner agreements between SaaS and IaaS/PaaS providers or by a large number of standardized SaaS agreements, each covering only a small area, with end users. It is obviously foreseeable that IaaS and PaaS providers will be able to survive solely by operating jointly with SaaS providers in partner models.

PaaS and IaaS service providers, above all in the area of computing services, must rigorously transform their traditional business models into partner models and simultaneously strive to secure cost leadership through strict automation, effects of scale, and high near-shore and offshore quotas.

5   Outlook

The greatest competitive intensity rules the computing service segment. Providers such as AWS[4] have long been offering a highly standardized, low-cost service in a fully automated provisioning function and are increasingly bold in launching a frontal attack on the traditional infrastructure providers. Surviving in this environment means implementing cloud brokerage and automation functions and building up partner structures with the most important cost leaders on the market before it is too late. Traditional outsourcing business is being replaced in more and more cases by XaaS models.

In the network service segment an effect similar to that will appear from software defined networks. In the future, it will be possible to provision software defined networks that are more fully automatic. Telcos can learn much from the difficulties and the survival strategies in the computing service segment where competition is substantially more intense. Similar fully automated provisioning is moving into the network service segment as well. Thanks to the Intercloud-based creation of global alliances by local cost leaders, the cutthroat competition will intensify. Strong Intercloud alliances will emerge as the victors when the dust clears.

The software and service segment, especially with regard to the subject of IoT, is dominated by a gold-rush atmosphere ("it's land-grabbing time"). The diversity of the players bustling around on the IoT and Industry 4.0 market is almost infinite. The most important are industry automation providers like Bosch and Siemens, software providers like SAP and Microsoft, OTTs like Google, and telecommunications providers like Vodafone, Deutsche Telekom, and British Telecom, not to mention a huge number of specialized providers.

In the long term, however, the winners in the business customer segment will be the software providers along with the providers who are able to protect their competitive position by means of intellectual property rights. Software providers today already have their anchors set most firmly in the verticals through their software solutions. The software solutions support many of the business processes of the customers. IoT solutions must integrate themselves into these processes. Telcos should turn more and more to sell-side partnerships with software providers because this is the approach that seems most likely to result in large-volume business. As long as there is so little standardization on the market, the direct distribution channel can be taken as a pathway to the customers, cooperating with customers to realize tailored solutions—to acquire the necessary know-how, if for no

---

[4]Amazon Web Services.

other reason. Large-volume business would appear to be a rare occurrence for telcos in this area, however; the possibilities for productization are limited. Still, if software providers are carefully selected for partnering, viable products can be realized.

Reference

Porter, M. E. (1980). *Competitive strategy*. New York: Free Press. isbn:0-684-84148-7.

Evolution Path: The Telco-centric Digital Ecosystem

Jan Grineisen and Giulia Rehme

1   Current Challenges: The Telco Evolution Path

Changes in the competitive environment and new requirements arising from customer behavior are compelling telecommunications companies (Telcos) to rethink their strategy. Telcos find themselves in direct competition with over-the-top (OTT) players who are substituting classic telecommunication services like voice calls, messaging etc. Such services are mostly integrated in a broader digital ecosystem with integration features (e.g. single sign-on, single invoice, common user interface). Since the rise of OTT services has triggered new market conditions and customer behavior, telcos have been facing challenges for the last 10 years. In order to keep up with the changing competitive environment, two general directions can be identified.

On the one hand, telcos accept the superior relevance of OTT services for customers and shift their focus from services towards connectivity. This scenario is often referred to as the "bit pipe", or negatively speaking "dumb pipe". Following Porters (1985) generic strategies this will be the move to the cost leadership. Some voices also call this scenario the extreme case with no return for telcos (Schneider 2013). Examples from telcos reach from T-Mobile USA to Telefonica and many MVNOs.

On the other hand, telcos try to anticipate the relevance of OTT services by either creating their own services besides the classic telco offer or by partnering with major OTT providers. The services include for example communication like messaging, TV and video on demand, music streaming, smart home, collaboration tools etc. Therefore telcos try to sell these services mostly by integrating them on top of their core offering of connectivity. By doing so they are also able to extend their portfolio. If the decision is between 'make' or 'partner' the better choice often depends on the competitiveness of inhouse developed services. Referring to Porter (1985) this

J. Grineisen · G. Rehme (✉)
Detecon International GmbH, Cologne, Germany
e-mail: Jan.Grineisen@detecon.com; Giulia.Rehme@detecon.com

© Springer International Publishing AG, part of Springer Nature 2019
P. Krüssel (ed.), *Future Telco*, Management for Professionals,
https://doi.org/10.1007/978-3-319-77724-5_27

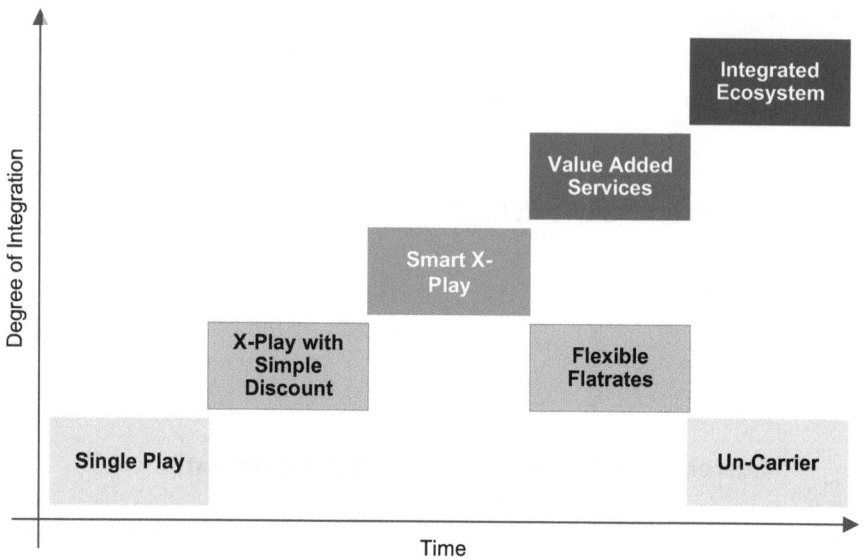

Fig. 1 Development path of telecommunications providers

scenario will be the differentiator. Examples from telcos anticipating this direction reach from Deutsche Telekom, Vodafone, Portugal Telecom to TDC.

Each direction cannot be achieved by leapfrogging the traditional business model of telcos and implementing the new target state in the short term. In fact telcos tend to follow an evolutionary path where each target state takes several years to be reached. As shown in Fig. 1 this evolution can be simplified in a few intermediary steps. The path is unified for integrated (fixed and mobile) and mobile only telcos but the existence of fixed assets will shape the decision making process about which direction to follow. In general both paths are suitable to both types of telcos.

In the first step of the evolution path telcos used to focus on individual products. The portfolio was made up of products such as fixed network lines for telephony and mobile lines for telephony, later supplemented by the data connections for each. The only differentiator from other players was the price and coverage but when the market became saturated margins were forced downwards.

The second stage consists of companies whose focus is on the convergence of fixed and mobile network products. In this stage, the primary interest is on the provision of fixed and mobile network services as a package at a discount price. With the upcoming of fixed-mobile convergence (FMC) offers customers are drawn to the highest discount for the offered bundle.

The third stage of evolution is based on the previous stage. Now, however, for the first time, telcos attempt to acquire customers by presenting arguments that are not based solely on price. Instead of a discount for the conclusion of a bundled contract, customers are offered a bonus in the form of additional services at no extra charge. This stage is often referred to as FMC+. Services that are hardly ever booked by customers and will be seen as hygiene factors in the future (e.g., unlimited calls from fixed to mobile networks, higher LTE data speeds, or an improved level of customer support) are particularly suitable for such offers.

The fourth evolutionary stage is characterized by telcos extending their bundles of FMC+ with further value-added services that move further away from the telco core business, e.g. including music streaming, video on demand, or smart home. The goal here is to cover a constantly expanding part of customers' digital lives, creating as a whole a higher added value for customers and simultaneously increasing the share of wallet for the telcos. When telcos begin offering these services, they enter into direct competition with other industries who have comparable services and products, especially OTTs. Alternatively, telcos can decide to focus on classic telecommunication services and offer flexible contracts with a variety of options regarding volume, contract period etc. to differentiate from new competitors with their core services.

In the fifth and final evolutionary stage, telcos decide to create a customer experience by building up an integrated digital ecosystem which goes beyond just offering FMC+ and additional value added services. In the contrary telco clould try to totally unbundle connectivity and make the offering highly flexible and low-cost. This is often referred to as un-carrier.

Based on our global discussion with telcos on the question how to win over OTTs, building up a telco-centric ecosystem is often considered to be a potential strategic move, especially for integrated telcos. Therefore in this article we will focus on the option of building up an ecosystem by telcos based on core services. The contrary option of an un-carrier strategy will not be discussed in this article.

2   The Creation of a Telco-centric Digital Ecosystem

2.1   The Rise of Digital Ecosystems Across Industries

A natural ecosystem is defined as a system of organisms occupying a habitat, together with those aspects of the environment with which they interact (The New Shorter Oxford English Dictionary 1993). The term 'business ecosystem' was defined by James Moore (1993). He states that companies were traditionally competing head-to-head for market shares, but nowadays new technologies will make this model obsolete. Companies will not be fighting alone in their vertical but will be part of a greater ecosystem of companies. Similar to the natural ecosystem, companies are highly dependent on the success and failure of the complete ecosystem (Iansiti and Levien 2004). In this context the concept of co-evolution is often used because of the interrelation of companies' success and failure. As today's business models are focusing on digital technology, business ecosystems are simultaneously referred to as digital ecosystems, even if they include physical parts like computer hardware, mobile phones etc.

As described above a number of technology companies have been highly successful with the establishment of an ecosystem with partners complementing their offerings and accelerating growth. Observing the emergence of digital ecosystems most of them seem to be triggered by one or a few companies and afterwards opened to more players. Apple laid the foundation for one of the first digital ecosystems by transforming the music industry from the analog world (e.g. Sony's Disc-Men) to the digital world with the release of the iPod in October 2001. The iPod itself cannot be considered as a digital ecosystem, but combined with the software iTunes it enables music producers and customers to interact on Apple's platform.

The most prominent and relevant digital ecosystems are the major app-stores of Apple iTunes, Google Play, Windows Store as well as the social media platforms of Facebook and LinkedIn. Looking at the offering across the app-stores roughly 1000 new apps are added every day already surpassing 2,000,000 apps offered so far (Statista 2017). The Economist (2014) states that such digital platforms will be an integral building block of our macroeconomic system and will increasingly aggregate digital products and services. But the above-mentioned companies carry out different roles in their ecosystem. In simple terms, Apple is the proprietary owner of the iTunes Store (and its iOS operating system) and decides independently from the rest of the ecosystem on future developments, access to the ecosystem etc. On the other hand Google owns the Google Play store but uses the Android operating system as an open source software. This position of power is often referred to as the ecosystem's orchestrator. There are many other examples of digital ecosystems which differ significantly regarding architecture, distribution of power, developments etc. Generally, there is no "one size fits all" when it comes to digital ecosystems.

However, what digital ecosystems have in common is that they are perceived as highly competitive and profitable. By now there is no common approach in order to isolate and measure the impact of digital ecosystems on a company's performance. But considering the stock performance of companies like Apple, Google and Facebook in comparison to adjacent companies and especially telcos, the hypothesis of a higher value creation is adequate. Especially an ecosystem orchestrator is capable of capturing a bigger stake of the generated value compared to small-scale entities like app-developers or content contributors (Dejager et al. 2014).

2.2 Designing a Telco-centric Digital Ecosystem

But what does this mean for telcos? As described in the evolution path, telcos should utilize the business model of a digital ecosystem to stay competitive, especially against OTT players. To leverage the existing assets of core telco services, telcos need to build an ecosystem around this nucleus of services, since this is both a lever to differentiate from OTTs as well as the "right to win" for telcos. As said there are many types of digital ecosystems and the perfect one for telcos is not yet defined. In general the underlying technologies Google, Apple etc. that are utilized to build up a digital ecosystem on top of the telco core service are nowadays in a high maturity level and available on the open market (Karhu et al. 2011).

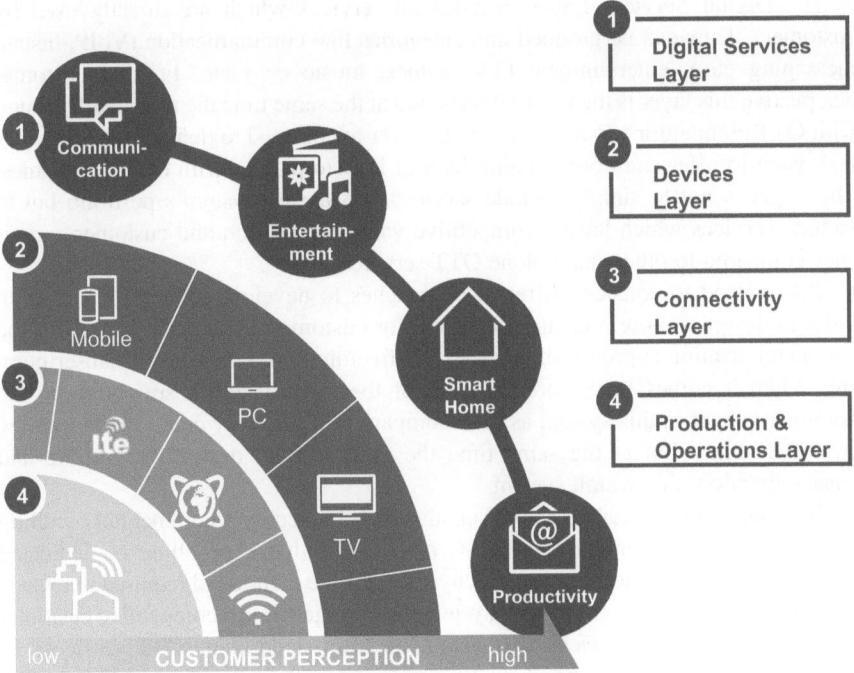

Fig. 2 The composition of a telco-centric digital ecosystem

Focusing on the design of the digital ecosystem, telcos should consider the system on four different layers (see Fig. 2) which need to be treated differently when it comes to sourcing and orchestration.

At the heart of the digital ecosystem is the Production & Operations Layer—the company itself—this is why we call it a telco-centric ecosystem. As the system is based on connectivity services the company's assets and capabilities build the very core. A solid foundation for the ecosystem is based on integrated networks, efficient IT-systems, customer centric operating model and corporate capabilities.

The Connectivity Layer includes all connectivity services for mobile and, if applicable, fixed network. This layer includes the core offering which makes a telco a strong differentiator against OTT competition and enables the "right to win" in the digital ecosystem game. This part of the value chain of digital services requires state of the art technology to enable the outer layers of the ecosystem.

The Devices Layer is about devices of any kind like mobile phones, tablets, TVs but also cars, smart home hardware etc. This layer makes all connectivity services tangible for customers and it is an essential part of the value chain for digital services. Telcos are traditionally experienced in reselling devices, especially mobile phones, and partnering with OEMs. In this layer telcos need to expand the scope to be more relevant by offering the last-mentioned device types.

The Digital Services Layer includes all services which are directly used by customers. They can be grouped into categories like communication (VoIP, instant messaging etc.), entertainment (TV, games, music etc.) etc. From a customer perspective this layer is the most relevant and at the same time the main battleground with OTT competitors. Especially for this layer telcos need to define which services they want to offer and how to build them or how to partner with other companies. The target is not to simply include services into the ecosystem's portfolio but to include services which have a competitive value proposition and customer experience compared to other stand-alone OTT services.

Telcos need to consider different approaches to develop, acquire or partner in order to design a more relevant ecosystem for customers. The same is true for the commercialization approach of services like freemium, pay-as-you-go, subscription etc. When it comes to the orchestration of the system, telcos are set to be the dominant player in this system as their company is at the very core. Therefore they have the ability but at the same time the obligation to design, orchestrate and constantly adopt the overall system.

The four layers described above can also be achieved by a commercial bundling approach, like it is common for FMC offerings with the upselling of dedicated additional services on top of connectivity. Besides the mentioned required relevance of the digital ecosystem, the overall system needs to generate a competitive customer experience across all services to differentiate from stand-alone OTT services. To achieve this the portfolio needs to be integrated with dedicated technology and commercial features to create a converged experience.

3 Key Features for a Converged Experience in Digital Ecosystems

The features needed to create a converged experience vary from many options depending on the market, customer needs and the design of the digital ecosystem. In general four features are essential to move the offering from a plain commercial bundle towards a digital ecosystem.

3.1 Single Sign-on

For many bundles, customers use different log-in credentials for all their digital services. This is especially the case for services which are delivered by a partner company. The log-in services of OTT players are already widely spread, especially from Facebook, Google and Amazon. In order to create a converged experience, telcos need to create their own log-in system which is used across all services and devices. This includes customer self-service portals, hot-spots, self-service terminals in stores and also all digital services like music streaming, smart home, instant messenger etc.

3.2   Unified User Interface and Self-service

As customers use a wide range of digital services, a digital ecosystem needs to offer a central user interface to create a converged experience. Such interface gives customers the ability to access and manage all services at a single place. In order to illustrate the service portfolio, a customer has an overview of all his services, sees relevant advertising for additional features and can access all his services. By selecting for example the video on demand service, customers will be redirected to the service's app. A self-service functionality for all digital services as well as the connectivity alongside a bill monitoring feature will give customers a central place to manage all their requirements.

3.3   Standardized Payment Method

Regardless of the chosen commercialization model—like subscription, pay-as-you-go or freemium—a digital ecosystem needs to establish a standardized payment method which can be used for all offered services. This payment method should be linked to the single sign-on feature in order to also enable single-click purchases and try-and-buy options. In contrast to OTT players, telcos have a huge customer base with existing contracts and bank collection agreements. This could be used as a baseline but needs to be extended especially to credit and debit cards. For this a wallet solution is necessary where telcos can either partner with the prominent players like PayPal, AliPay etc. or use the opportunity to promote an own payment solution. These solutions can also be scaled outside the digital ecosystem.

3.4   Centralized Customer Support

To finally create a converged experience, customers also need to get a centralized customer support for all services in the digital ecosystem. This also includes partner services. Utilizing the unified user-interface, customers should be able to select a customer support service directly from the service they are using. As the customer service cannot handle every customer request, this centralized support should be designed as first level support. Especially for partner services the support request needs to be forwarded if required. New channels and technologies need to be utilized in order to mitigate the new increased scope of support requested. Customer-to-customer forums, chat bots and artificial intelligence solutions will decrease the workload for the work force and increase the quality.

4   Key Enabler for Telco Ecosystems

The developments described above have substantial implications not only for marketing, but also for the organizational structure and IT. A change in the strategy seems to be inevitable and there are four major elements that executives should consider: the development of a modular portfolio based on valuable partnerships, more agile processes that can support a fast product development and partner integration and finally a new way of managing the great number of customer data.

On the customer side telcos should build a modular product portfolio to allow transparency over the offered services and quick adjustments to customer wishes. A modular and software-based production ensures fast time to provide and flexible contracts.

But a new portfolio strategy will not be enough to cope with the demanding and fast changing consumer needs if the organization is not changed accordingly. In order to keep pace with changing consumer needs telcos need to speed up product development by moving from waterfall methods to agile processes that allow adaptation and consider customer needs. Furthermore, the organizational structure and processes should support a fast integration of third party services such as music on demand services or third party payment solutions. To guarantee a seamless customer experience, services need to be integrated along the whole way of the customer journey from purchasing the different services in one marketplace until the customer service via one contact center. A well-considered partner management ensures a smooth communication and clear defined accounting models to avoid delays in launch dates. But customer-centric practices that go beyond the organization require a strong information infrastructure based on a rich customer database.

And here is where one key challenge for telcos lies. Although telcos have always had strong data assets, they have been relatively cautious when it comes to using those data for marketing purposes or product development. One reason for their restraint lies in their concerns about privacy, since this is their competitive advantage compared to OTTs. However, studies show that consumers indeed support the usage of customer data in exchange for a value added product or service or special offers (Heilmann and Liegl 2013). As a starting point telcos need to build uniform IT systems that support data transparency and usage of data in real-time across all processes. Furthermore, it needs to be assured that the IT systems of the ecosystem partners can easily be linked. This will be one crucial success factor for digital ecosystems to not just offer bundled products but to also provide a seamless experience.

5   Business Benefits of Building an Ecosystem

But why doing all this effort and investments to build up a digital ecosystem? The answer is simple when highlighting the potential benefits. According to the 2017 CIO Agenda report published by Gartner (2017), 79% of top-performing digital companies are being involved in digital ecosystems. Therefore, being part of a digital ecosystem is increasingly becoming an essential factor that differentiates high profitable organizations able to deliver customer value from lower-performing ones.

Most of the benefits that arise from the development of digital ecosystems result in higher revenues through an increase in wallet share, a higher customer willingness to buy or an avoidance of loss by keeping up with the competition. Three underlying benefit drivers can be defined: the enhancement of collaborative innovation, the growth of the business into new markets and customer segments and the development of new marketing capabilities. All three benefits are thought to become important elements in driving the profitability of businesses in the next few years and they are the result of a greater performance achieved by businesses operating within an ecosystem.

5.1   Collaborative Innovation

As already analyzed in the previous sections, key drivers of an effective digital ecosystem are the management of valuable partnerships and a good synergy between the different players of the ecosystem. As a direct result, several different businesses with different resources, capabilities and different key competencies find themselves working together towards a common goal. This collaboration between partners with complementary expertise can result in the unlocking of unexpected value that potentially gives rise to innovative products and services. Digital ecosystems enhance the collaboration between partners and the different skills and inputs foster innovation and give rise to new disruptive revenue growth.

5.2   New Markets and Customer Segments

Highly connected to the increasing number of partnerships and collaborations between businesses coming from different industries and often targeting different customer segments, through digital ecosystems companies have the possibility to expand towards new markets. As Gartner Research confirms in their report, "digital ecosystems enable you to interact with customers, partners, adjacent industries—and even your competition". Therefore, the focus of businesses will not only be their traditional customers anymore, but they will need to consider all potential new users coming from the sharing of a digital ecosystem with other players. This can be a great chance for companies that will be able to take advantage of this event in order to broaden their influence and brand image towards new and previously unattainable opportunities.

5.3   Enhanced Marketing Capabilities

The management and processing of the massive amount of data produced by the usage of different products and services were identified as important key enablers for telcos. If implemented successfully, this may result in improved marketing capabilities due to the broader information base that allows the creation of more precise customer profiles. As a result, telcos will not just be able to reach a better market segmentation with a more specific address of users in their social context, but to also have valuable insights about future trends and changing customer needs that will be the base for product development.

These extended capabilities will lead to a higher customer interaction and an enhanced customer experience increasing stickiness and willingness to buy.

What initial measures should telcos now initiate to successfully build up a digital ecosystem to set themselves apart long-term from OTTs and other telcos? The first step is to define their vision and their target role in the digital ecosystem. This includes the nature and structure of the ecosystem, deciding what they want to stand for in the eyes of their customers. In a next step it is important to start the transformation of the organization including a shift in skills and mindset in order to reach the full potential of a digital ecosystems.

References

Dejager, L., Vandevyver, S., Petta, I., & Libert, C. (2014). Dominance of the strongest: Inflammatory cytokines versus glucocorticoids. *Cytokine & Growth Factor Reviews, 25*(1), 21–33.

Gartner. (2017). *The 2017 CIO agenda: Seize the digital ecosystem opportunity.*

Heilmann, D., & Liegl, T. (2013). *Big data und Datenschutz.* Düsseldorf: Handelsblatt Research Institute.

Iansiti, M., & Levien, R. (2004). Strategy as ecology. *Harvard Business Review.*

Karhu, K., Botero, A., & Vihavainen, S. (2011). A digital ecosystem for co-creating business with people. *Journal of Emerging Technologies in Web Intelligence, 3*(3). https://doi.org/10.4304/jetwi.3.3.197-205.

Moore, J. F. (1993). *The death of competition: Leadership and strategy in the age of business ecosystems.* Harper Paperbacks.

Porter, M. E. (1985). *Competitive advantage.* New York: Free Press.

Schneider, M. (2013). *Management von Medienunternehmen. Digitale Innovationen – crossmediale Strategien.* Wiesbaden: Springer-Gabler.

Statista. (2017). *TOP 100 telecommunications companies (Global).* Online. Available at https://de.statista.com/statistik/studie/id/42747/dokument/top-100-telekommunikationsunternehmen-global/

The Economist. (2014). *Something to stand on.* Online. Available at https://www.economist.com/news/special-report/21593583-proliferating-digital-platforms-will-be-heart-tomorrows-economy-and-even

Pushing the Right Buttons: How the Internet of Things Simplifies the Customer Journey

Nikolai Nölle and Frank Wisselink

1 A New Destination: How Affordable Internet of Things Will Change Customer Experiences in Future

When it comes to customer journey mapping, the usual approach is to arrange the existing touch points in order to optimize the different phases of the customer life cycle. To reach the customer in different environments, these touch points are spread across different channels. Hence, Customer Experience Management (CEM) used to be incremental, focusing on improving present Customer Journeys, enhancing Moments of Truth or aligning existing touch points. New touch points are rarely introduced, new technologies are hardly considered.

The introduction of affordable Internet of Things (IoT) has the potential to transform the customer journey design and its individual touch points considerably. By this, customer needs can be addressed more adequately, individually and certain in the customer's favor. "Smart" IoT devices can now be included into the customer journey improving customer experience in the pre- as well as the after sales phase. This development could potentially lead Customer Experience Management on a new journey to its next paramount.

Even in the age of online shopping a customer journey has many detours. On closer examination each step within the journey means a point of decision (illustrated detours in Fig. 1). In the pre-sales phase already, a costumer has to choose whether to buy online or offline, select one out of a multitude of retailers, compare several brands, products and their variations or flavors, considering price differences or special offerings before it finally comes to a purchase decision. Buying online may even increase this complexity already within the pre-sales phase: After the selection of a favored product, the electronic shopping cart has to be checked-out. During this final stage even more steps are introduced: sign-in or in case of an initial purchase

N. Nölle (✉) · F. Wisselink
Detecon International GmbH, Cologne, Germany
e-mail: nikolai.noelle@detecon.com; frank.wisselink@detecon.com

© Springer International Publishing AG, part of Springer Nature 2019
P. Krüssel (ed.), *Future Telco*, Management for Professionals,
https://doi.org/10.1007/978-3-319-77724-5_28

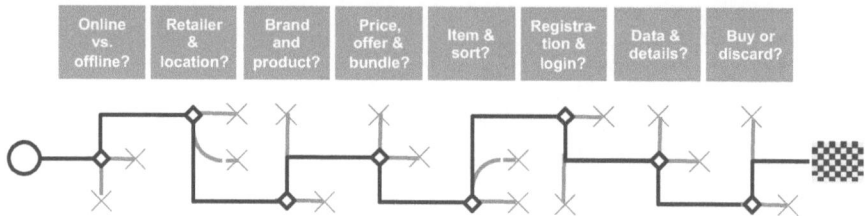

Fig. 1 Each point of decision within a customer journey represents a potential exit

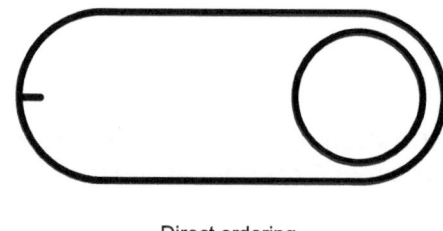

Direct ordering
at a push of a button

Fig. 2 IoT-Devices like the Amazon Dash Button have the potential to cut customers' detours

register, enter the entire personal confidential data including address and bank details before ultimately committing to purchase. And then waiting for delivery before it comes to utilization of the product or return.

Each point of decision within the customer journey may lead to a refusal of a product or brand. This results in customers going astray from the planned customer journey. And the path may start again with the next purchase, even more at consumer goods with high turnover rates and low customer involvement. All this leads to smaller conversion rates and results in uncertainties concerning the customer relationship on both the customer's and a company's side. Usually without insights when and why the customer left the journey.

IoT has the potential to circumnavigate these detours. A well-known example: the Amazon Dash Button. With one device, Amazon simplifies and encapsulates customers' decisions regarding Point of Purchase (online vs. offline), retailer & location, brand and product selection, price, offer and bundle, as well as item by a push of a button (see Fig. 2). Also registration, login and indication of confidential data needs to be entered just once. This increased ease of use prevents retailer, product and price comparison and lead customers to a convenient repurchase and steady retention.

In fact, implementing IoT devices into the customer journey may lead to less points of decision and thus less exit points. Consequently, more customers will run through the defined customer journey, passing the complete pre- and after-sales phase. That means an increase in the conversion of initial purchases and repurchases. Moreover the Internet of Things offers possibilities to redefine a customer journey which can be summarized in six dimensions.

2   Route Planning: Six Dimensions for IoT Driven Re-design of Customer Journeys

When a customer leaves a pre-defined customer journey detours are not just occurring in a straight rational decision making process. Following Forrester customer experience is, "how customers perceive their interactions with your company" (Forrester Research 2010). Hence, customer experience is rather a composition of factual and emotional perception in many layers across all involved touch points. The IoT delivers expansion and control about both, layers and touch points along the customer journey. In order to shape these layers and touch points structurally, Detecon defined six dimensions the IoT offers new opportunities in: learn, interact, innovate, sell, support and simplify (inspired by Groopman 2015) (see Fig. 3).

Learn Due to the large amount of devices and versatile applications, the IoT is able to gather and transport data from manifold resources at almost any time and any place. Therefore, almost every IoT-device is equipped with at least one sensor. The provided internet connection makes this data available afar and without any intermediary. Exerting this data will bring new customer insights and clear dark spots in a company's knowledge base: How do customers feel? How do they act in detail? What is there current situation? This knowledge can be applied in customer experience management within the further dimensions.

Fig. 3 The IoT has the potential to enhance customer experiences in six dimensions

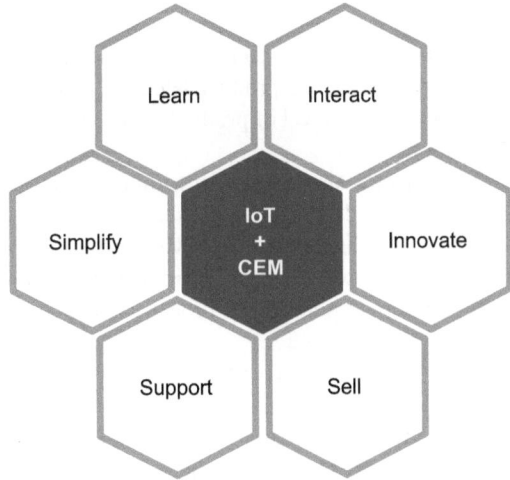

Interact The IoT enables to gain insights into a customer's personal situation and needs. Since communication via IoT devices is not necessarily one directional, the IoT enables instant and direct interaction with the customer. Knowing a customer's actions enables inference concerning his emotions and thoughts. Knowing these emotions and thoughts enables to act accordingly when it comes to interaction.

In addition, communication can happen directly between a company and the customer without any intermediary. A direct line to the customer can be established, overriding retailers, sales agents etc. For producing companies, this facilitates a new way of direct interaction and communication with the customer. Operated in the right way, brand values and culture can be transmitted to strengthen the customer relationship. Using the right technology, like Artificial Intelligence this interaction can also be approached predictively or in advance.

Innovate IoT-based information means also a new basis of information for product and business development. Additionally to or even in place of market research, up-to-date unfiltered customer insights will help to improve current products, develop new features or expand or emphasize single product benefits as needed. By this, products may change or develop in a new as yet unexpected direction.

Potentially, the new access to customer insights can also be used to refine or rethink fundamental business models. New data based services may gain importance and could amend current products or even replace them.

Sell Beside a customer's situation, IoT devices may also be able to detect a customer's geographical position. This enables situation- and location-based offers during a customer's shopping tour.

Knowing the customers profile and the dedicated preferences and needs, this information can also be utilized within a sales interaction. By this, the sales agent is able to present a suitable offer for a certain customer matching their needs. Implementing new IoT-based shopping devices has also the potential to create new methods of instore and online or remote sales approaches e.g. by sales robots.

Support IoT devices can support customer processes. For example, decision making processes like product selection can be guided by smart devices. Or search processes like retrieval of product locations within a store can be assisted by in-store navigation.

The IoT may also be applied within the after sales phase of a customer journey. In case of an incident, gathered user and device data can be used to offer direct or automated service and support. Monitoring the customer's usage patterns in combination with knowledge about a products limitations or critical operational thresholds offers additional potential in proactive or predictive services.

Simplify Since the IoT offers a direct communication channel to a customer it offers also potential in reducing complexity of customer or client processes. Thereby, it could increase convenience by replacing uncomfortable or cumbersome but yet necessary touch points such as order forms. Due to remote interaction it could also digitize several stationary touch points. By this, not just process cost on the client side but also on a company's side can be reduced due to straight processes safe process costs for both a company and its customers (e.g. the Amazon Dash Button).

Transferring these dimensions into practice, most IoT driven customer journeys have the potential to combine several of the mentioned dimensions. Combining simplified processes with additional IoT-based support and situational interaction may impact the value of the Customer Experience considerably. By this, the integrated combination may reduce the total amount of a company's touch points and lead to a more convenient experience on a customer's side. In fact, some few IoT-based touch points might be able to turn a customer journey into a new direction. In the matter of digitization, companies and also Telcos are already deploying IoT devices as touch points. How the IoT is applied within customer journeys already is described in the next section.

3   On the Trail: IoT-Based Touch Points in the Wild

Many companies strive to digitize their customer experience. Hence, several IoT-based touch points are already up and running. Since IoT-Applications are usually operating with mobile networks, Telcos have a certain interest in different perspectives. Sometimes Telcos are acting as network operator trying to build and utilize their infrastructure. In other cases, Telcos are in the role of a technology partner, co-creating new services and solutions. And sometimes Telcos are applying smart touch points within the customer journey of their own core business. However, the use cases are operating at least one of the six dimensions. The actual application is described in detail for the use cases Gillette Box, Paketbutler and the CarConnect.

3.1   Repurchase Made Easy: The Gillette Box

In 2015 Procter & Gamble's brand Gillette introduced in cooperation with the e-tailer Perfect Shave the Gillette Box: a razor holder equipped with an M2M-Module of Deutsche Telekom (Computerwoche 2015 and Perfect Shave n.d.). A push of a button triggers a repurchase of suitable razor blades. The connectivity of the M2M-Module is used to transmit the purchase data to the e-tailer. The delivery is promised within 24 h right to the door. M2M connectivity means a plug-and-play application with neither need for installation nor pairing and by this an increase in convenience.

The Gillette-Box offers the consumer a new way to repurchase razor blades straight from the basin. By this, Gillette offers a solution that reduces complexity within the decisions making process and thus has a share in the dimension *Simplify*. Due to the clear focus on repurchases the Gillette-Box is also contributing to the dimension *Sell*.

3.2   Heading for New Businesses: Deutsche Telekom's Paketbutler

In cooperation with DHL Telekom Deutschland released in 2017 the Paketbutler. Paketbutler is a container to receive parcels in absence (Telekom Deutschland GmbH 2017a and feldsechs service Gesellschaft mbH 2017). The container of robust and fire-resistant material is attached to the house, flat or office door of the addressee. In case of absence, the parcel carrier can place parcels in the Paketbutler. To avoid undesirable charging or unauthorized withdrawals, the Paketbutler is equipped with a combination lock. After delivery, the addressee receives a notification via smartphone app. By this, failed delivery and parcel pick-ups at neighbors or pick-up-stations should be avoided. Therefore, Deutsche Telekom promises the customer to become "independent from delivery periods and opening hours" (Telekom Deutschland GmbH 2017a).

The Paketbutler operates a NB-IoT-Module and is thereby a plug-and-play application with neither need for installation nor pairing. From the perspective of the cooperation partner DHL the reduced amount of failed delivery attempts means a reduction of unnecessary tours and hence, costs.

Due to the ease of parcel retrieval, the customer decreases efforts for pick-ups. In addition, he gains security for his deliveries. By this, the Paketbutler means a contribution to the dimension of *Simplify*. In addition, the costumer gains security for his deliveries what may result in an increase of delivery orders. By this, the Paketbutler is also active in the dimension *Sell* from the perspective of DHL. As original Telco provider, Telekom Deutschland is entering a new area of business: logistics. Hence, the Paketbutler means a share in the dimension *Innovate*.

3.3   Internet Access & Services for Cars: CarConnect by Deutsche Telekom

In September 2017 Deutsche Telekom announced the release of CarConnect (Deutsche Telekom AG 2017). CarConnect is a retro-fit device for cars which offers mobile internet access for passengers. Connecting the device via the current standard interface for on-board diagnostics (OBD) grants also access to localization and further vehicle data. In fact, CarConnect is not just offering on-board connectivity via mobile radio telephone services but also additional services (Telekom Deutschland GmbH 2017b). Via App, the current position and navigated routes are accessible in real time via GPS. The registered user is notified when the car is moved in his absence or leaves respectively enters a predefined area. Based on this information, Telekom Deutschland promotes an increase in security in case of parking dents, towing or theft. Over OBD the device is also able to gather data about the vehicle status.

Accessing the internet within the car means a reduced complexity in customer processes. Being off-line or arranging tethering with other mobile devices means efforts resulting in inconvenience. Furthermore, the customer gains security for his vehicle. Thus, CarConnect contributes to the dimension *Simplify*.

With its diverse functionalities, CarConnect creates new features for existing products. Given that cars are not part of the original core business of Deutsche Telekom, CarConnect is a new touch point in a different market and hence, supporting the dimension *Innovate*.

Developed further, each use case has the potential to cover additional dimensions. For example, all of them are able to contribute to the dimensions *Learn* or *Interact* by gathering deep customer insights (*Learn*) and initiate correspondent communication measures (*Interact*). But new developments need investments in R&D, IT, infrastructure, knowledge development etc. to be successful. At this point, due to the starting situation and partially initiated activities within the industry, Telcos are in a head start position (further reading in chapter "The Artificial Intelligence Challenge: How Telcos can obtain a Grand Prix for Insights Monetization"). In addition, Marketing and CRM, usually the home of Customer Experience Management are cost driven. Its value creation is based rather on 'soft' than on 'hard' factors and thereby difficult to measure. But IoT-driven customer journeys are still able to add value for a Telco. How this works is described in the next paragraph.

4   Towards Summit: Value Creation with the IoT

Besides the contributions to the six dimensions, each IoT-based touch point offers a common opportunity: The IoT-inherent access to the internet. By this, each connected device offers the opportunity to gather and transmit data. "The more IoT Solutions scale, the greater the incentive will be to leverage this data" is stated by the GSM Association (GSMA 2015). Thereby, the IoT becomes an enabler of Big Data.

Following the Oxford Dictionary data are "things known or assumed as facts, making the basis of reasoning or calculation" (Oxford Dictionary n.d.). Hence, more or deeper data means a broader basis for decision making. But further processing is needed to generate value with insights: First, processing data can increase the informational value. Hence, it will lead to 'better information'. But 'better information' does not create added value on its own. Applying the better information could lead to better decisions ('decision agility') (Wisselink et al. 2016). Following this, the added value of big data is the difference between the outcomes of a decision made without better information and a decision utilizing 'decision agility' based on 'better information' (Alpar et al. 2016).

Condition of value creation is trustworthy handling of customer information. Besides legal limitations, customers must be aware of the application and dealing with their confidential information within a company. Unconsenting or unbeknown use of customer information will destruct customers' trust. The aftermath: customer churn and a damaged image. Both with a lasting effect on the entire company. By this, any value creation will be abolished or inverted. Hence, IoT-driven value

Risk:
Churn and image loss
due to false handling of
customer information

Value Creation :
Added value based on
trustworthy handling of
customer information

Fig. 4 IoT-based value creation must prevail potential risks

creation should always prevail the potential risk (as illustrated in the balance in Fig. 4) and strictly respect customers' privacy.

Especially Telcos are regarded as trustful companies predominantly (Institut für Demoskopie Allensbach 2016). Hence, factoring the risk is important to retain this status. Considering the risk, prospective use cases must be reasonably selected to benefit from the potential value creation. This is valid for both, information-driven applications outside of and within Customer Experience Management. How the definition of value creation is applicable for Customer Experience Management is specified in the following section.

5 Panoramic View: The Meaning of "Learning" in Artificial Customer Intelligence

Transferring the definition of value creation to Customer Experience Management the IoT generates value based on better customer information and its application for agile decision making within CEM. Within the Six Dimensions 'better information' is represented by the dimension of *Learning*. This 'better information' should be used for decisions in regards to the other dimensions to benefit from the 'decision agility'. Therefore, *Learning* is a key element and basis for the value creation of IoT-based Customer Experience Management.

Examining present approaches in Customer Management, CRM is operating within a limited frame of retrospective data and data quality. Current CRM-Systems are gathering customer information across several channels at a certain point of time with occasional updates. In Customer Experience Management, this information can be used along the customer journey as basis for customer interaction. Therefore, current situation, emotions or events are unknown and cannot always be addressed adequately. Up-to-date changes at the customer side or situational conditions are not considered. In particular, "speed" in fast responses and a quick retrieve of information are crucial for an ideal Customer Experience (Li et al. 2017).

The IoT means a new ecosystem to gain customer information continuously and near real-time. This includes deep customer knowledge based on real customer information instead of punctual broad inference. In fact, this information can be used as market research and deliver real-time customer insights. Considering the six dimensions, this will bring new chances for Customer Experience Management.

But gaining the most value out of 'better information' and 'decision agility', actions have to be taken immediately to be responsive to the customer's current situation. And there is a challenge occurring: Continuous and direct reaction on current customer situations is not manually manageable. Therefore, new solutions are needed which must be able to analyze available customer information near real-time. Based on the analysis, the solution must be able to make decisions concerning the customer treatment automatically and trigger appropriate actions. Here, Artificial Intelligence (AI) comes into play.

Artificial intelligence describes computer science applications that aim for intelligent behavior (Bitkom 2017). Therefore, certain core capabilities are needed: *perceive, understand, act* and *learn*. In comparison to conventional information processing systems, the elements of *understand* and *learn* are novel. Conventional systems are following predefined, fixed programmed rules. In contrast, modern AI-systems are able to train their processing and by this to learn to reach better results or decisions.

Applying Artificial Intelligence in Customer Experience Management the dimension of *Learning* is basis to understand customers and act accordingly. This means direct and individual customer interaction corresponding to their current needs along the entire Customer Journey. Precondition: IoT-based customer journey design and touch points factoring the six dimensions. But further steps have to be taken to create new experiences.

6 Optimize Your Ascend: The Path to IoT-Based Customer Experience

The way to IoT-based Customer Experience and smart touch points is a progress affected by technological innovations and societal developments. In general, the following steps are providing a frame for the trial of new use cases and applications:

1. Rethink your Customer Journey
2. Consider IoT as element of your Customer Journey
3. Factor the six dimensions in touch point design
4. Emphasize the dimension of Learning for value creation
5. Optimize Customer Experiences by customer insights
6. Deliver immediate customer experience by Artificial Intelligence

All these steps have to be taken under one basic principle: always respect customers' privacy.

The race for the paramount is already started and competitors are pulling in into the track. Telcos need to embrace the Grand Prix and accelerate now to achieve the winners' rostrum.

References

Alpar, P., Alt, R., Bensberg, F., Grob, H. L., Weinmann, P., & Winter, R. (2016). *Anwendungsorientierte Wirtschaftsinformatik: Strategische Planung, Entwicklung und Nutzung von Informationssystemen*. Wiesbaden: Springer Vieweg.

Bitkom. (2017). *Entscheidungsunterstützung mit Künstlicher Intelligenz, Bundesverband Informationswirtschaft*. Berlin: Telekommunikation und neue Medien e. V.

Computerwoche. (2015). Gillette-Box bestellt Ersatzklingen per Knopfdruck. Accessed September 19, 2017, from https://www.computerwoche.de/a/angetestet-gillette-box-bestellt-ersatzklingen-per-knopfdruck,3091981

Deutsche Telekom AG. (2017). CarConnect – small adapter makes any car smart. Accessed September 20, 2017, from https://www.telekom.com/en/media/media-information/archive/carconnect-501374

feldsechs service Gesellschaft mbH. (2017). Accessed September 20, 2017, from Paketbutler.com, https://www.paketbutler.com/

Forrester Research. (2010, November 23) Drive revenue with CX. Retrieved from Customer Experience Defined. Accessed September 22, 2017, https://go.forrester.com/blogs/10-11-23-customer_experience_defined/

Groopman, J. (2015). *Customer experience in the internet of things: five ways brands can use sensors to build better customer relationships*. San Francisco, CA: Altimeter.

GSMA. (2015). *Unlocking the value of IoT through big data*. London: GSM Association.

Institut für Demoskopie Allensbach. (2016). *Sicherheitsreport Bevölkerung 2016*. Bonn: Deutsche Telekom/T-Systems.

Li, C., Littleton, A., & Akhtar, O. (2017). *Experience strategy: Connecting customer experience to business strategy*. San Francisco, CA: Altimeter.

Oxford Dictionary. (n.d.) British & World English. Retrieved from Data. Accessed September 21, 2017, https://en.oxforddictionaries.com/definition/data

Perfect Shave. (n.d.). Gillette-box: Rasieren bekommt eine neue Dimension. Accessed on September 20, 2017, from http://www.perfect-shave.de/gillette-box

Telekom Deutschland GmbH. (2017a). Der PaketButler. Accessed September 20, 2017, from http://www.telekom.de/paketbutler

Telekom Deutschland GmbH. (2017b). Telekom CarConnect Adapter. Accessed September 20, 2017, from https://www.telekom.de/unterwegs/telekom/telekom-carconnect-adapter

Wisselink, D. F., Meinberg, D. R., Horn, T., Obeloer, J., & Ujhelyiová, D. (2016). *Der Wert von Big Data für Telekommunikationsunternehmen: Schatzkiste oder Büchse der Pandora?* Cologne: Detecon International GmbH.

The Artificial Intelligence Challenge: How Telcos Can Obtain a Grand Prix for Insights Monetization

Frank Wisselink and Dominik Schneider

1 Artificial Intelligence and the Internet of Things Emergence in Many Households

Artificial Intelligence (AI) emerges in many households fueled by the physical speech assistants (Amazon 2017) and other Internet of Things (IoT) devices. Physical speech assistants are systems that enable household residents to use web services conveniently by voice control. For example Amazon Echo users currently can use more than 3000 web services via voice control (Manager Magazin 2016). It allows its users amongst others to order goods, request information, stream music or control the heating. After Amazon has launched the first physical speech assistant on the market in June 2015 Google, Microsoft and Apple were following suit. The reason why they followed isn't because of the expected earnings from direct sales only. The assistants offer a new possibility to collect new kinds of data in households concerning the behavior of the residents in the home. AI enables the OTTs hereby improving their existing and supporting new insights driven business models.

In the domain of households, Telcos have data with a high level of uniqueness for years as they have provided fixed line and more recently TV products. In case of fixed mobile integration also knowledge about other household members than the principle resident is available. The reliability of the data is often very high since Telcos have ID validated data about at least the contract holder, mostly the principal resident of the household. Using the data for generating insights is in many countries restricted due to regulation or due to issues of customer trust and only feasible when care about privacy concerns is taken.

Recently, many Telcos started extending their portfolio with *Smart Home* introducing smart *things* like light bulbs, electronic switches or for instance energy measurement devices in the home. As an example, the German Smart Home market

F. Wisselink (✉) · D. Schneider
Detecon International GmbH, Cologne, Germany
e-mail: frank.wisselink@detecon.com; dominik.schneider@detecon.com

© Springer International Publishing AG, part of Springer Nature 2019
P. Krüssel (ed.), *Future Telco*, Management for Professionals,
https://doi.org/10.1007/978-3-319-77724-5_29

comprised around 2 million households in 2015 (Statista 2016). In 2017 this is predicted to be already 4.4 million households in Germany that are using Smart Home solutions supporting a broad portfolio of energy management, home automation, building security, home entertainment and ambient assisted living (Statista 2016). A forecast suggests even that this triples until 2021 (Statista 2016). Some Smart Home solutions even cooperate with OTT speech assistants. Since the Smart Home market increases, this recent portfolio extension can assist Telcos enhancing their footprint in the domain of households and offering a basis for AI driven business models.

Having a look on the starting lineup there are OTTs and Telcos qualified for the race about the households. But who is in pole position?

2   Concerning AI and IoT the OTTs Are in Pole Position in the Race About the Households

After outrunning Telcos in the World Wide Web (WWW) the next race is on. This next challenge concerns a domain which was previously solely the domain of the Telco: the home and being more precisely, its residents. The OTTs would like to understand household members even better than by solely analyzing their behavior in the WWW.

So far, OTTs succeed better in entering households and analyzing its residents' behavior due to physical speech assistants. Actually the amount of Smart Home devices in households might be higher than the amount of physical speech assistants, but currently Smart Home devices are predominantly used to control the household remotely and not to collect all kinds of data. Limitations by law and by terms and conditions restrict Telcos very heavily on use of data whereas limitations for OTTs and therefore for physical speech assistants are not that heavy. For example a smart heating meter only collects data which are necessary for the specific use case (temperature of heating, energy consumption or regulation of heating). Physical speech assistants on the contrary store everything they get asked.

Using AI, OTTs can synchronize the collected data with their data lakes, find correlations and patterns and take automated decisions (Bitkom 2017). Automated decisions are based on algorithms that determine action alternatives, evaluate them and put them into practice (Bitkom 2017). Thus, algorithms create value by transforming data into information and by preparing the information for targeted applications in society (Bitkom 2017). Two parameters are significant when talking about value creation of algorithms (Fig. 1):

- The first parameter is the improved knowledge of a situation. But only improved knowledge does not create added value (Wisselink et al. 2015).
- The second parameter is decision agility. Decision agility means that the improved knowledge is used for decisions. Before knowledge is used, added value cannot be created (Wisselink et al. 2015).

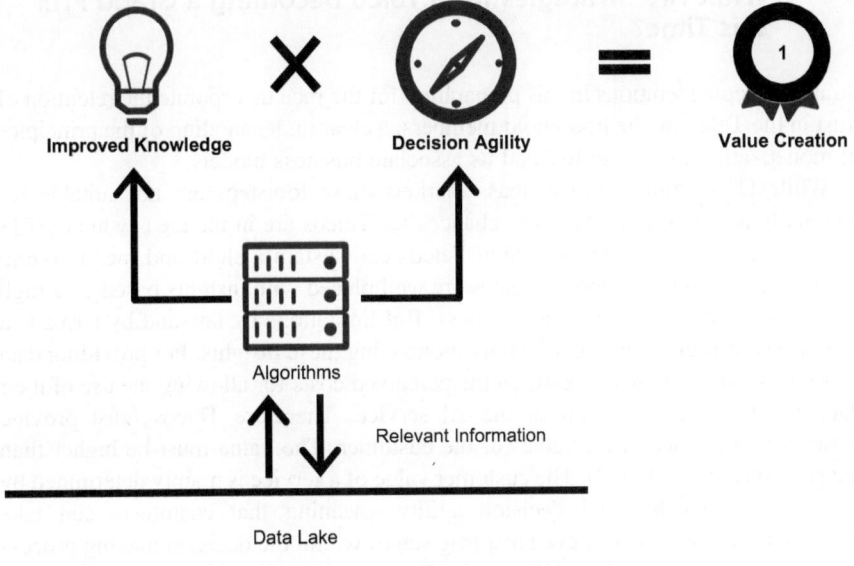

Fig. 1 The interplay of data and algorithms create value through improved knowledge and decision agility (Bitkom 2017)

For example OTTs can generate improved knowledge based on household data to optimize the performance of physical speech assistants to be able to answer residents' requests more precisely.

Although OTTs are in pole position in the race about the households, Telcos are not without a chance since they are fueled by data quality and the trust relationship with customers (see Nölle/Wisselink—Pushing the Right Buttons).

Data quality is an important requirement for AI, because like human decision-makers also algorithms can come to wrong decisions due to incomplete or false data. Similar to human decision making the trust relationship with the data provider is an important prerequisite for the correctness of delivered information. An asset of Telcos is that they have unique data sources validated through identification procedures. The use of physical speech assistants and in general the use of services from OTTs do not require an explicit identification of the user. OTTs rely on correct data entries and possibly additional information by the user.

To overtake OTTs and become a Grand Prix this time, Telcos have to react fast and act tactically smart.

3   What Are Strategies for a Telco Becoming a Grand Prix this Time?

Strategical considerations in this preparation for the race incorporate the retention of trust in the Telco by the household members, a clear understanding of the principles of monetization of AI and IoT and its associate business models.

While OTTs aim at global mass markets these footsteps are not suitable for nationally acting Telcos. The best chances for Telcos are in the areas which OTTs currently do not cover and in which Telcos can sustain a short and medium-term position. As mentioned above, Telcos are well placed with insights based on a high quality and a certain level of uniqueness. But limitations by law and by terms and conditions restrict Telcos very heavily monetizing these insights. For providing data customers make a tradeoff between the perceived costs for allowing the use of their data and the value creation of the AI service. Therefore Telcos must provide attractive AI services with value for the customer. The value must be higher than the perceived costs (Fig. 2). The customer value of a service is mainly determined by improved knowledge and decision agility, meaning that customers can take decisions better or faster or even leapfrog stages within the decision making process due to the service (see Nölle/Wisselink—Pushing the Right Buttons).

Chat bots like Tinka are an example for this strategy. Tinka has been introduced by T-Mobile Austria and allows customers to interact with an AI not distinguishable from a human counterpart (T-Mobile Austria 2017) answering customer service questions. To make use of Tinka's support, customers must approve the use of their data enabling customers' individual and personal questions to be answered by Tinka.

When services offered generating more customer value than the perceived cost for providing the customer will give his consent (1) for providing data. If this is the case Telcos are allowed to generate insights and monetize the insights in a

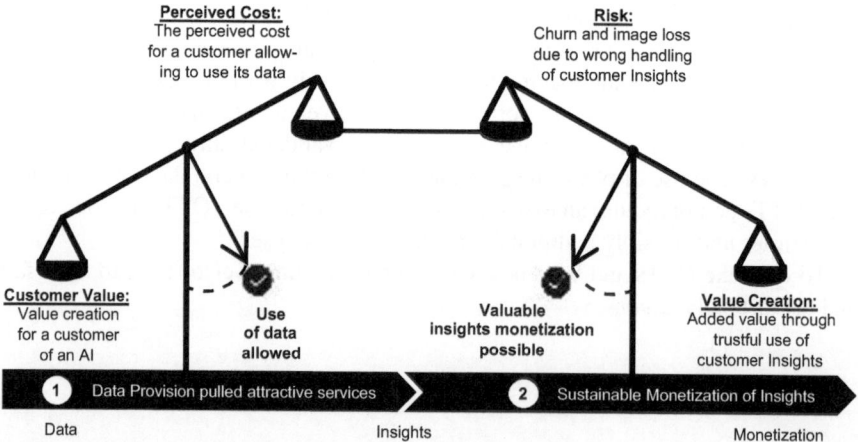

Fig. 2 Tradeoff for a customer of an AI service

sustainable manner (2). It is to be considered that insights monetization is a very sensitive issue with many risks. For a Telco customer trust is of upmost importance (Wisselink et al. 2016). Since the loss of customer trust affects the total revenue of a Telco, it should minimize risks and assure that value creation is sustainable.

If this is the case AI creates value both for the customer and telco and a *win-win* situation is achieved. If balance between risk and value creation however shifts, also the perceived cost for the customer increases and a *lose-lose* situation occurs.

By keeping the trust relationship with customers high, Telcos can get speed enough to overtake OTTs. To increase the lead on the finishing straight Telcos must understand another parameter that defines the value of insights: The uniqueness of insights and their market value. It is key to understand this parameter since it is part of principles of insights monetization.

4   Understanding of these Principles of Monetization of AI and IoT Is Key

To ensure that insight monetization, insights must not only provide improved knowledge and increased decision agility, but must also be unique in the market. Thus, insights are only valuable if they cannot be offered from other providers or if they are not offered with a similar quality. The model of qualitative value creation (Fig. 1) therefore needs to be complemented with the factor uniqueness to be able to take market effects into account (Fig. 3).

As mentioned before Telcos can generate unique insights which are of a high quality. OTTs basically offer services based on data that are not validated and therefore a certain data quality is not guaranteed. The uniqueness and high quality of Telcos' insights are major asset that must be played off against OTTs.

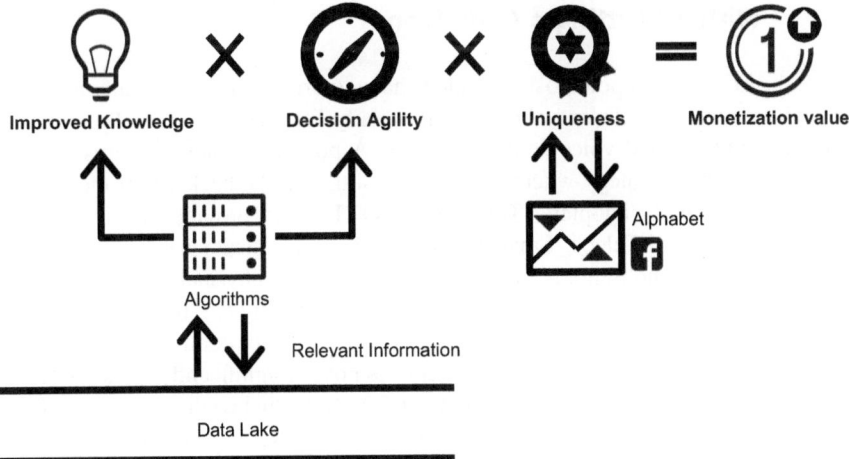

Fig. 3 The monetization of insights is defined by their uniqueness

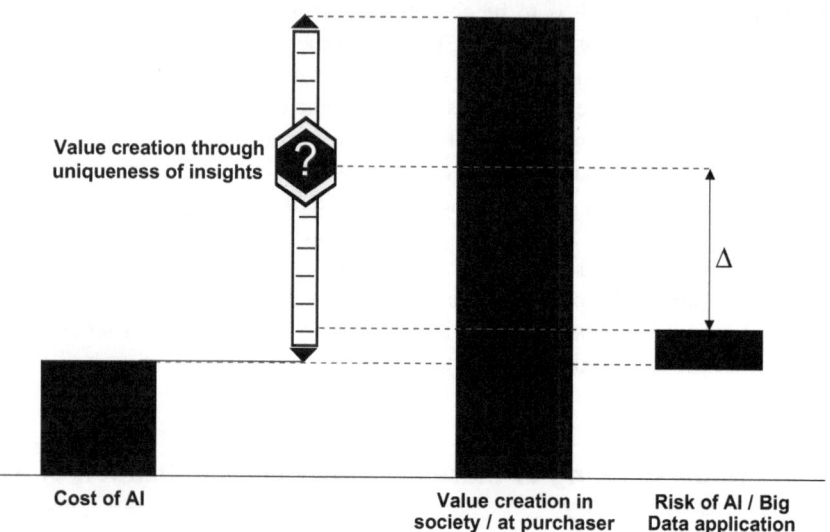

Fig. 4 The value creation of insights for the provider is defined by their uniqueness. The risk must not be higher than this value creation

The uniqueness is key for determining the value creation for a Telco as shown in Fig. 4. In case of high uniqueness a large high share of the value creation can be captured by a Telco (Δ) provided that a Telco acts trustfully and takes little risk.

Knowing these strategical considerations how should a telco tactically act to overhaul the OTT's?

5   Telcos Must Drive Tactically Smart to Overhaul the OTTs Without Hitting a Curb Stone

Since the OTTs are in pole position Telcos have to drive fast yet tactically act smart not to hit a curb stone. To enter households quickly Telcos need to offer Smart Home solutions with AI and voice control services. A possible strategy for this purpose could be a hybrid strategy where the Telco is sitting in the backseat of the OTT in order to serve early adopters. Thereby the OTT could provide physical speech assistants to control Telcos' Smart Home solutions. Physical speech assistants are currently entering households quickly and the trend of physical speech assistants seems to be becoming accepted. While partnering with an OTT the customer value of a Telco's Smart Home solution is increased through voice control, as illustrated in Fig. 5. The perceived cost for the customer does not get significantly higher, because Telco's trustworthiness is transferred to the OTT. Nevertheless the risk for the Telco becomes higher and the branding gets worse since privacy issues at the OTT are partly transferred to the Telco. As indicated in Fig. 5 the Telco risk is decoupled and not controllable anymore because data generated are not transferred to the Telco's

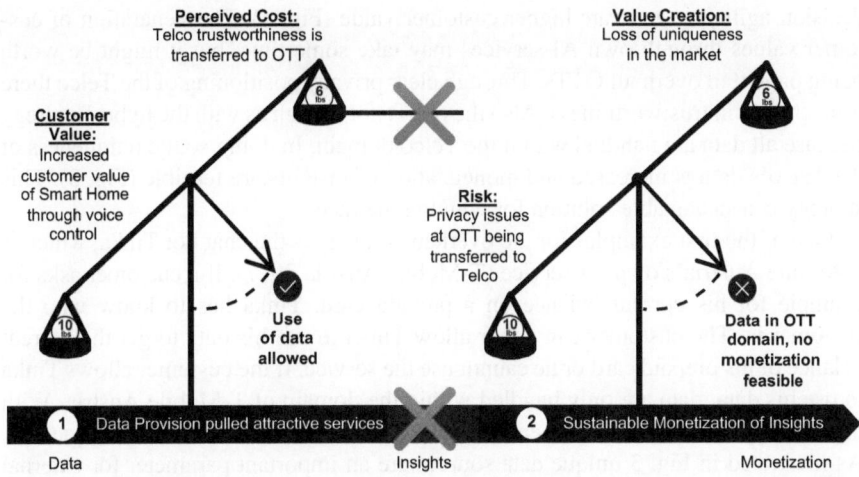

Fig. 5 Hybrid strategy: Telco sitting in the backseat of the OTT

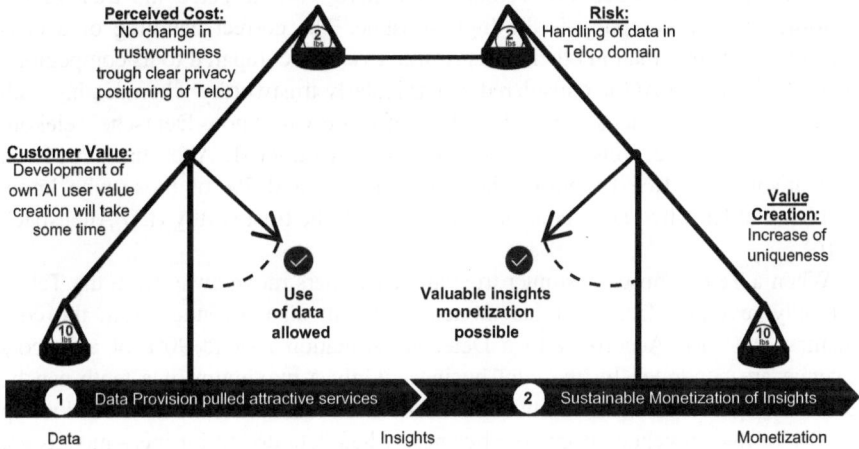

Fig. 6 Overhaul strategy: Telco takes the wheel

and are under the OTT control. Since no significant data is transferred to the Telco, the value creation for the Telco decreases due to the loss of uniqueness. Although Telcos are able to enter households faster by using OTT's voice control solutions, sitting in the backseat will result in dropping out of the race or hitting a curb stone in the long run. It is only an option to catch innovators and should therefore only be used temporarily.

Telcos are better advised to follow the overhaul strategy taking the wheel themselves, which is illustrated in Fig. 6. Telcos should develop and integrate their own attractive AI services also in the Smart Home solutions focusing on customer trust. With own attractive AI services they can provide better information and increased decision agility to generate higher customer value (Fig. 1). The generation of customer values through own AI services may take some time, but it might be worth being patient to overhaul OTTs. Through clear privacy positioning of the Telco there is no change in trustworthiness. Also the risk is not as high as with the hybrid strategy because all data are handled within the Telco domain. In doing so the uniqueness of the Telco's data is increased and monetization of insights are feasible (Fig. 3). This strategy is a sustainable solution for winning the race.

One of the first examples for the overhaul strategy is the chat bot Tinka, which is T-Mobile Austria's own AI service (T-Mobile Austria 2017). If a customer asks for example for his current balance on a prepaid card, Tinka has to know who the customer is. The customer can either allow Tinka to use his data to get the current balance of his prepaid card or he cannot use the service. If the customer allows Tinka to use his data, data are only handled within the domain of T-Mobile Austria. With this strategy, Telcos are able to enrich their data lakes and preserve their uniqueness. As illustrated in Fig. 3 unique data sources are an important parameter for external insights monetization.

External monetization of data insights can only be done sustainably when customer trust is maintained (Wisselink et al. 2016). For a Telco the trust of the customers in the company is the highest value. The correct handling of data is even more sensible than in other industries. In example compared to its competitors, Deutsche Telekom AG is considered as particularly trustworthy when dealing with personal data (IfD Allensbach 2016). 47% of those who know Deutsche Telekom AG at least by name, consider Telekom to be trustworthy (IfD Allensbach 2016). In comparison only 18% of those who know Google and 7% of those who know Facebook at least by name, consider these OTTs to be trustworthy (IfD Allensbach 2016).

When a Telco abuses customer trust and customers move away from the Telco, not only revenues from insights monetization but also revenues from the core business get lost. According to a Detecon estimation around 90% of a Telco's revenue are generated by the core business. Only a maximum of a tenth can be generated by Telco data driven business models. That means that Telcos can lose up to 90% of their revenues from core business when data driven business models are not conducted carefully with regards to customer trust and privacy issues.

To prevent the loss of customer trust and churn from happening, Telcos should work out guidelines regarding the governance. An example for these guidelines offering a sound basis are the guidelines of Deutsche Telekom AG, which cover major topics with regards to data protection and compliance (Deutsche Telekom AG 2018).

References

Amazon. (2017). *Magenta smart home*. Accessed September 15, 2017, from https://www.amazon. de/Deutsche-Telekom-AG-Magenta-SmartHome/dp/B06X1BR9H1

Bitkom. (2017). *Entscheidungsunterstützung mit Künstlicher Intelligenz – Wirtschaftliche Bedeutung, gesellschaftliche Herausforderungen, menschliche Verantwortung*. Bitkom, Berlin, pp. 66–77.

Deutsche Telekom AG. (2018). *Deutsche Telekom's guidelines artificial intelligence*. Accessed June 5, 2018, from https://www.telekom.com/en/company/digital-responsibility/details/artifi cial-intelligence-ai-guideline-524366

Institut für Demoskopie (IfD) Allensbach. (2016). *Sicherheitsreport Bevölkerung 2016*. Bonn, Deutsche Telekom/T-Systems.

Manager Magazin. (2016). *Warum auch Google und Apple künftig mithören wollen*. Accessed August 30, 2017, from http://www.manager-magazin.de/unternehmen/it/nach-amazon-echo-erfolg-ziehen-apple-und-google-nach-a-1114024.html

Statista. (2016). *Smart homes*. Accessed August 30, 2017, from https://de.statista.com/outlook/279/ 137/smart-home/deutschland#market-smarthomes

T-Mobile Austria. (2017). *Tinka Lernt*. Accessed September 15, 2017, from http://www.t-mobile. at/tinka-lernt/abgerufen

Wisselink, F., Meinberg, R., Horn, T., Oberloer, J., & Ujhelyiova, D. (2015). *The value of big data for a Telco: Treasure Trove or Pandora's Box, Future Telco Reloaded*, Detecon 2015.

Wisselink, F., Meinberg, R., & Obeloer, J. (2016). *Vertrauensvoll Mehrwert für Kunden schaffen*. Accessed October 4, 2017, from https://www.detecon.com/de/Publikationen/vertrauensvoll-mehrwert-fuer-kunden-schaffen

Part VII
Internal Enabler

Enabling Digital Excellence Through Business Process Management and Process Frameworks

Chin-Gi Hong and Christian Dietze

1 The Impact of Digital Technologies on Business Processes

The rise of new digitization technologies impacts business processes of organizations across all industries. Such technologies do not only pave the way for business innovations and unlock new revenue streams but also carry the potential for organizations to drive automation and unlock tremendous efficiency and effectiveness gains. Especially in the context of customer-facing core processes, the impact through digitization is even greater as the customer experience itself is directly impacted.

Some of the most promising technologies that are becoming more and more ubiquitously available and create much traction across all industries include Artificial Intelligence (AI), Big Data, as well as Internet of Things (IoT) and Machine-to-Machine Communication (M2M). The benefit potentials of these technologies as relevant examples in the context of digitization on business processes of an atypical organization are briefly described in the following section through practical use cases.

1.1 Artificial Intelligence

AI is widely regarded as the one of the key technologies of the digital age and comes with a tremendous potential to drive process automation. It is defined as a

C.-G. Hong (✉)
Detecon International GmbH, Frankfurt am Main, Germany
e-mail: Chin-Gi.Hong@detecon.com

C. Dietze
Detecon Consulting FZ-LLC, Abu Dhabi, UAE
e-mail: Christian.Dietze@detecon.com

© Springer International Publishing AG, part of Springer Nature 2019
P. Krüssel (ed.), *Future Telco*, Management for Professionals,
https://doi.org/10.1007/978-3-319-77724-5_30

"technology that appears to emulate human performance typically by learning, coming to its own conclusions, appearing to understand complex content" (Gartner 2017a).

The overall objective of AI is not only to emulate to human behavior or intelligence, for example based on simple decision-tree-based approaches, but to replicate human behavior across the entire information processing and decision-making process.

One key application domain of AI is the Smart Agents technology based on avatars which are able to interact with other objects within an environment. In telecommunications, AI is consequently often applied in the complaint management area, for instance via chat-based agents handling customer complaints.

1.2 Big Data

According to Gartner, Big Data is defined as a "high-volume, high-velocity and/or high-variety information assets that demand cost-effective, innovative forms of information processing that enable enhanced insight, decision making, and process automation" (Gartner 2017b). As this definition suggests, process automation inherent to the concept of Big Data as one of the key drivers for according initiatives and programs as they are created across all industries.

The areas of application and use cases are manifold and sheer endless. In the telco context, Big Data is often applied in the area of customer relationship management (CRM), sales and marketing. The collection and structured analytics of customer data allows for highly personalized, targeted marketing measures. Besides these rather obvious areas of application, Big Data can be also often found in finance, Supply Chain Management (SCM) and operations.

1.3 Machine-to-Machine Communication and Internet of Things

As traditional revenue streams have been declining in recent years, many telcos have been pursuing to unlock new revenue stream through providing IoT and M2M services. Both technology concepts are closely interrelated to one another. Whereas M2M describes the concept of machines being interconnected to one another to communication and transmit data which are gathered through embedded sensors or actuators, the concept of IoT describes the holistic system of interconnected devices as a whole.

In the telco context, M2M/IoT can be used to improve means of remote infrastructure management. This field in particular has been fully based on manual activities requiring on-site labor at the device location for installation and maintenance. With the capabilities provided through a M2M-platform, monitoring can be done completely remotely with strong efficiency gains for incident resolutions based on improved means for remote diagnosis. Figure 1 provides an overview on the use cases described.

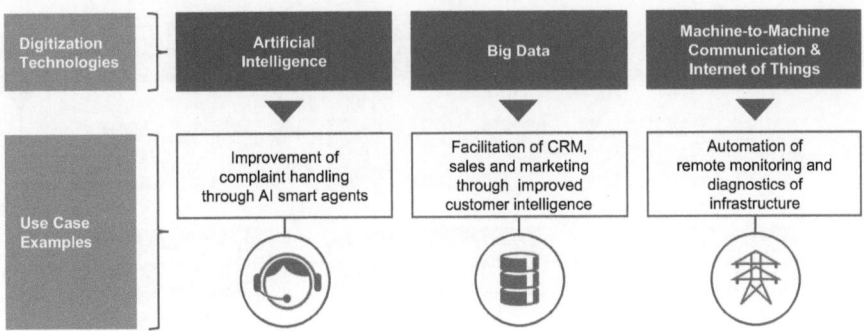

Fig. 1 Digitization technologies and use case examples

The three technology examples above clearly illustrate the automation potentials that digitization provides. Whereas each technology of its own provides for opportunities for increased process automation, it is in the combination of all in which the greatest potentials for automation can be realized. AI requires the capabilities of Big Data which itself requires the capabilities provided through M2M/IoT.

However, many digitization endeavors are still purely technology-driven, while often the business and operational perspectives are not sufficiently considered. In order to be able to fully capitalize on the opportunities that such digital technologies provide, a well-structured and comprehensive approach is required. Making use of an integrated BPM framework and process reference frameworks can help to approach such transformation endeavors towards digital excellence in a well-structured manner. The following sections describe the facets of such integrated BPM lifecycle approach.

2 BPM Lifecycle for Agility and Digital Transformation

As illustrated in the previous section, there is an increasing penetration of digital technologies in organizations, such as AI, Big Data and M2M/IoT. The selection and implementation of those technologies has a strong impact on the business processes of the telecommunications operator.

Business processes are an essential driver for agility in organizations—especially when it comes to the development and implementation of new products and services. Hence, the management of business process has to be ensured through a dedicated Business Process Management (BPM) function in an organization. The BPM function can take advantage of an existing and comprehensive BPM Lifecycle that covers all relevant elements for successfully managing processes.

The comprehensive BPM Lifecycle is introduced in this Section. Over the last decade, it has successfully been applied in various project assignments with international telecommunications operators and it led to tangible improvements. The BPM Lifecycle consists of nine dedicated elements that are described in Sects. 2.1–2.9.

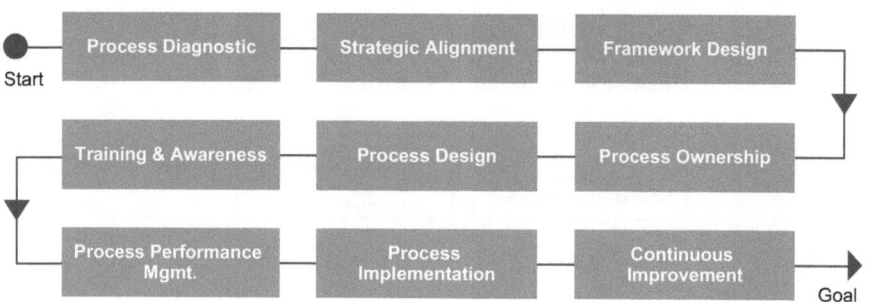

Fig. 2 BPM lifecycle

Figure 2 provides an illustrative overview of the BPM Lifecycle and its nine elements. The concept of the BPM Lifecycle is flexible and only some elements might be required for a specific project. In addition, the order of the elements does not necessarily have to follow the flow of elements illustrated in Fig. 2.

The nine elements of the BPM Lifecycle can be used to drive digital excellence across different business processes of a telecommunications operator.

2.1   Process Diagnostics

The objective of the Process Diagnostics element is the identification of existing gaps in the as-is processes. In this respect, interviews and workshops with experts of the different process areas are important in order to get transparency about the strengths and weaknesses. In addition, available process documents providing further details are reviewed. In this context, it is important to differentiate between the implemented as-is processes and the existing process documents.

In the traditional BPM context, the outcome of the Process Diagnostics element is a list of existing gaps related to the as-is processes compared to best practices and leading industry standards like TM Forum's enhanced Telecom Operations Map (eTOM). In the context of process digitization, the focus is put on the identification of automation potentials for existing processes. Digital processes are executed without manual interaction and are highly automated through maximum IT-usage.

2.2   Strategic Alignment

The Strategic Alignment element of the BPM Lifecycle focuses on the alignment of the strategic targets of the organization and the BPM approach itself. The strategy of an organization is defined through its vision, mission as well as short-term and long-term goals. The strategic targets of the organization are translated into different process targets.

For example, an organization with the goal of becoming more customer-focused will probably have the strategic target to optimize the customer-centric processes. This strategic target has an immediate impact on the process framework of the organization.

Any organization that finds itself on a journey towards digital transformation will most likely have the strategic target to digitize processes for enabling the transformation. The digital transformation target of the organization has a direct impact on the desired automation degree of processes.

2.3   Process Framework

The Framework Design element provides the basis and the starting point for the detailed design of the processes. It is recommended to distinguish between the content view and the methodical view of the Framework Design. The content view of the Framework Design defines the process domains and the most relevant processes of the process framework.

In general, the Framework Design element is applicable for manual but also for digital processes. However, the degree of digitization might require a customization of the process framework. Additional domains could for example be introduced that accommodate especially those processes that are supposed to be highly automated.

2.4   Process Ownership

The Process Ownership element defines clear roles and responsibilities in the organization for analyzing, designing, implementing and continuously improving the processes.

The responsible top management representatives of an organization should nominate one Process Owner for each end-to-end process of the process framework. In the context of digitization, it is important that process owners have in-depth knowledge about the latest technologies and also IT-systems that allow the automation of processes. Process Owners have the role of driving and facilitating process digitization at the same time.

2.5   Process Design

Process owners are accountable and functional experts are responsible for the detailed design of their respective end-to-end process as defined in the Process Design element. It is recommended to follow a top-down approach for designing the details of each end-to-end process and to introduce different process design levels.

Today process design is more than just defining process descriptions and drawing process flows. While the digitization of processes is becoming more important, the identification, the selection, and the deployment of the right technologies has to be an integral part of any process development activity. The desired degree of digitization can only be realized by using the right technologies to automate processes.

2.6   Training and Awareness

Trainings and awareness sessions need to be planned, prepared and conducted for various target groups that will be either involved in or affected by the later process implementation. Those training and awareness sessions related to the BPM Lifecycle can range from strategic alignment, process framework development, process ownership conception, process design, and process implementation to process improvement.

For digital processes the focus of the training and awareness campaigns is mainly on the introduction of new technologies and IT-systems that are executing the automated processes. On the other hand, emphasizing the importance of automating processes along the BPM lifecycle to facilitate digital transformation of an organization should be an essential part of the Training and Awareness element.

2.7   Process Performance Management

The Process Performance Management element of the BPM Lifecycle links the processes with the performance view. In this respect, KPIs are defined and associated with the processes. Monitoring of process performance is the prerequisite for continuous improvement activities. The digitization of processes should also include an automated monitoring of KPIs. This automation can only be realized through the usage of performance management tools and advanced technologies.

Process targets are either defined in Service Level Agreements (SLAs) between the organization and external parties or in Operational Level Agreements (OLAs) between different units within the organization. Each process target should be associated with KPIs that are linked to end-to-end processes.

2.8   Process Implementation

Based on the detailed design of the end-to-end processes the implementation of the processes has to be planned and executed. For this, the BPM Lifecycle foresees a dedicated Process Implementation element. The extensive experience from different BPM related projects suggests that the complexity and the effort of implementing end-to-end processes should not be underestimated.

There is an increased complexity especially if digital processes have to be implemented. The implementation of digital processes requires either the introduction of new IT-systems or the change of existing IT-systems. Time-consuming vendor selections, IT-system changes, and system tests are part of digital process implementations.

2.9 Continuous Improvement

Once processes are implemented, the monitoring and improvement of the implemented processes is a continuous endeavor. Especially in the context of digital processes the identification of existing pain points can be realized through the execution of continuous software tests. In this case the frequent elimination of pain points in the existing software that automates the process can be considered as the continuous improvement.

The selection of a continuous improvement approach depends on the overall strategy. A stepwise approach including standardized actions like process analysis, pain point identification, optimization measure development, and optimization measure realization has proven to be successful in real-life projects.

3 The Practical Usage of Reference Process Frameworks

Section 2 has concentrated on the introduction and description of a comprehensive BPM Lifecycle that can be utilized as guideline and tool by a dedicated BPM function in an organization. In addition, a BPM function is also in the possibility to make the use of existing reference process framework. TM Forum's Business Process Framework (eTOM) (see Sect. 2) is an example for a very advanced reference process framework that has been developed by leading organizations in the telecommunications industry and it is widely used (Czarnecki 2009).

Reference process frameworks such as eTOM provide a comprehensive and consistent view on process domains and functional business processes that are relevant for a telecommunications operator. eTOM can be utilized as guideline for process design activities and it provides a common process terminology for (project) managers, experts, system integrators, researchers, and consultants. In the context of digital technologies, eTOM can also be utilized for establishing an ecosystem of partners for dedicated process areas.

As shown in Fig. 3, eTOM provides on its top level three process group that are namely *Operations*; *Strategy, Infrastructure and Products (SIP)*; and *Enterprise Management*.

In eTOM, processes are further decomposed in a hierarchical manner. The industry standard is a hierarchical collection of business processes and it includes definitions for each process. eTOM can serve as guideline and reference for business process design activities. It provides a comprehensive set of process groups with process steps for each group.

Initially, TM Forum's Business Process Framework (eTOM) did not provide an end-to-end perspective for the different process groups. Therefore, eTOM has been enhanced by a customer-centric Business Process Framework that provides the

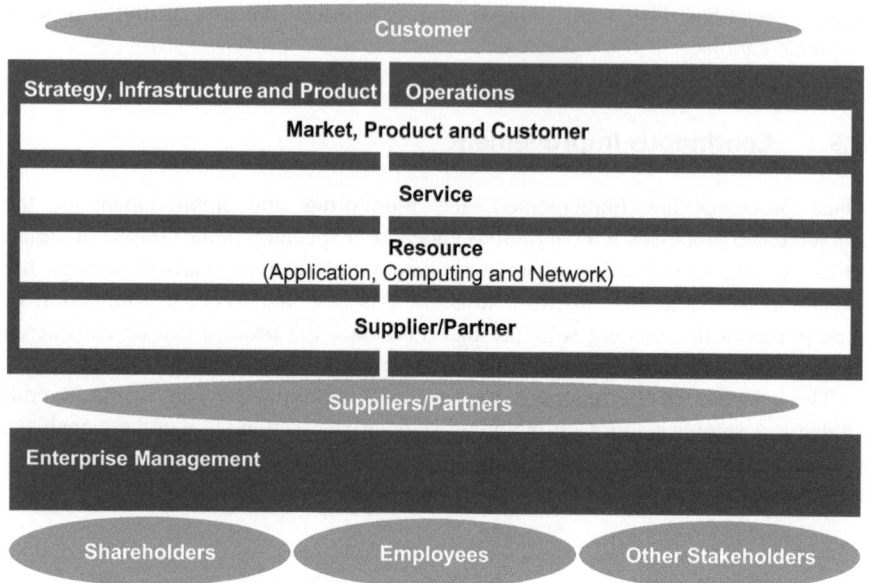

Fig. 3 Structure of the business process framework—eTOM (reproduced from Czarnecki and Dietze 2017, p. 81)

desired end-to-end flow for processes. , This customer-centric Business Process Framework is compliant to eTOM and it was published as part of Addendum GB921E (TM Forum 2015).

References

Czarnecki, C. (2009). Gestaltung von customer relationship management über die Grenzen von Telekommunikationsunternehmen hinweg. In T. Eymann (Ed.), *Bayreuther Arbeitspapiere Zur Wirtschaftsinformatik—Tagungsband Zum Doctoral Consortium Der WI 2009* (pp. 11–23). Bayreuth: Universität Bayreuth.

Czarnecki, C., & Dietze, C. (2017). *Reference architecture for the telecommunications industry: Transformation of strategy, organization, processes, data, and applications.* Berlin: Springer.

Gartner. (2017a). Gartner IT glossary. *Artificial intelligence.* Retrieved October 12, 2017, from https://www.gartner.com/it-glossary/artificial-intelligence/

Gartner. (2017b). Gartner IT glossary. *Big data.* Retrieved October 12, 2017, from https://www.gartner.com/it-glossary/big-data/

TM Forum. (2015). *Business process framework (eTOM): Concepts and principles (GB921 CP). Version 15.0.0. ed.*

Smart Automation as Enabler of Digitalization? A Review of RPA/AI Potential and Barriers to Its Realization

Manfred Schmitz, Christoph Stummer, and Michael Gerke

1 What Do We Mean by "Smart Automation"?

Smart automation and related fields such as robotic process automation (RPA), virtual assistants, chatbots, artificial intelligence (AI), and machine learning (ML) are currently the subjects of intense discussion in the press and in numerous enterprises. But what do we mean by "smart automation," and how do all these different aspects like RPA, AI/ML etc. relate to one another?

Smart automation refers to the use of software, with or without AI, and machine learning capabilities to perform repetitive tasks and manipulate data, work that has always been handled by employees. If we want to understand these aspects and how they interact with one another, we will need to view them within the context of an overall framework. Let us start with a classification in four categories at varying levels of maturity:

1. Stand-alone automation
2. Robot process automation (RPA)
3. Artificial Intelligence (AI)/Machine learning (ML)
4. Natural language processing (NLP)

Stand-alone automation has been on the market for quite a long time and encompasses client-based programs/plugins for specific applications running on a single machine. It is intended to take care of monotonous and rule-based tasks by following simple algorithms. Typical examples are macros, mouse/keyboard emulation, or office program automation.

M. Schmitz · C. Stummer (✉) · M. Gerke
Detecon International GmbH, Cologne, Germany
e-mail: Manfred.Schmitz@detecon.com; Christoph.Stummer@detecon.com;
Michael.Gerke@detecon.com

© Springer International Publishing AG, part of Springer Nature 2019
P. Krüssel (ed.), *Future Telco*, Management for Professionals,
https://doi.org/10.1007/978-3-319-77724-5_31

Robotic process automation (RPA) basically takes this automation to an enterprise-grade level. It typically runs on a server, allowing more complex automation of process tasks. In contrast to stand-alone automation, it comes with APIs, quality checkpoints, and reporting capabilities for professional use. Unlike business process management (BPM), it does not require deep IT integration and is often implemented by business departments themselves as solutions to specific situations, sometimes bypassing the IT department altogether. RPA is viable solely for rule-based, routine, and predictable tasks in combination with structured and stable data. This typically restricts the use of RPA to the automation of sub-tasks which meet specific task and data criteria rather than entire processes.

Artificial intelligence (AI)/machine learning (ML) capabilities enhance these simpler functions with the addition of pattern recognition (images, objects, people) and the automation of algorithms. They have the ability to learn and can draw on the huge amounts of data available to the system to adapt to a specific situation instead of mindlessly applying fixed and inflexible rules. Complete processes, including dynamic judgements and ad-hoc decisions, can be automated, and the data to be processed can be unstructured and more volatile. In this sense, AI/ML can be viewed as a natural extension of the RPA automation. There are many more AI use cases going far beyond the type of automation described here (autonomous driving is one such application), but they do not fall within the scope of this article.

Understanding and responding to a natural human language is the objective of the fourth category, natural language processing (NLP). It seeks to generate appropriate output in response to oral or written input. Typical applications are virtual personal assistants (e.g., Apple Siri, Amazon Echo, etc.), customer service chatbots, or machine translations.

This article examines a typical application in this field: the introduction of an RPA solution, potentially enhanced by AI/ML or chatbot functionality, resulting in smart automation.

2   What Are the Benefits of Smart Automation?

The purpose of smart automation is the customization of solutions to digitalize processes and deliver sustainable and significant value within a short time frame. Organizations use smart automation to improve efficiency, reduce costs and human error, and secure regulatory compliance. Some of the potential benefits are shown below:

1. Increased customer satisfaction from higher quality and greater consistency without human intervention, available 24/7
2. Greater profitability from significant gains in productivity; fewer full-time equivalents (FTEs) are tied up with the performance of repetitive tasks; easier scaling based on demand fluctuations
3. Better process compliance from increased processing speed, and improved auditability and transaction visibility for monitoring of integrated processes

4. Higher staff satisfaction because monotonous tasks are eliminated, allowing employees to focus on activities that are more substantive for the task at hand

Technology is becoming increasingly sophisticated and offers enterprises a way to integrate systems, rapidly scale and deliver complex services securely, and adapt business processes with greater control. Smart automation does not eliminate the need for business process management, but provides a valuable option for the automation of complex and costly processes.

3   How Big Is the Smart Automation Market?

Process automation is now widely accepted as the path of the future and a first step on the digitalization journey. Although many organizations have embraced smart automation and tested the technology through limited proof-of-concepts (POCs), very few have been able to scale them into meaningful operational impact. This is reflected in projections for the global process automation market. It will develop from a niche technology with a global market volume of US$70 million to US$80m in 2014 into a market worth billions by 2020. Various analysts predict a market volume in 2020 of between US$3 billion and US$5 billion based on a CAGR of 60%–80% (Nasscom; Transparency Market Research; WinterGreen Research).

Initial implementations have frequently been realized in North America and Western Europe. Now awareness of process automation software is growing in regions such as Asia-Pacific, Latin America, and MEA, where its deployment in offshore setups has been shown to cut the cost of a full-time employee in half. The implementation of smart automation requires performance of a broad range of consulting and integration services to ensure full understanding of the demands and processes on the client's side. The rising need for smart automation adoption is expected to increase the demand for professional services among large enterprises in particular.

In offshore setups, smart automation technology has proved it can cut the cost of a full-time employee in half. Due to higher FTE deployment costs, smart automation implementations in Western Europe have even achieved a positive return on investment (ROI) in less than three months with savings in the range of 60%–80% (based Detecon project results).

4   What Are the Typical Use Cases for RPA?

RPA is a kind of outsourcing of repetitive tasks to the computer. Any activity that a person performs with mouse and keyboard can be carried out by a software robot. Typical indications for use cases are digitalization gaps in processes, i.e., the manual bridging of interfaces between IT systems by humans. These so-called "swivel chair" activities frequently involve the transfer of information from one application to another by means of mouse or keyboard actions.

4.1   Scenario 1 "Order2Cash"

In an Order2Cash process, there is an interface that order management operates manually. Orders that are ready for production are generated as PDF documents in upstream IT systems, then manually transferred to downstream systems (web front end and IT entry masks) for the contracting of preliminary services. This process requires about 20 min for the two partial orders for each of 14,000 transactions a year, excluding time for entry errors.

In future, this activity will be handled by a software robot as a part of smart automation. The PDF files are read out by OCR and entered into the IT fields or web mask by a copy and paste operation. The software robot uses mouse and keyboard movements to accomplish this, just like a human employee would—but faster and with greater precision, i.e., without any errors. The outsourcing of this activity to the robot relieves order management of its onerous work. Errors are avoided and peak loads are smoothed out by the scalability of the robots.

Only 6 weeks passed between concept and operation of the software robots. Implementation on the basis of legacy IT would have taken at least a year.

4.2   Scenario 2 "Sales Representatives"

The sales representative must perform a functional test at the client's site when the project is completed. This requires contact to the sales office via cellphone so that the staff can initiate the functional test in an IT system. The result must then be conveyed by the sales office to the sales representative by phone or chat. The involved parties are required to devote their attention to this transaction for a minimum of 10–15 min. During this period, the sales representative must also "keep quiet" while at the client's site so that he/she is reachable. The situation does not offer an especially high experience value for the people involved. This activity is supported by a software robot today.

The sales representative sends a test order to the robot via an app, and the robot forwards the order directly to the IT in the sales office. The robot waits for the report and sends the test results back to the sales representative via the app. The transaction takes at most one minute, and the sales representative can take care of the client's needs in the meantime.

This is another case in which the implementation based on the legacy IT would have required substantial additional time and effort. Thanks to the RPA solution, the rollout for the approximately 3,000 sales representatives, including third-party providers, was possible after only 8 weeks. Sales office and representatives were able to accomplish more and do it faster, and the experience value for the client was enhanced directly.

4.3   How Do We Arrive at these Results?

The goal is to reach the "go live" phase within 6–8 weeks. The following steps are required:

1. Identification of digitalization gaps by an end2end process analysis
2. Determination of the effects (quantity × time × error index) and selection of the prototype
3. Preparation of the click instructions (storyboard) with the users
4. Programming of the software robot ("must have" features)
5. Testing and fine tuning
6. Release and go live
7. Expansion of the activities to include other scenarios

5   Going Digital: But How?

"From the establishment of an RPA factory to the responsibility for robots."

Unlike the traditional release planning by IT and management, high speed is necessary for the implementation of RPA. The technology quickly closes automation gaps and relieves the burden of performing tasks that do not generate added value. Implementation in the "passing lane" is therefore required during "smart automation." Budget processes, decision-making paths, and the required expertise must be organized accordingly. A second, agile organization alongside the legacy IT is typically required for "smart automation." An "RPA factory" of this type must be equipped with the appropriate processes and structures so that the advantages of "IT with 2 speeds" can be exploited fully. The objective is the supplement, and relief of the burdens, between release- and Scrum-driven IT, not cannibalization.

5.1   Success Factor "RPA Expertise"

When RPA and AI are used, the client's needs and business-procedural activities are at the forefront. The implementation speed and the "exactness of the fit" for the user or the target group represent the major added value. The requirements management, operation, and service for RPA solutions necessitate the corresponding expertise—new payment models such as "pay per user" to "pay per use" force new ways of thinking.

During the search for experts, it is not enough to look for people with "new" talents or to integrate "externals"; internal resources must also be enabled to develop, establish, and operate systems based on RPA and AI. Otherwise, it will not be possible to integrate RPA long-term as a key signpost on the digitalization road map.

5.2   Success Factor "Understanding of RPA Role" (Business vs. IT)

Software robots, the "new co-workers", must be assigned to the business departments where they will be active. This includes the professional responsibility, the functional capability, and the commercial mirroring.

The business units are drivers and owners of the automation projects. The RPA factory takes over the role of coordinator so that inconsistencies and redundant solutions are avoided and is responsible for secure and stable operations. The RPA factory must always become involved in new RPA projects. It provides the specifications and sets the framework conditions. Once the framework has been set, process automation activities can be carried out by the employees in the business departments without any further support from the IT department.

This is made possible by the implementation on the basis of simple instructions or click dummies that the business departments prepare autonomously. No more than brief training is required. RPA projects are possible with minimum IT involvement while still securing good automation results. IT resources are conserved.

Since the same guidelines apply to the instructions to the software robots as to human workers and the same user interfaces are used, compliance with the guidelines stipulated by the IT department is largely assured from the very beginning. The risk of major or irreparable damage from the RPA is virtually zero.

This simplification, including the positive effects on costs, lowers the threshold for the realization of automation projects.

5.3   Success Factor "Change Management"

"The software robot must be integrated as a new colleague."

RPA as outsourcing relieves employees of monotonous, boring activities. Employees are free to devote their attention more intensively to specific customer needs. Subjectively, their jobs take on a higher quality and become more exciting. It is a question of attractiveness with regard to the newly created opportunities for assignment that enhances the argumentation and motivation for what is "new" among employees. It must be a good match!

Just as with all process changes and reorganization projects, the social partners must become involved at an early stage. The innovations must be accompanied by adequate, open communication measures that "take away people's fear."

The new colleague, the software robot, must be seen as a friend and supporter. It gives recommendations, asks questions, irons out errors, and warns of manipulation or risks (see Computerwoche 2017). It does not replace human jobs—it changes them.

5.4 Success Factor "Agile Project Management"

The use of agile project management and development methods is the best approach for the development of RPA. Agility means "the cyclical further development of the applications." It starts with the definition of "must-have functions" and extends to the determination of optional functions. The testing should be carried out directly by the business department after every cycle so that faulty functions or misunderstandings can be corrected early. Methods such as Design Thinking and rapid prototyping are a help.

6 An Embarrassment of Riches

The numbers of RPA and AI products on the market are growing, but differ in what they have to offer. For this reason, the requirements for the specific use case must be mirrored in the specific characteristics of the various products. Useful criteria for assessment in this sense:

- Billing model (by transaction, by license, or other)
- Housing focus (who sets it up, who operates it and where)
- Implementation costs and effort
- Business-unit friendliness (e.g., learning time)
- Max. complexity (simple, decisions, cognitive intelligence)
- Industry focus
- Function focus (e.g., CRM, purchasing, documentation management)
- Performance indicator ($\varnothing$ gain in effectiveness)
- Platform compatibility

7 Technologies for Rule-Based Robots

The biggest advantage of robotic process automation is its simplification of process automation. Instead of managing a complex system integration project, you can set up a separate stand-alone system which acts in the same way as humans do, a virtual assistant. Employees generally sit in front of their notebooks, type on the keyboard, move the mouse around, click on data fields and enter information, select drop-down menus, and press "Enter." All these tasks can be done by a virtual employee or robot. But what technologies allow RPA to take over human tasks very quickly and simply?

People couldn't work without a mouse and keyboard. Virtual assistants or robots are just as dependent on these basic enablers. RPA systems steer the absolute or relative mouse position and automate key taps. A robot can run any application without regard for the operating system; it is irrelevant whether the application is running on legacy, stand-alone, cloud, or antique mainframe systems. RPAs are flexible and can automate processes end-to-end or only selected sections.

Many RPA systems provide image and optical character recognition (OCR) technology. The system searches for a small object or image and the cursor is moved to this position. Taking this as its orientation point, the system positions the cursor 100 pixels down and 200 pixels to left, for instance. It selects the correct data field, enters data, or clicks on the button. Important considerations when using these technologies are display resolution and the appropriate language settings in the automation script.

More and more applications today are web-based. Employees open a browser and click through HTML documents within a document object model (DOM) structure. RPA systems identify the relevant UI ID or element on the screen and select it for subsequent operations. When the element has been identified, it can be modified or controlled automatically. The presence of an HTML, CSS, and JavaScript expert on the team who can deal with special cases and automation challenges is vital for this type of automation.

Every company works with Windows applications on both the client and server sides. The application is embedded in the Windows UI, and the RPA system must be able to control this interface. All functions based on Windows and desktop (e.g., Windows OS, Excel, Word, Outlook) can be automated with this technology. For instance, the virtual robot opens an Excel spreadsheet, selects a cell, enters data, and clicks on "Save".

Many shared service teams work with applications installed on a Citrix platform. In this case, the controls are not available directly on the client side, and there are two options for robotic process automation. One is the use of screen scraper technology; the other, the installation of the virtual agent on the server side. The server has access to all UI controls of the application. The latter is the recommended approach for process automation.

There are applications (such as SAP) that provide content in a mixed format on both the web UI and iFrames—a browser uses JavaScript for its software and this runs in a dedicated iFrame. JavaScript is typically used for dynamic content that in fact is not simple to control. There are two options for automation in this instance: screen scraping or falling back on the individual GUI scripting function of the software.

There are pros and cons as well as security concerns for the use of SAP scripting. Should you mix old and new SAP installations, a bug in the user authorizations might show up, allowing more users than actually assigned to run automation scripts. One workaround is to configure the system manually. An intensive testing phase should be completed before SAP scripting is allowed. Another security gap in SAP script is the lack of encryption. Moreover, it cannot store passwords for automatic login procedures.

8   The Intelligent Part of Smart Process Automation

Artificial intelligence is the top search item in the digital community. Customers who hear the phrase "robotic process automation" expect some degree of intelligence. RPA providers understood this demand and have added intelligence, albeit of a differing nature. But what part of the game of robotic or smart process automation is intelligent?

The standard answer is that any unstructured input such as the human voice, pictures, or free text that enters into the system can be recast as structured output so that subsequent automatic processing becomes possible. In this sense, intelligence means the transfer of human knowledge to a machine. Today, there are many processes that cannot be executed without human knowledge, but in the future, the ability of machines to learn will open the door to the automation of a far greater range of processes. The following use case will help to understand this better.

Although the legal content of invoices is stipulated, the variety of formats used to convey this content is infinite. Accountants processing the invoices must manually extract the information relevant for the posting (customer ID, bank routing number, account number, value, and, if an international company, the currency) from the format of the specific invoice before it can be released and payment made. Experienced accountants have acquired the requisite knowledge allowing them to determine the correct answers to the key questions such as: "Which of these data elements could be the correct ID?" or "Which of these elements could be the value of the invoice?" They are then able to classify the data correctly. If there is any ambiguity, the accountant calls the appropriate contact and asks for clarification. Mistakes happen, and accuracy is never 100%.

The technique for automation of this process is slightly different from the one for rule-based process automation familiar to companies because the system needs to acquire the knowledge currently found in the accountants' minds. The system must learn to separate relevant from irrelevant information and to identify numbers and values. The learning phase takes place before the process can run automatically. Analogously to the human world, the machine learns from examples known as training data. In this particular use case, the many different invoice formats that may possibly be received are entered in the system, which learns from its examination of the formats and stores the results. Once a certain critical mass has been reached, the program is itself able to distinguish between the customer ID and the invoice number in a document regardless of its format.

The above use case started with a clear vision: "What data should be analyzed?" and "What is to be accomplished?" Such clear guidelines cannot always be defined. Companies in every industry have dead data storage that could potentially be used for financial controlling, customer relationship management, or other purposes. Just think about all the emails stored in a free-text format in the company. Can you imagine how much knowledge could be gleaned from these data? Normally, data scientists are presented with a business problem before they start to clean and analyze data. It is worth thinking about.

Once the data have been cleaned and analyzed, the next step is the selection of the proper algorithm. Algorithms are the essential elements of machine learning. For most people, algorithms are mysterious oracles with the ability to predict the future. This is generally good enough, but it is important to clarify how algorithms can be applied correctly and what degree of accuracy is to be expected.

Let's return to our real-life use case, invoice processing. The company achieved both objectives: a reduction in processing time from 23 min to only one minute and accuracy of between 58% and 91% for all recognized cases. The system was first trained with a dataset comprising 650 documents, then tested with 100 documents. Substantial improvements could be achieved with 300 additional training documents.

Smart automation with the inclusion of RPA and AI/ML practically offers unlimited possibilities for digitalization and automation. Software Robots are enablers for almost all kind of business processes and are considered as one element of the digital automation strategy. While RPA focuses on quick rule-based automation relying on structured data, AI/ML solutions delivers mid-term benefits. Our very obvious recommendation is to apply RPA immediately to achieve a better production, analyze the data architecture and skill up your resources with AI/ML capabilities. The opportunities of AI/ML are so huge that companies have to work out their own strategy with individual use cases.

References

Computerwoche. (2017). KI am Wendepunkt.
Nasscom: Seizing the robotic process automation (RPA) market opportunity.
Transparency Market Research: IT robotic automation market (RPA Tools and RPA Services) – Global industry analysis, size, share, growth, trends and forecast 2014–2020.
WinterGreen Research: Business process management, cloud, mobile and patterns: Market shares, strategies and forecasts, worldwide, 2014–2020.

Technology Transformation: Reinvent the Network and Change the Way of Working, Now!

Jörg Borowski and Birinder Singh Khurana

1   Magic of the Future Network

"Network Is King" is all the truer the more the network turns into software and connects all devices in an Internet of Things world with 5G, opening up doors to other business verticals. With softwarization and cloudification and 5G, technology organizations can emancipate themselves from being the sheer recipient of demand to becoming the driver for innovative customer propositions. This will open up new sources of revenue and carrier's attractiveness for the capital market will rise.

Today, customers adapt to the network. By and large, standard best-effort services are produced. They are not differentiated and do not guarantee quality. Tomorrow with 5G, the future network will adapt to the customers and offer them tailored solutions that are infinitely scalable and can be provided within a very short time—network slicing is a concept for this. Slicing provides tailored capabilities for industry or specific client demands, e.g. slices for autonomous driving or smart grids for energy suppliers, which have significantly different requirements for features like latency, bandwidth, availability and security. The configurability of the network according to customer requirements will enable efficient network operation and thus turn previously unprofitable businesses cases into profitable business cases.

Edge Computing is another technology with potential to be a game changer. Edge Computing enables the introduction of innovative products and services, like augmented reality glasses using computing power from the network. Customer devices like intelligent glasses will become light weight and easy to handle because the power consuming computing power is not build in the device itself, but "outsourced"

J. Borowski
Detecon International GmbH, Cologne, Germany
e-mail: Joerg.Borowski@detecon.com

B. S. Khurana (✉)
Deutsche Telekom AG, Bonn, Germany
e-mail: Birinder.Khurana@telekom.de

© Springer International Publishing AG, part of Springer Nature 2019
P. Krüssel (ed.), *Future Telco*, Management for Professionals,
https://doi.org/10.1007/978-3-319-77724-5_32

to the network. This is enabled by a network topology with a large number of small data centers close to the edge of the network to ensure the realization of the needed low latency—a competitive advantage for telco operators who control their own network and already have sites in the area like their base stations for cellular network.

Network function virtualization (NFV) and software defined network (SDN) will constitute the underlying structure and bring cloudification respectively softwarization to the network. The target is to operate a network factory that is highly flexible and fully automated. Principles from the IT application production will be transferred into the carrier's network world. The way to realize such a production model for networks in a carrier grade environment is long—but the first encouraging steps are made.

The true value of 5G, Edge Computing, SDN and NFV lies less in their isolated features, but more in their combination along with other emerging technologies, paving the way to the true magic of the network: SDN/NFV opens Network Technology (NT) the doors to the software world. New business models are thinkable, because Application Programming Interfaces will enable an interconnection to other game changing technologies like Data Analytics and Virtual Reality with the network at the heart of it all. Why not to combine Edge Computing with a 5G mobile connectivity slice—both provided from an operator—with the device intelligence from a partner, to produce Augmented Reality glasses?

This technology also opens the door to a highly performant scaling environment and brings Telcos to a level playing field with hyper scaling companies like Amazon or Facebook.

The scaling effects in an ecosystem in which the network is virtually mapped and managed from a central position are all the greater when the use of network capacity is further optimized by joint utilization. The creation of the largest possible networks through cooperative activities is the logical step for making the most of the scaling advantages. This will result in new business models based on cooperation among network operators. Models of international alliances, such as those common since long in the aviation industry, can be transferred to the world of telecommunications.

One example for this is the Pan-Net of Deutsche Telekom: A common infrastructure in a central data center virtualizes the production of services for multiple national companies. Another example is the Next Generation Enterprise Network Alliance (NGENA): A global alliance of telecommunications providers, Ngena offers telecommunications services such as VPN access or WAN connectivity of locations to international business customers requiring communications across national borders. A joint production platform is established in an independent company and links the transport resources of its members in a cloud so that various national access services can be used.

The above mentioned technologies have something in common. They enable the setup of a new technology architecture based on the following three design principles:

- **Keep it simple—Standardization**: Proprietary systems will be pushed aside by standardized solutions. The diversity of the components in use will be reduced. The components themselves will become more homogeneous. Open source principles will replace closed solutions in the manufacturers' ecosystems. Interfaces will be standardized, complexity reduced, costs lowered.

- **I am the center of attention—Centralization**: Decoupling of traffic and control will enable a centralization of functions and network elements such as data centers and network operation centers (NOC). The more extensive the centralization, the greater the scaling effects that can be expected. This will make it possible to offer a homogeneous line of services. A standard portfolio that transcends all of the boundaries between national companies and holdings in the corporate group can be created.

- **Code rules—Convergence of IT and NT**: Everything that can turn from hardware to software will turn to software. Software-based solutions will be used more and more frequently and will replace solutions that are hard-wired in the hardware. Basic principles from the world of IT will find ever-broader utilization in the network domain. While NT and processes of proprietary systems closely related to the networks were controlled by only a few providers, IT services have developed in a highly competitive world. Flexibility and efficiency, familiar elements for IT organizations, will take their place in the NT world.

2   Design Principles for the New Operating Model

The isolated implementation of the new technologies—incremental change—is the easiest part of the whole journey. The issue is that Telcos will never be successful in unfolding the magic of the network—operating the new technologies and leveraging the advantages, which stem from an ability to combine and evolve, unless structures that evolved around the old way of working are overcome and a complete new operating model is established. It would be the same as asking a McDonald's restaurant to serve food like the Osteria Francescana in Modena.

A holistic approach must be taken using an operating model, which describes how resources are organized in order to create value according to the laid out strategy.

2.1   Challenges of the Current Operating Model

Companies, especially those in the game since the introduction of 2G, are all struggling with the same issue: The network architecture and its implementations have grown fastly, like the Amazon jungle instead of the well maintained flower garden Keukenhof, with each evolution cycle e.g. 3G and LTE.

The negative effect of this wild evolution has been enforced through failure of retiring the majority of legacy systems. Consequently, the current operating model has developed into a complex interlinked rainforest. It has become nearly impossible to understand the end-to-end value chain in terms of true value creating steps and overhead activities. The lack of transparency chokes any breath of innovation, because no one wants to change the running system. A culture based on consensus and preservation of vested rights has prevailed. On top, managers define their value by the amount of employees they have and by the amount of decisions that are made by them.

Change is necessary! A transparent operating model with a culture is called for that allows the courageous combination of technologies and enables the proliferation of best practices beyond organizational seams.

The challenge will be to manage the balancing act between exploiting the legacy technology securing today's profit on the one hand and exploring the possibilities of the new technologies building up tomorrow's profit on the other hand. This ambidexterity is at the core of dynamic capabilities that firms must have (O'Reilly and Tushman 2011). This is especially true and challenging for today's telco technology departments.

Specifically, in the Telco context, the current cash cows need to continue giving milk. Connectivity and telephony services, the backbone of the fixed and mobile services offering of Telcos need continuous improvement. Telcos are familiar with this efficiency oriented and deterministic planning—the old world. However, planning the future based on only lessons learned from the past is literally in the past when it comes to innovation in a digitalized world.

Companies must learn to look ahead and manage uncertainty throughout the organization with an agile approach, both in terms of team setup and ways of working, especially in the technology domain. Additionally, the interplay of efficiency and agility must be learned. Building up both capabilities is nothing less than an open-heart surgery with the heart of the company being the culture. Enduring the tension that the coexistence of both efficiency and agility cause will be the main challenge.

The operating model approach helps sort out the implications of ambidexterity for a firm. In the following chapter, the operating model for the future will be described by focusing on the exploratory capabilities of the operating model.

2.2   Design Principles for the Target Operating Model

The target operating model for the technology department of a Telco company must be tailored to the specific strategy of the company. However, there are three guiding principles, which are universal for the technology department of every Telco company that want to avoid becoming the dumb bit pipe in the future, but rather want to unfold the magic of the network. This is especially true for integrated companies offering both fixed and mobile access having a history in providing voice and internet connectivity in saturated markets.

2.2.1   New Way of Working: Grow Empathic Leaders and Empower Employees

Today's organizations make decisions too high up in the hierarchy, at positions that are at least two floors above the operating business that is ultimately affected by the decisions. A large number of decision-making bodies and levels make it an easy matter to dodge accountability and leave decisions up to others. The consequences of such behavior are bottlenecks and congestions as pending decisions pile up. Ultimately, employees are hesitant to try out technological and business innovations.

Managers must truly learn to delegate, shift the volume of operational decisions closer to the employees at the operating business and empower employees to make decisions within a budget and resource range. Managers can then spend the newly won time to guide their employees with truly strategic decisions and be able to develop them professionally and personally. This is done more with empathy than with ego. Empathy will be a key qualification for pressing the reset button, in order to embrace innovative impulses (Jennewein 2017). Consequently, the number of coordination and cooperation programs in the organization can be reduced to a minimum.

This kind of leadership will not be every managers and employees cup of tea. There will be both managers and employees that will fail to live up to new role models and change their behavior accordingly. However, the transition is necessary and will need to be managed as smoothly as possible with respect to expectations from employees, shareholders and customers.

2.2.2 Convergence: Break Down Silos

Technology organizations of Telco companies have developed organizational silos along different dimensions. The advantage of such organizations is the specialization and professionalization within each area and this perfectly fits to the old legacy world. The disadvantage is that end-to-end relationships are difficult to recognize, implementation of changes and innovations is slow, and knowledge transfer is restricted. For the target operating model these silos need to be broken down, in order to pave the way for convergence, especially along the dimensions value chain, network architecture and NT & IT interworking.

Historically, technology organizations have been aligned along the "plan, build and run" **value chain**. The chain has a critical defined breaking point: the handover of products, services and platforms from development to operations. Today the value creating activities have a "wall" at the border lines, which manifests itself in a weak team work between the development and operations department. Operational requirements are considered at a very late stage of the value chain. If enterprises act faster and more flexibly on the market and, at the same time, reduce costs for the development of new products, the value creating activities for development and operations must be meshed more tightly with one another. The most extreme form this can take is described by the DevOps (development and operations) concept that has proved its value in software development and pursues the goal of developing innovations flexibly and quickly. Intensive communication and close cooperation among the employees in development and operations enable recognizing and eliminating mistakes at an early stage of the value chain. DevOps is successfully used by OTTs such as Facebook or Amazon. A couple of release launches every year does not suffice for these companies. Quite the contrary—releases are continuously rolled out. While a full fledge adaptation to DevOps is challenging, technology organizations can start by carefully reviewing, which characteristics of DevOps can be implemented quickly and with low effort, in order to reap benefits early and design a Telco tailored DevOps value chain.

The **network architecture** is the second dimension to consider. Organizational silos have also occurred due to design in-line with the network architecture access, transport, aggregation and support systems. In the access arena there's still the differentiation between mobile and fixed that has to be overcome at least from the point of view of the end customer, who doesn't care whether mobile or fixed technology enables him to use services.

In the face of technological trends like NFV and SDN the future network architecture will differ from the as-is version and will need adaptation. The organization must be laid out in such a way that it reflects the functions to be organized in a software defined network instead of the classic elements of a network architecture.

Network Technology and IT started out as two separate business functions—resulting in the third dimension of silo building. New IT and NT architecture principles make it possible to decouple services from the resources, to manage the network centrally, and to separate the operation of the network from the traffic. The assignment of tasks to IT and NT organizations will become increasingly difficult because the functions will also blend into one another. Cloud-based networks make it possible to fuse the control over IT and NT in their operating system. The organization must follow this lead. IT and NT organizations must be tightly meshed and grow together in the future and learn from each other.

2.2.3   Innovation: Fail Fast and Fail Cheap

Reinventing the innovation culture means building an environment that enables a "fail fast, fail cheap" way of working. New services can be rolled out within a brief period. Due to close customer interaction, the company will quickly know if the new products and services are meeting customer needs. A high percentage of flops is deliberately accepted because the expenditures for development and rollout are very low. This is because technology departments will increasingly apply economical aspects using a high degree of automation and centralization.

2.3   Operating Model Blueprint for a Successful Technology Organization

The three design principles, firstly the people orientation of the new way of working, secondly the structure related principle of convergence and thirdly the new business orientated innovation must be applied to the new operating model. In this article, we suggest a technology operating model that consists of eight building blocks (Fig. 1). It serves as the blueprint for future technology organizations:

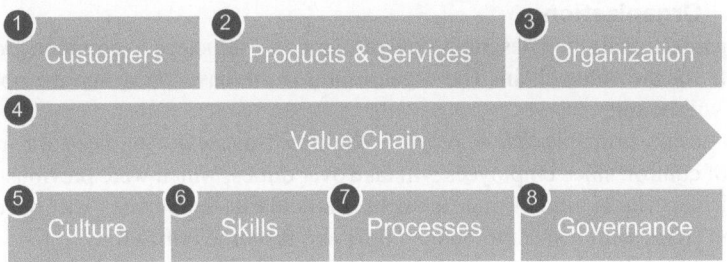

Fig. 1 Blueprint for technology operating model

2.3.1 Customers

The organization maintains a clear view on its customer base. Customers for technology organizations are typically the marketing department and the IT department.

The design principle "Innovation: fail fast and fail cheap", will require employees to keep their customer needs and demands in mind, which will be the decisive compass when it comes to product and service life cycle management. Furthermore, gold standard solutions that serve their own purpose will be of the past, because the incentive will be to produce the product and service that best fits the customer's need and demand at the lowest cost. Moreover, clarity will enable and fuel dialogue between the technology organization and its customers, be it internal customers like the marketing department or external customers like other companies.

The opened up dialogue between marketing and technology will allow both push and pull solutions. In particular, the pull will enable technology organizations to transform from a strict cost center to an impulse generator for new sources of revenue.

2.3.2 Products and Services

A clear view about the products and services portfolio, and the underlying platform assets are the basis for running the organization. Closely linked with the client's perspective all actions should focus on the values of the products and services that are produced in the network. An intensive communication between demand and supply parties should be established in the new operating model.

The "Convergence" principle implies that the new organization must handle an emerging portfolio of products and services like apps for Edge Computing users, slices for 5G and converged products like seamless access. This will result in a more differentiated and heterogeneous products and service landscape than before.

To be innovative with products and service the usage of modular design principles, helps to provide the outcome cheaply and quickly.

2.3.3 Organization

The organizational setup describes how resources are allocated to common groups of interests for the value chain. The organization is the basis on which the operating model will be run.

The design principle "New Way of Working", dissolves the need for multiple layers of control, since employees will take over duties, which were previously done by their superiors—middle management. Particularly, employees will have more decision space with respect to budget, time and resource utilization.

The second design principle "Convergence", will cause a gravitational shift from line work to project work. Consequently, the organization will be sorted more according to skills like "cloud engineer" than to a particular technology function like "layer 1 network signaling mobile design engineer".

The third design principle "Innovation", implies a flexible access to employees and skills that could be realized by total workforce management. In order to handle the uncertainty of future work demand with respect to duration and expertise, it is expected that the share of project work will gain much more significance than regular line work. Hence, there will be a pool of experts with different areas of focus. The separation line between internal and external work force will blur out. Total Workforce Management concepts enable the resource steering of internal and external resources. A pre-condition for this to work is that the respective national legislation design the labor market to be transparent and liberal.

2.3.4 Value Chain

The value chain describes the value creating areas and activities of the business.

The "New Way of Working" principle has the potential to improve supplier selection when faced with a buy decision and it has the potential to improve partner selection. This is because all employees and especially the engineers are empowered to decide new sourcing and partnership models. The classical role separation between business department and procurement will be dissolved.

The design principle "Convergence" resulting in the softwarization of the network will bring in completely new technology suppliers and partners into the consideration set. On the one hand, this increases the flexibility and the speed to change suppliers, but on the other hand, it reopens the core competency question box once again: what are the core tasks for operators and what tasks are better done by partners. New business models result from this and the role and tasks of procurement departments have to be adjusted.

The design principle "Innovation" will also affect the value chain. "Fail fast, fail cheap" results in smaller teams that are staffed cross-functionally. The small team size will speed up decision-making. Cross-functional staffing, will enable front-loading. This means experts from the test and operations department will be involved early on in the value chain, namely the plan and build phase. Consequently, implementation or even design errors can be detected early and costly bug-fixing in the live network will be reduced.

2.3.5 Culture

Culture will become the core asset of the future technology organization and the actual engine of the whole operating model. Culture will attract talents and experts, but it will also be able to keep employees and ultimately make them brand ambassadors.

On the one hand applying the first design principle "New Way of Working" will translate into much more freedom to make operative business decisions, on the other hand, it will also translate into much more freedom with respect to one's own professional progress and one's development of value for the company. For example, a cloud engineer can decide to move to a role as a presales consultant, because he would like to put more emphasis on his business skills than his engineering skills.

The second principle "Convergence", will amplify such a bold career move. This is because the present silo among planning, building and running will blur out with agile ways of working in small-sized and cross-professionally staffed teams working in open space environments. Consequently, the barriers of switching will be lower than before. Additionally, job rotations will be encouraged. Especially, employees in leadership positions will be encouraged to switch jobs after a period of e.g. about 5 years—sufficient time to create impact and just enough time to not get comfortable. This will improve their ability to understand the business from an E2E-perspective.

The third principle "Innovation" will have many practical implications in order to enable a fail fast and fail cheap business operation. People must learn to act as an entrepreneur and to take over more responsibility. The culture must award courage and avoid punishing failure. In addition, the culture must be flexible and enable employees to work more efficiently according to their specific needs by giving them the opportunity to work from home, or any other remote location. Usage of social media platforms will not compensate for face-to-face meetings, but they will enrich the traditional means for information exchange and discussions by providing a new platform. However, communication through this platform needs to be learned just like etiquettes need to be learned when going for an interview.

This change in culture comes at a price that the employees must pay: reduced plannability and greater uncertainty in their actions. The reward is that the performance of employees will have significantly more substantive content and meaning. The bottom line of this approach will profit most employees—enhancing their motivation.

2.3.6 Skills

Skills are the capabilities that are required by employees to create value.

The first design principle "New Way of Working" requires employees, both managers and experts, to interact with each other more than before. This is because the newly won freedom of employees will require them instead of their managers to find solutions to changing customer demands or technological developments. Agile methods like Scrum are designed for this way of working.

"Convergence", the second principle, will require adjustment of the skillset. E.g. network technology engineers must understand coding and handle new agile methods. It will become more and more important to understand the complete value chain.

The skills to master a method like "Design Thinking" takes a different approach to solving issues then the traditional systematic way. The customer, both internal and external, is put at the center of the matter. The status quo is questioned. Solutions are found quickly, developed in a team and tested immediately. During these phases, a team spirit is established and mutual respect develops. A willingness to embrace new and uncertain ways emerges and sparks of innovation are ignited.

The third principle "Innovation", enables fail fast and fail cheap by trying new things. Employees that endeavor to create value for the customer by enriching their work with topics outside of their comfort zone must be appreciated for their courage and effort to learn something new and combine it with existing knowledge. This includes the capability to reflect upon one's own skills and to recognize the skills, which are required to run the technology organization in the future. Trying out something new to create value for the customer or making existing value creating activities more efficient must be rewarded. Employees must not be conditioned to go back to their familiar cage of expertise, even if they fail.

Another aspect of "innovation" is the requirement to develop solutions economically. Most engineers know little about this. Therefore, there is frequently extensive potential for economic efficiency in technology organizations. An example for a procedure that is suitable for achieving an economic optimum is the design-to-cost and -value method that seeks the most efficient design. The rigorous application of design-to-cost can enhance the strictly technologically oriented performance model by economic efficiency and customer centricity.

2.3.7 Processes

Processes are a framework for doing repetitively occurring tasks in the most efficient manner.

Design Principle, "New Way of Working", calls for as less process overhead as possible. Consequently, employees can focus on value creating activities like the programming of new products and services rather than on the preparation of management slides for approval of a new product.

Design Principle "Innovation" implies cutting costs of existing processes through the usage of automation. In contrast to "New Way of Working", this requires an even more detailed process description than before, in order to use automation.

The challenge for the area of processes will be to decide when processes are needed at all. However, when they are needed, the aim would be to use best-practices and apply automation as much as possible.

2.3.8 Governance

This block describes how crucial business decisions are made impacting human resources and budget utilization.

The design principle "New Way of Working" will reduce the number of management alignment meetings, because many decisions will be made by the cross-functional expert teams. Only issues with out of tolerance budget, time and resource situations or strategic topics will need guidance from management.

The target system must ensure that teamwork is enhanced and individual optimizations is avoided. This could be achieved by switching from individual targets, still dominant in traditional tech organizations, to common team targets.

Another impact to the target system results from the design principle "Convergence", which results in a new SDN & NFV architecture. The steering of the system no longer focuses on hardware based network elements. The orchestration of SDN/NFV architecture is dominated by software driven services, which will be reflected in a new governance with adequate targets and KPIs.

3   Conclusion

Why—Technological changes like 5G, Edge Computing, softwarization and cloudification will change the name of the game for the technology departments of telecommunication companies. These changes enable a transformation of technology departments from the mere cost center to the driver of innovative business models.

What—However, in order to enable this transformation not only the pure technological development is necessary, but the operating model of the technology organization must evolve as well, in particular the culture, skills and value chain as we know it must change, in order to allow the transformation.

How—Detecon suggests three key design principles that will make future technology organizations fit for the future. While these three key design principles apply to all telco companies, their implementation needs to be tailored to the particular situation e.g. current culture and core competencies of the company at hand.

It's a big chance for all carriers to unfold the magic of the network. It's not enough to set up "just another" engineering programm. Success is dependend on changing the way of working summarized in a new target operation model. Not starting the transformation journey is a risk; therefore the right time to start is NOW!

References

Jennewein, W. (2017). Harvard business manager Edition 11/2010, Die Distanz reduzieren, Manager Magazin Verlagsgesellschaft mbH (in press).

O'Reilly, C. A., III, & Tushman, M. L. (2011). Organizational ambidexterity in action: How managers explore and exploit. *California Management Review, 53*(5), 5–22.

Generating Value Through Digitalization: Simple and Digital

Andreas Lischka, Michaela Wolfering-Zoerner, and Eva Faust

1 Introduction

As digitalization advances, it inevitably leads to in-depth changes in many areas of society. Everyone is affected, whether business owners who find themselves facing new service and business models or consumers whose everyday lives, from online banking to health-care services, are shifting more and more into the digital world. Even the role of geographical location has become less significant and is today more or less irrelevant .

Politics acknowledge this structural transformation, and the Ministry of Economics has issued the "Digital Strategy 2025", a clear 10-point digital transformation agenda for Germany. The National IT Summit has been renamed to Digital Summit. This move intended to reflect digitalization in its full breadth and considers the provider and user sides as well as Industry 4.0 right up to cultural and creative industries. The provider side in particular has to deal with digital transformation challenges like handling the complexity of organizational structures and systems (Bundesregierung 2017). Today, the main demand is for innovative solutions and not only about merely selling products to customers.

Digital transformation must be driven forward so that it can successfully match the changes in customer behavior. The status of digitalization in companies has changed from nice-to-have to a fundamental requirement for maintaining competitiveness in the market. Companies will lose their competitiveness, when they don't drive the digital transformation of their core business. Examples of challenging traditional concepts are the taxi and hotel industries, where companies such as Uber, Lyft and Airbnb are bringing down the existing market structures. Another sector that is struggling with digital business models is the travel industry. More and

A. Lischka · M. Wolfering-Zoerner · E. Faust (✉)
Telekom Deutschland GmbH, Bonn, Germany
e-mail: Andreas.Lischka@telekom.de; Michaela.Wolfering-Zoerner@telekom.de;
E.Faust@telekom.de

© Springer International Publishing AG, part of Springer Nature 2019
P. Krüssel (ed.), *Future Telco*, Management for Professionals,
https://doi.org/10.1007/978-3-319-77724-5_33

more travelers are booking their vacation in online portals like Booking.com, an Alphabet Company. As a consequence, the number of travel agents operating in the USA declined from 132,000 in 1990 to 74,000 in 2014 and is still dropping.

Digitalization is opening many new doors for the telecommunication industry. Many companies have committed to trusting in digitalization during the interaction with their customers and for management of their networks. The added value is generated here by improving the customer experience, developing new business fields, and reducing costs. According to the McKinsey analysis "How Telecom Companies Can Win in the Digital Revolution" from 2016, customer journeys can be significantly improved by digitalization and data analysis.

2   Customers and Market in Digital Transformation: Opportunities of Digitalization

Digitalization is drastically changing the ways customers and companies interact with each other. In this sense, omnichannel management aims to increase the customer contact (see Fig. 1). One characteristic feature is the transition of traditional contact points such as shops and customer service towards integrated digital platforms. The ultimate objective is for the whole customer journey to be completely and seamlessly integrated. This shall include the entire customer journey starting with the search for the appropriate tariff, the registration for services and options through to contract cancellation.

According to a study on user behavior conducted by Nice Systems, a leading provider of call center IT, the success of an omnichannel approach with regard to customer services can vary significantly from company to company. Among other factors, success is dependent on the organization and culture of the company. The introduction of omnichannel models must overcome channel conflicts in sales and service by consistently taking the customer viewpoint across all touch points. This is highly significant because the study determines that customers today frequently use more than one channel to find solutions for their requests.

On average, a customer will use 5.8 of 10 available channels to get in contact with companies. Moreover, the study showed that one out of three customers will also use self-service opportunities in an attempt to find solutions. If this is not possible because of inefficiency or dissatisfaction, they will move to a different channel. Self-services and their quality are becoming increasingly valuable criteria for customers and their satisfaction (Mustica 2015).

Deutsche Telekom also relies on the realization of omnichannel management— the interplay of sales and service across all channels. All sales and service channels are bundled into one service unit so that the customer experience will be permanently improved. In consequence, customers can be offered a consistent seamless service experience throughout the entire customer journey.

Telecommunication companies have the opportunity to improve the customer experience by increasing the availability of (digital) customer touch points in various sales & service channels and process steps.

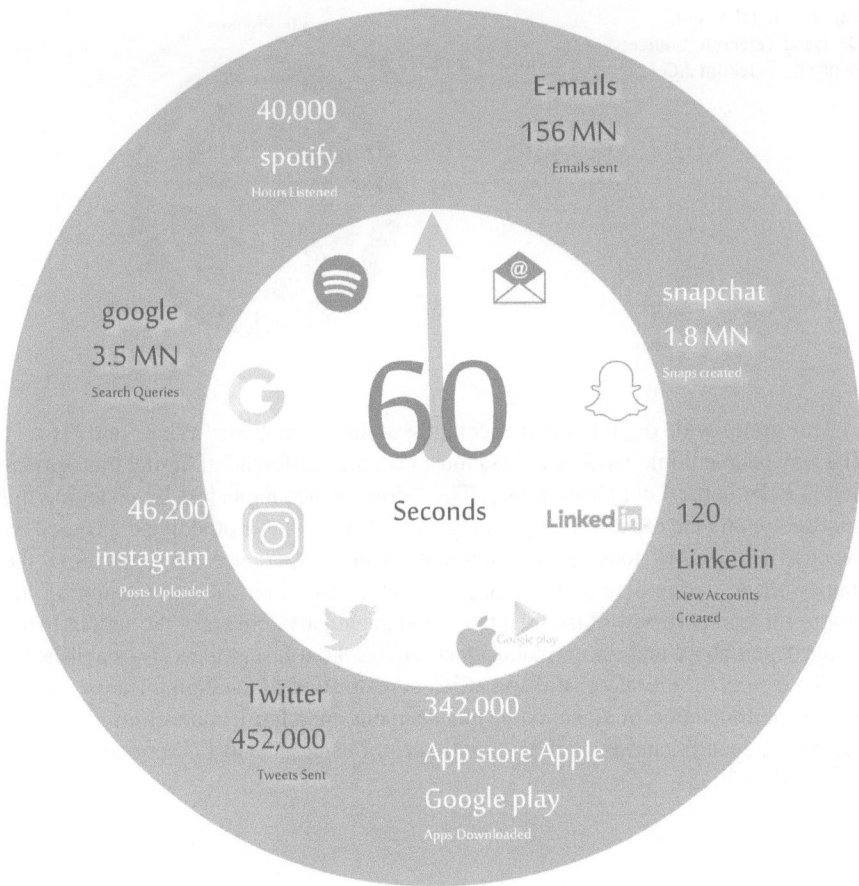

Fig. 1 What happens online in 60 s. Source: businessinsider (2016)

The integration of apps can offer added value. Application-based interaction ensures customers to benefit from an interactive experience with the company. This personalized experience can contribute to a strengthening of customer loyalty.

3 The Challenge of Digitalization to Deutsche Telekom

Telekom has developed a digital vision valid throughout the corporate group to keep up with market developments and changes in customer expectations for cross-channel, digital interaction. It includes the creation of the position of a CDO in the German market unit as well as in the European business segment. The digital vision will be implemented by 2020 and is driven by the program "Be Digital".

Fig. 2 Digital Vision
Deutsche Telekom. Source:
Deutsche Telekom AG

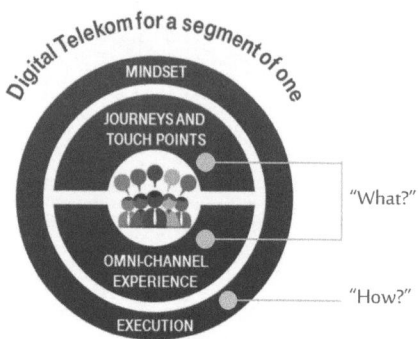

This group-wide digital vision encompasses the dimensions "What" and "How". The way people think, work, and lead must become "different" if digital transformation is to be carried out successfully. The "How" is accomplished by changing the mindset for and the execution of projects. The "What", on the other hand, is based on the customer journey, touch points, and omnichannel experience. This constitutes the framework to derive actions for change processes (see Fig. 2). For monitoring the progress a regular review of the qualitative and quantitative effects of the actions to the management board and senior management of Telekom Deutschland is established.

The factors for the successful and efficient realization of the digitalization agenda are the centralization of stakeholder management and clarification of conflicts in a functional department, as well as "multi-speed business" and IT Services (Dörner and Meffert 2015).

3.1   The Change Dimension "What" in Telekom's Digital Vision

The goals of the "What" (Fig. 2) along the digital roadmap are individualized customer journeys, consolidated interactions, and customer experience that is both consistent and intuitive. Individualized customer journeys generate a competitive advantage when a 360° view is provided on all customer contact points, whether online or offline, and across all sales and service channels—obviously with due regard for the security of customer data.

If we take Netflix as the benchmark, we note in particular that this company understands not only how to gather and analyze relevant customer data, but also how to use the insights as a foundation for addressing specifically customer needs that make the relevant difference. The best practices for individualization at Netflix illustrate to us, that a definition of the relevant seamless customer journeys across all channels, including shops, and the identification of a set of measures for individualization at the digital contact points along these customer journeys are necessary. The goal is to define actions for individualization at the digital touch points along the customer journey that contribute to centralization of customer information.

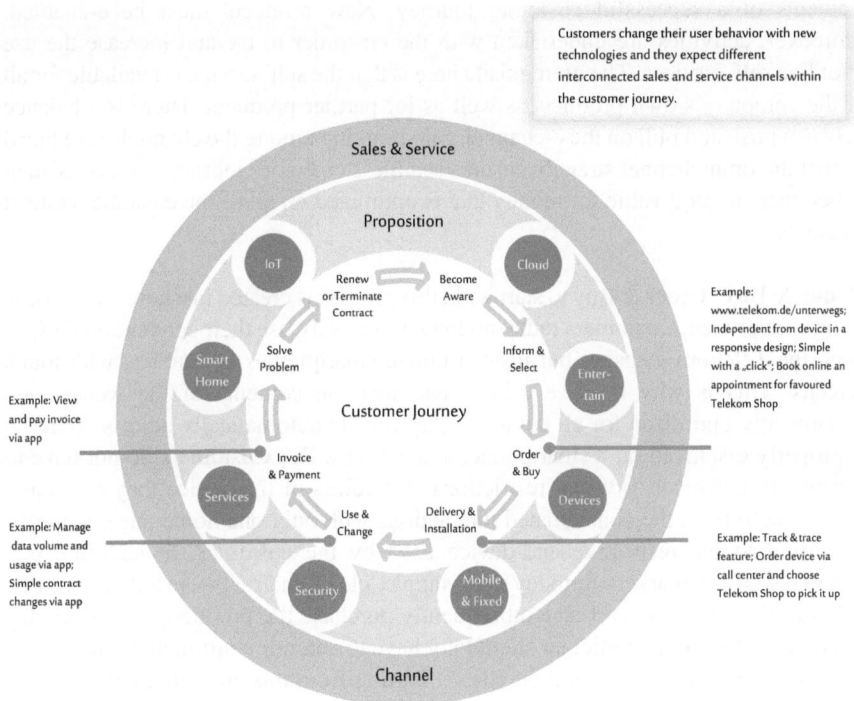

Fig. 3 Customer journey and a seamless service experience. Source: Telekom Deutschland GmbH

This form of individualization represents an opportunity for telecommunication companies to guarantee a consistent customer experience. A function of this type for the German market was launched on a personalized website telekom.de (see Fig. 3).

Drawing on this source, the customer journeys are being systematically examined and used as blueprints at Telekom Deutschland. This procedure (Fig. 3) is being followed to analyze where customers today abort the customer journey and experience problems. Interviews with experts and assessments are used as a basis to determine where customers find themselves blocked on a "digital channel" or e-channel or do not receive adequate information. These are the points that prompt customers to call the hotline. From the customer journey perspective in the e-channel, we find three points through which we seek to improve the customer experience:

1. Order & Buy
2. Delivery & Installation
3. Invoice & Payment

This is where the cross-channel customer experience (omnichannel) and the prioritization of the digital or mobile customer contact channels are indispensable elements of a successful customer journey. New products must be e-enabled. Moreover, activities are undertaken with the customer to try and increase the use of online self-services. The prerequisite here is that the self-service is available for all of the company's own products as well as for partner products. There is a balance between push and pull on the e-channel. Silo thinking among the channels is reduced so that the omnichannel strategy ensures seamless customer journeys and consistent experience of good value for money that is optimized on all of the customer contact channels.

Order & Buy Order & Buy to start with, this process is created in a new "responsive design" approach. Customers today no longer primarily use their conventional PC to open the Telekom website, but instead utilize smartphones and tablets with touch screens. This is why it is very important that the contents of the website are dynamically optimized for all devices—the website automatically adjusts so that it is properly displayed on a smartphone, a tablet, or a PC. Customers do not have to accept any technical or visual restrictions, regardless of the device they are using. They benefit from the consolidated interactions with only one home page, rendered according to the respective end-device used by the customer. In addition, buy processes on the market are today very simple, clear, and fast—customers must be able to fill a shopping cart and subsequently purchase the products with a "click". Existing customers in particular should not have to enter or confirm any information again right before their second purchase. Starting from this and other premises, we have designed and implemented a simple, clear, and fast purchase experience (see www.telekom.de/unterwegs).

Another example is the possibility to book appointments in a shop via online channel. The customer can set up this appointment at the agreed time in a Telekom shop of his choice.

Delivery & Installation In addition, we note what customers appreciate when receiving information about their order such as the simple review of their past orders and a track and trace function that makes it possible for customers to locate the position of the order at any specific time. These functions of viewing recent orders and tracking & tracing will in the future be displayed in an optimized layout on the e-channel. If desired, customers can obtain assistance for the setup of a new smartphone via an app by reading community articles, viewing help-videos, or looking at Q&As. Another flexibility is to order the chosen devices online and collect them in the shop.

Invoice & Payment The need is to satisfy customers' wishes for a simple and clear presentation of the invoice. An app and a website function that provide simple access to invoices for customers have already been implemented to meet this customer expectation. Customers will be given a simple invoice view in the future that also allows the display of "jumps" in the amounts of invoices, giving more detailed information. This can be used to ward off payment difficulties.

The examples demonstrate that the generation of consistent and intuitive customer experiences can create a new simplicity and the "same look and feel" across all customer contact channels. The application of a general "DT Design Guideline" and compliance with the guideline ensures the consistency of this customer experience across all of the contact channels.

3.2   The "How" as a Change Dimension in Telekom's Digital Vision

The "How" along the digital road map is aimed at the mindset and execution as main influencing factors of realization. Digitalization through leadership from the very top means that management's leadership is visibly digital and that it motivates the organization and resolutely provides it with the appropriate resources by continuously defining and incentivizing digitalization as a priority in the company's business operations.

The centralization of digital transformation activities in the purview of a CDO who reports directly to the CEO firmly establishes responsibility and clearly sets digitalization high on the company agenda. Core elements are the organizational structure and culture, supporting incentives, and a lead set of key performance indicators (KPI) that pay into the digital value contribution. The changes resulting from digitalization should become visible throughout the entire organization. One key digital lead KPI as an internal goal and one KPI as an external goal in the direction of the capital market aid communication, whereby the CEO also has a major role to play as communication carrier.

Besides the key KPI, the CDO is also to be seen as change manager of the organization and culture who reports on digital goals in the corporation and anchors and supports digital "evangelists" as promoters within the organization. Incentives aligned to the digital vision are assured consistently across all channels and management levels. Fast learning from customers is incentivized in particular. Design Thinking methods promote and encourage the taking of the customer viewpoint in all processes. The incentive can, for example, break down silos existing in middle management in the sales and service divisions and promote cross-functional, agile teams to solve operational issues. These agile teams are already supported at Telekom Deutschland, e.g., by Scrum teams.

Fast learning from our customers is possible only if customer insights are analyzed quickly and precisely and utilized according to relevance for customers by means of a continuous and iterative improvement process. From the starting point of initial customer hypotheses and minimally produced test products, it is possible to learn from customers and no longer to prepare a pre-defined, detailed product planning that is from the beginning defined in depth and stretching far into the future. Instead, the test products and their releases are quickly improved and strictly prioritized by cross-functional teams in response to customer feedback. Buying decisions are analyzed using IT and standard solutions are implemented. Initial initiatives for fast learning have already been launched in Germany.

Telekom Deutschland already maintains digital analysis units that exchange synergies and findings that are closely meshed with the CDO teams. The analysis teams essentially use three layers. In the front end, data about all types of customer interactions are recorded and collected in an integrated database, in turn analyzed and assessed by specialist teams, and returned iteratively to the data recording layer. The analysis results automatically flow into operations—for instance, into churn prevention in the current account management. Telekom Deutschland is already using customer engagement at digital customer touch points so that specific advertising with personalized touch points can be created. Customers read an article about music and see the appropriate personalized advertising from Deutsche Telekom and Spotify, including a personalized website with the focus on Spotify products.

The consolidated user behavior across all of the customer touch points on an integrated data management platform enables an advanced analysis so that users and use cases can be addressed in e-sales and e-service. These user information are utilized at all of the online touch points for personalization and advertising directed at specific target groups. In this way, advertising directed at specific target groups and the personalized address of the customer lead to greater customer satisfaction.

Multi-speed business and IT, supported by precise customer analysis using big data, create another prerequisite for customer-centric improvements. So-called multi-speed business and IT become possible when front-end and back-end systems are decoupled and processed parallel to each other. A central API (application programming interface) management team can be set up to ensure that APIs are free of any redundant services and instable performance. Besides this decoupling, a structure that enables digital work at the best-in-class level and in which new processes and methods are developed that enable multi-speed work for Telekom must be created. Cross-functional teams, along with business and IT, support the fast realization, which ideally should be brought together for all digital products and services aimed at customers.

This decoupling also allows multi-speed in development. One example is Telekom Deutschland's new mobile portal. Along with the releases in the planning process, shorter release cycles with only a few, but flexible functions are created.

Improvements for customers in the direction of simplicity, efficiency, and speed-driven benefits also drive the automation of sales and service transactions. This automation must be driven forward so that digitalization can be made possible. The advantages are shorter and faster order & service processes and real-time data generation (e.g., for individualization), marketing specific to target groups, and efficiency where manual work can be reduced. Complexity is reduced by focusing on existing products and processes. Products and processes are the primary drivers of complexity and costs. The reduction of complexity simplifies and drives automation. Moreover, automation should always be done together with a complexity reduction. One example of automated self-service solutions is the disruption assistant at Telekom Deutschland. The virtual assistant supports customers via chat, phone or mail, who are having technical problems with the Internet, TV and voice service.

4 Success Factors for Digitalization at Telekom

Various critical factors must be given careful consideration if the challenges of digitalization are to be mastered successfully.

Digitalization must be lived top-down. The topic is a key component of group and company strategy, and the implementation must be steered by a central authority (at the beginning, at least). Cross-functional teams are required for strategic development and its implementation because digitalization impacts all levels of the value-chain. Digitally proficient users can enhance the learning effect across sector and company boundaries. Another necessity is the intense further development of the company's own employees so that the major "capabilities" become available within the company. The objective is to convey relevant capabilities such as data analytics. This is supplemented at the top level with specific investments in technologies and experience for the further advancement of the company (e.g., through the provision of risk capital).

The digitalization of products and services also demands a change in the way of thinking. Agility in product and service development along with continuous involvement of customers in development and further development of products and services help the company to offer defect-free products and services tailored to customers. Moreover, adaptations can be made in shorter cycles.

The share of customers expecting to be able to contact providers digitally will continue to rise. Accordingly, digital contact channels are already of high importance today, and they will become even more essential in the future. Sales and service must adapt to this change in customer behavior. Customers will decide themselves which channel they want to use for their issues—advice in a shop, information from the Internet, or product reviews in social media, for instance. They will use different channels, and that means that the same information must be available to customers at all touch points, whether it is information about a product or service or personalized offers. This is where the use of advanced analytics of existing stored customer information can play a decisive role.

Digitalization also means driving further development in the operating units. Sales and service units will become increasingly digital—this applies to customer-facing units as well as internal units. The relevant customer-facing touch points must be addressed. Customer concerns must run through the company's processes automatically, however. What good is a "beautiful" website if the customer's order does not move through the process chain successfully in the customer's eyes? Business and IT must work together on this further development. Agile evolvement of the systems, automation, and ongoing trials with customers are required.

Culturally speaking, digitalization means setting a good example. Managers must be role models to their employees, living the culture of an open and flexible approach to change. A culture of constant learning must be established in the company. Time and resources are allocated to promote the ongoing further development of the workforce—without any fears of possible failures and always with a focus on market and customers.

Some companies already have large reserves of capabilities just waiting to be used. A culture of trust encourages employees to find their own creative solutions to problems. New technologies for communication and collaboration—social media, cloud computing, and all the rest—generate tremendous potential for companies to simplify and accelerate processes. These new, disruptive opportunities afforded by digitalization should be exploited as they can aid in offering the customer the best possible customer journey.

References

Bundesregierung. (2017). Accessed December 20, 2017, from https://www.bundesregierung.de/Content/DE/Artikel/2017/06/2017-06-12-digital-gipfel-2017.html?nn=694676

Businessinsider. (2016). Das alles passiert innerhalb einer Minute im Internet. Accessed December 14, 2017, from http://www.businessinsider.de/das-passiert-in-einer-minute-im-internet-2016-11

Dörner, K., & Meffert, J. (2015). McKinsey, article: "Nine Questions to Help You Get Your Digital Transformation Right".

Mustica, S. (2015). Kunden glücklich machen. *Acquisa, 15*(62), 68–69.

Challenges for HR
in the Telecommunications Industry:
Competence Transformation Follows
Market Change

Steffen Roos and Björn Menden

1 Elementary Changes on the Telecommunications Market

Anyone speaking about the future of telecommunications must take care not to forget about the people working in the telecommunications industry. Even today, one of the key questions for carriers revolves around what employees will perform what tasks in the near future, what future skills will be required—and what skills will not be needed.

The innovation and transformation process on the ICT market requires a corresponding, ongoing update of the role and skills portfolio. The changes are complex and simultaneously fast and furious; the required skills are in some cases highly specific, rare, and very difficult to find; many of the details in the shaping of the new roles are often unforeseeable. All of these factors make the shift in roles and skills one of the greatest challenges confronting carriers, and it must be managed actively and mastered.

The drivers of change on the telecommunications market are numerous and varied, and their simultaneous occurrence generates a "storm of transformation" for carriers. We regard the following factors (besides the drivers resulting from the current competitive position of the telcos) to be of the greatest relevance for our deliberations:

- Telecommunications companies are following internationalization strategies as a means of expanding their presence as well as of providing cross-border solutions in contexts such as machine-to-machine and Internet of Things. English skills are becoming increasingly important.
- Customer expectations and customer interfaces are changing; digital channels and social media are growing in significance. But "brick and mortar stores" are

S. Roos (✉) · B. Menden
Detecon International GmbH, Cologne, Germany
e-mail: steffen.roos@detecon.com; bjoern.menden@detecon.com

experiencing their own renaissance by introducing new, fresh concepts. Operators must employ a multichannel approach to develop into genuine omnichannel companies, overcome channel egoism, and address and serve customers coherently at many different touch points.

- The convergence of mobile and fixed ICT continues to advance, making it possible to offer significantly more complex solutions and new services while at the same time exploiting efficiency effects in the long term.
- The fundamental change of the technology in Next Generation Networks will open the door to new services and customer offers in the shift of the operators' business model toward data services, from telephone providers to data transport companies. In addition, completely new, centralized network operations will appear as part of the all-IP conversion of the networks. At the same time, the network standards developed by IT will mean increasing relevance of the software with respect to hardware. Telecommunications is going through a process of wide-area "softwareization".
- Products, services, and entire business models built on data can be modified and adapted very rapidly. Comprehensive analyses permit ad-hoc and real-time responses. Micro services for special and specific target groups can theoretically be realized very simply and quickly; management decisions are backed up more strongly with data material. Telecommunications companies are becoming "big data-driven companies".
- Thanks to the increase in data quantities, touch points, and interactions with customers, trust will become a core element of an operator's brand. Data security and data protection are inseparably integrated into the DNA of a successful telecommunications company.
- The automation rate of processes and repetitive activities will increase substantially in the telecommunications industry. This can lead to a lessening of the burdens on the workforce and free up the resources for creative, innovative activities.
- Last, but not least, the technical opportunities for support of virtual collaboration and digital information management will continue to develop enormously. Modern digital working environments will enable much tighter networking of spatially separated, cross-functional, and subject-oriented teams, cooperation across physical boundaries, and collaboration on projects. Managers will be called upon to lead teams whose members are at different physical locations. Orientation to results and performance culture based on trust will of necessity gradually replace the previous "command and control" leadership style.

The consequences from the changes in the business models, the new working world, and the leaps forward in technology also found in the telecommunications sector (and especially here) are far-reaching.

Table 1 Changes in business and resulting changes in skills requirements

	Past	Future
Business demand	Telecommunications technology and hardware as a basis for the business model	Services as a competitive factor
	Non-specific mass business	Micro-segmented target groups/mass customizing
	Mastery of mass processes	Flexible process and service design
	Securing of basic service	Partner management
	Presence culture and command and control leadership	Customer service as a competitive factor
Skill requirements	Technology affinity with in-depth, detailed specialist knowledge	Customer and service orientation
	Reliable execution	Performance flexibility
		Creativity for high innovation rate
		Willingness to learn, development of personality
		Project management
		Self-determination of daily business

1.1   Paradigm Change

The movement in the direction of all-IP networks and software solutions that more and more are replacing the traditional infrastructure using dedicated telecommunications hardware stands for more than just a change in the technical infrastructure; it is actually representative for a fundamental paradigm change in the telecommunications industry. It is certainly legitimate to ask whether the occupational profile of the telecommunications engineer actually has a viable future and whether it would not be more advisable to seek training in the IT field as the basis for a career on the technical side of the telecommunications business.

An especially important objective must be to transform the basic know-how about voice and data operations as well as switching and routing from hardware-oriented network planning into IT management expertise. Telecommunications engineers must have the ability "to simulate" with software (on standard hardware) the underlying processes of the telecommunications industry and to translate specifications for network hardware step by step into requirements for software companies or for their own developments. Table 1 gives an exemplary overview of needed skills in the future.

What does this now mean for established telecommunications companies, for the incumbents on the saturated markets in Europe, America, or Asia? Obviously dramatic changes of this scope must lead to adaptations in the organization. HR must (within the framework of this transformation) develop a fundamental and solid understanding of the technological changes that are taking place and the impact they will have.

2   New Roles and Patience in the Transformation

If telecommunications companies are to meet the changed demands related to technology, market, and business strategy, they will have to do more than just think about new skills. They will have to go far beyond this and describe and implement new roles. The organization of the future will be much more strongly project-driven, more flexible, networked, and fluid than the line organizations currently in place. Expert roles with specialized and specific know-how will be handling assignments in various fields and different projects. Projects and activities will become increasingly international, foreign language skills and intercultural cooperation will increase in significance. Knowledge of project and program management as well as IT skills will become more essential. Obviously the transformation will not sweep the old world away with a single stroke. There is still a long road ahead until the legacy world in ICT systems, processes, procedural models, and organization has been set aside. Telecommunications companies suffer intensely from the complexity arising from over-adapted ICT systems, overloaded processes, restrictive requirements, rigid line organizations, silo thinking, and departmental egoism, and they have no one to blame for this but themselves. This legacy world will be subjected to increasing efficiency pressures while the new world is still in the parallel construction phase.

But what are the key roles that must be filled as the first priority? Transformation is demanding changes in all positions, but three areas will be highlighted here as representative of all.

2.1   Chief Digital Officer (CDO)

Current discussion consistently points to one key role critical for success that must urgently be filled with a highly competent candidate: the chief digital officer. CDOs will be expected to use their central position to drive the organization's digital transformation and move it forward. But what is the instrument set of CDOs? Where are they anchored? Are they a part of top management, or do they report to the CIOs? What muscle power will be given to CDOs so that they can achieve their goals and overcome resistance?

We are convinced that ideal CDOs must be brought in from outside the organization and must, if at all possible, be a member of top management. Moreover, they need these important traits: they must be committed drivers of digitalization and, in this sense, be drivers of transformation, networkers in their ideas and actions, motivators of the organization, able to convince and carry employees and management alike with them; and they must feel beholden to the rollout of a new, digital operating system for the company.

2.2   Data Managers

As the significance of data in the telecommunications business model continues to grow, the roles related to the management of data will become more relevant as well. Among such roles are the profiles that take care of the collection, analysis, steward-ship, security, and protection of data as well as product developers who generate services and products from data. A carrier is well advised to develop fully and fill the role of chief data officer. Finding candidates on the outside market to fill any of the roles revolving around data management is difficult even today. Major efforts will be required additionally in the areas of training and ongoing training of employees and their further training and qualification in these roles.

2.3   Solution Architects

Telecommunications companies have little to no positive tradition in the manage-ment of software development. The OSS/BSS landscape currently found at almost all carriers is a sad testimonial to over-adaptation, a lack of standardization, inade-quate documentation, and poor interface and data management. In the future, software will play a significantly greater role than it does even today in networks, interaction with partners and OTTs, and at the customer interface. That is why it will be necessary to develop capable solution architects who—with end-to-end responsi-bility—design, coordinate, and implement complex solutions at the interface between the business side, IT, technology, and external service providers. In addi-tion, this role will have the task of defining, rolling out, and continuously evolving documentation and development standards. This role will continue to grow in relevance for business and success in view of the greater impact of software in critical areas.

3   Consequences and Recommendations for HR

As a rule, there are a number of consequences whose management will present a massive, but solvable challenge.

The first consequence is the necessity to transform the skills and capabilities of the workforce. Both technological transformation and the rising pressure for greater efficiency will leave companies no other choice. The second consequence is the achievement of flexibility and agility such as found among OTT players. Carriers are facing the question of how and where they can utilize as well as develop these capabilities. How do you combine agile product development with efficiency-oriented production? What strategy should be pursued in dealing with the OTT players, and what competencies are required to accomplish this?

HR departments and strategists in many companies are struggling with the question of what future competence models in their business will look like—and in some cases, even what the business itself will look like. Although it is possible to discern the development of strategy patterns, many telecommunications companies are still uncertain about the business models they want to follow as they steer their operations into the future.

In this article, we take a look at workforce management as well as employment or transfer units as possible instruments to guide and shape this transformation. Both of these instruments can develop enormous leverage within the framework of a massive transformation. They are not, however, the only instruments, but merely an excerpt from the complete portfolio.

3.1 Workforce Management

As a concept, total workforce management is nothing new. There are challenges and opportunities inherent in this area because of digitalization that have never previously appeared in this form. Generally speaking, total workforce management means the assumption by the HR department of a significantly more analytical approach to personnel planning and management that goes beyond the HR department to encompass the personnel management tasks of each and every executive. Analysis, skills matching, development planning, filling of positions in the line, and the provision of capacities for projects will turn more and more into real-time tasks that can no longer be pressed into the mold of annual planning cycles. At the same time, however, planning and development of the concept portfolio demand a long-term view in harmony with strategy that attempts to predict supply and demand so that this important basis for business can be actively shaped with the appropriate HR instruments.

This gives rise in turn to further demands on subjects such as performance management, skills management, training, and advanced development. New learning methods are necessary so that the changes in skills and capabilities in the workforce can be mapped—including training catalogs that are not based on traditional courses. There will be more advanced development on the job itself, but it will by no means be arbitrary; it will be directed by the appropriate platforms and through the use of technology.

The composition of the workforce will change. A broad range of software solutions in the area of skills management and skills matching will appear, and they will serve as the basis for generating optimal staffing for projects in alignment with profiles—including projections for project success and critical points in various scenarios of personnel assignments and staffing. This is certainly not a new problem. The fact, however, that there will more and more frequently be a mixture of regular workforce and personnel acquired for a specific project indicates that there will be quite a fundamental change in the working world. The issue here concerns highly qualified specialists who are in demand on the market as well as any number of activities that can be handled at lower cost in this way. The setting is always that HR must assure and develop further the availability of certain core skills.

The subject of leadership must also be reconsidered. What do all of these topics mean for the demands on executives in this environment? The various dimensions—some of which contradict one another—must be managed, and this must take place in a situation in which the executives are also struggling with the changes in demands made on them and with a growing sense of uncertainty.

3.2   Employment and Transfer Companies

Employment and transfer companies are vehicles in competence transformation to the extent that deliberations go beyond the simple outsourcing of personnel who are no longer required. The mechanism employed here is the removal from the organization of personnel who do not have the required skills and capabilities. Skills and capabilities that are no longer required in the original added-value chain are eliminated, opening up the space for optimization of the added-value chain analogously to the challenges of the competition that is strongly driven by efficiency.

Two aspects are important in this context: one is that the block of personnel expenses that puts pressure on competitiveness must not be reduced to zero; the other is that a company has a social responsibility to its employees, many of whom have made contributions to the company's success over a period of many years. This social responsibility should be given consideration in each and every case.

Features of an employment and transfer company as a central instrument for the transformation of skills and capabilities:

- Assessment existing skills & capabilities, evaluation development possibilities
- Training and advanced training
- Support of the access to the outside labor market or outplacement
- Creation of an internal labor market
- Supporting and consulting function related to measures effective for employment in the company or the corporate group

Depending on the size of the transfer unit, a decision must be made whether a consulting institution can be set up within the HR department or whether it would be better to view it as a separate unit with its own business purposes.

4   HR Organization as Driver

But who will drive these changes forward in the organization? We are convinced that, when it comes to changes, it is absolutely mandatory—especially for larger telecommunications companies—to take advantage of the HR function as a decisive driver for digital transformation and the new business models. HR will then be able to make use of the appropriate instruments as described above not merely to guide, but actively to steer the comprehensive changes in the requirements for skills and capabilities of the organization and its employees. In this sense, the role of HR will also undergo a transformation and must take its leave of administrative support. Virtualization and digitalization must not be allowed to stop at the door to the HR department—above all because it will be virtually impossible to realize anything in the overall organization if HR keeps itself aloof from the same processes. More than ever before, HR will be required to integrate its services more tightly into the company's business. Only if this is understood, only if the department thinks holistically and focuses on the customer as the focal point will the HR position be able to exercise a decisive influence in a digital world. To achieve this, HR must replace the internal customer with the external customer and see itself as an integral element that works jointly with the "business side" to find solutions that benefit customers.

We have examined only two instruments or fields here that can play a major role in the transformation. They are, however, simply elements in a changing HR portfolio. It is important to have an overall philosophy in the area of competence and, in no small degree, talent management that the HR department must use to meet this massive transformation in requirements head-on. True, the technology aspect is very much at the forefront because of its significance for efficiency and successful business models. The most critical resource, however, is people; they must be capable of implementing technologies in organizations, processes, or business models. Employees who are able to do this are in demand—and not only by telecommunications companies, but also by attractive OTT players, startups, or the automotive industry in the segments Connected Car and Mobility Platforms.

A well-considered talent strategy is consequently the key to the solution for the HR department. This talent strategy should be viewed as more than just the fostering of high potentials; it is a holistic approach that can be broken down to the general skills and workforce management (see Fig. 1).

One must not forget that the HR department is asking the same questions as the rest of the company. Where can we find the skills and capabilities needed to prepare our employees digitally and to ensure that they can meet the new demands? And what is to be done with the skills and capabilities that we no longer need?

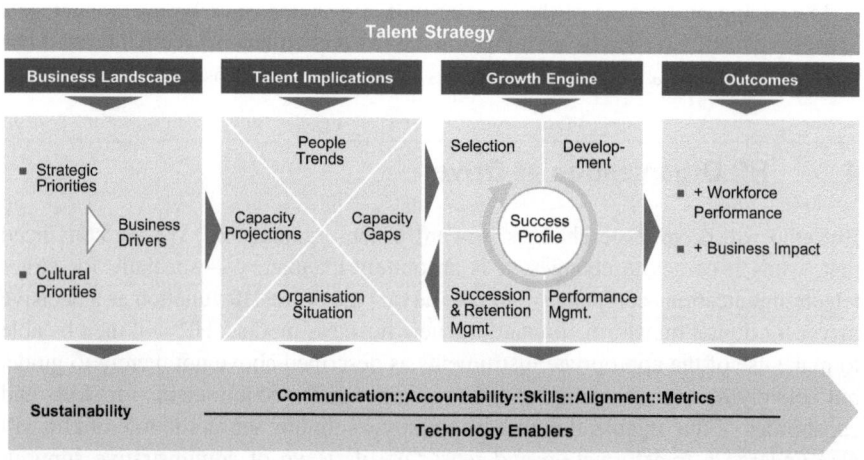

Fig. 1 Talent strategy

Field Report of a CDO

Interview with Mario Pieper, Chief Digital Officer, BSH Hausgeräte GmbH

Mario Pieper

BSH Hausgeräte GmbH is one of the world's leading companies in its industry and the largest manufacturer of household appliances in Europe. Employing a workforce of more than 58,000 worldwide, BSH was able to realize sales of about €13.1 billion in 2016. BSH manufactures the full range of modern household appliances in about 40 factories around the world. Digitalization, the Internet of Things, M2M-based services, the ever tighter interconnection of humans and things will radically change society and business—that is the consensus of most experts. BSH Hausgeräte GmbH is also preparing to take advantage of the opportunities and meet the challenges that will result.

Do you share the view that digitalization will bring about a fundamental transformation of the conditions of competition, in production, and in market approaches for all possible industries and that this will also be true for the manufacturers of so-called "white goods"?

M. Pieper: The bad news is that this appraisal also includes the manufacturers of "white goods," even though we are not the ones on the digital S curve, like the media and travel sectors, who were first affected or have been affected for years. Even the telco industry is undoubtedly ahead of us with respect to digital transformation. I would say that attention today is sharply focused on the automotive and retail sectors. We take these changes very seriously. We see ourselves as following close behind.

The good news for us is this: the positive course of our business allows us to approach digitalization from the side of the customer interfaces. We do not have to ask ourselves every day what efficiency advantages digitalization will create for us. On the contrary, our focus is on growth. For us, the question regarding the

M. Pieper (✉)
BSH Hausgeräte GmbH, Munich, Germany
e-mail: Mario.Pieper@bshg.com

© Springer International Publishing AG, part of Springer Nature 2019   389
P. Krüssel (ed.), *Future Telco*, Management for Professionals,
https://doi.org/10.1007/978-3-319-77724-5_35

increase in effectiveness of addressing customers plays a major role, i.e., the development and driving of new business models, the improvement of customer service, and the interconnection of our products.

Speaking concretely for BSH, what importance does BSH place on digitalization? Where do you see the primary opportunities and threats for BSH Group and its employees?

M. Pieper: Digital transformation is of high strategic importance for BSH and is one of the key strategic fields of action. In organizational terms, it is situated in my CDO role. When it comes to digitalization, we believe the opportunities are greater than the risks. Our efforts are aimed at aligning the development of our products and our distribution and service channels closely with digital opportunities. We want to position ourselves in greater proximity to our customers and understand the customer journey better.

Despite all our optimism, there are definitely risks. Specific capabilities are required to interconnect products. And in this case, the software engineers rather than the hardware engineers are in demand. This is not a skill that is broadly available within our company. In other words: we have skill gaps. This is no less the case for the marketing departments. In future, customer relationship management will become even more important. One example: As it stands today, we do not have established customer relationships, by which I mean we have very few data about our customers. But perhaps we will need precisely these data about the customers in tandem with analytical skills. If we do not succeed in closing this gap in good time, we could find ourselves with serious problems on our hands.

What is BSH doing to face these developments? How are the implications of digitalization taken into account in the innovation and product development processes, the customer contact points, the production process, employee development, organizational structures, or perhaps in concrete digitalization projects or programs as well?

M. Pieper: We have just spoken about digitalization as one of the strategic fields of action and, what is more, as being represented in its own organizational unit. Because of our business model, this unit takes the form of a center of excellence. In this sense, it seeks to function as an incubator for digital topics in the organizational structure, including its employees trained specifically for this purpose. To this extent, the digitalization process is one that begins right in the company and then leads to a type of standard organization. The digital topics also find expression in a digital program that is consistently managed and produces concrete road maps and IT prioritization. In this way, all digital topics are prioritized via a digital coffer. All digital topics are carried out by specific project managers, and the bottom-line target is for digital transformation to be carried out as a process in the company. The question of how I approach this is related to the specific corporate structure. At BSH, digitalization has its own governance in the company. It is for the most part separate from standard governance, but has key points of interaction, such as in budget planning. The digitalization governance operates at a different speed from most of the other governance processes.

What concrete examples of "digitalized" products are there at BSH? And has there already been verifiable success, either in terms of additional revenues or gains in efficiency?

M. Pieper: The best example is our core product: the increasingly tightly connected household appliance that we are putting on the market. There is software that, for instance, makes remote control possible along with the hardware. This remote control is intended to help customers to access new digital services. One example from everyday life: your refrigerator automatically takes inventory of its contents and places orders as needed (milk, for instance) without any further action on your part. Or imagine a dishwasher that autonomously sends a request for dishwasher tablets. Shopping by online delivery is a good example of digitalized services. The starting point is an investment in a software platform (at BSH: HomeConnect). Over time, this will also generate new income in new added-value categories and gains in efficiency through my development of standardized household appliances that communicate with one another, for example. Standardization in the software will secure gains in efficiency in production and will enhance the time to market. Clearly discernible benefits are already noticeable, as are the first signs of additional sales. We are aware, however, that household appliances are longer-term consumer goods.

Is digitalization changing the competitive environment for BSH? Are there new players entering the market who previously played no role and who are now seizing parts of the traditional added value as well as creating new added-value elements (e.g., in the service sector) where BSH must first develop its own position? Do you see here any examples of best practices from highly promising new players/products or services that will make a new, special customer experience in the environment of the household appliance sector possible?

M. Pieper: If we look at the global picture, we can already see initial trends in this direction. China, for instance, is a country where digitalization is organized quite centrally. The government strongly urges players to collaborate with one another. This leads to the fast establishment of ecosystems and new revenue streams. This development gives rise to a number of questions. Who generates what value and who heads the ecosystem? We clearly see here that major players such as Alibaba and Tencent with WeChat appear to be disruptive for the market for interconnected devices, including platforms. But other trends are also discernible. What is the connecting point for a customer? The recipe? The shopping transaction? The chances to obtain a share of future growth are good for us as household appliance manufacturers as well because we know better than anyone else what individual customers have in their refrigerators. And, subject to the consent of the customers, we will create contact points between them and the suppliers. The issue here is the linking of new contact chains with the chance to establish new revenue streams.

How is BSH positioning itself in this dynamic competitive environment? Are you taking the path of cooperation or of your own innovation with a corresponding added-value depth?

M. Pieper: The answer is this: both. One constant in the digitalization process is speed. Our attention is certainly on cooperation so that we can quickly come up with marketable solutions. Our network platform HomeConnect is a good example. Since the market launch, we have concluded cooperation agreements with 30 international partners, including Amazon, Kochhaus, and HelloFresh, who along with us generate added value for customers. In this sense, we certainly have a major focus in the area of cooperation. We are also investing in solutions in the fields of robotics and cognitive solutions because this is the future. Both our own innovation and cooperation are represented in a mixed form because we cannot generate all of the components for new solutions ourselves.

Let's talk about your role as CDO, a position you have held for 3 years. What prompted BSH to create this position in the first place, and how is this role brought to life? What goals, tasks, and competencies have been assigned to it?

M. Pieper: From my perspective, a fundamental element is the strategic realization that our industry will no more be able to avoid the impact of digitalization than others. The issue here is a response in good time. We must actively drive this subject from a position of strength. BSH took a bold step in this direction in 2014. We established a separate organizational unit that today employs more than 150 people and serves as an in-house incubator for digitalization solutions within the company. My task encompasses the digitalization of BSH and the transformation that goes along with this process. Every solution that is implemented brings new capabilities to the company. The exploitation of these capabilities is a part of the transformation and change they bring about. The role of the CDO is given clear prominence, reporting directly to the CEO. My tasks have a global, cross-departmental scope, and I have broad authority for deciding what is to be done because the capabilities of digitalization in the company have not yet put down very deep roots.

Looking back, what do you see as the key success factors in the performance of this role? What general conditions, what preconditions must be created, in general and in terms of the specific situation at BSH?

M. Pieper: The most important prerequisite is the awareness that a role of this nature is necessary and that it is simultaneously a part of the CEO agenda. The second point is that it must be clear to us that projects will not be able to function without digital heroes and specialists. The third consideration: digital subjects should be driven in collaboration with the people who actually understand them.

Companies do not usually have so many specialists available to them. As a company, I must ask myself the question: How do I integrate the right people into the discussion and work process? The most crucial success factor is my finding a way to plant digital capabilities in the company as quickly as possible. Change has many different facets. For instance, digitally talented people must also be found on the employer branding side. Our factories are rarely located in the digital centers, the places where the talented people from the software development sector can be found. But how do I draw these talented people to my locations? How do the new and old employees work together in the various fields of specialization? At BSH, we discussed intensely our vision and strategy right at the beginning. You must not back off during the systematic implementation; it is essential to create the opportunities and the fertile ground that digitalization must have to grow.

Large telecommunications companies are confronted by two challenges. On the one hand, they must realize digital transformation for their own business model and core business while, on the other hand, they must also act as a platform or enabler of digital transformation for other industries. What are your expectations with respect to telecommunications companies? What concrete services or solutions do you expect from telecommunications companies?

M. Pieper: The most essential condition of all is the availability of a full-coverage broadband infrastructure, including Wi-Fi. Our architecture for networking relies on a plug and play Wi-Fi solution. We can control our appliances remotely via the Internet. I hope that the infrastructure, which has not been adequately established everywhere, will systematically improve. Moreover, we do not see ourselves as the parties who serve the full integration of the smart home because we do not produce either consumer electronics or security systems. We are therefore delighted to find partners such as Qivicon in the case of Telekom who offer solutions for the complete networking of the home and who can integrate our solution into the network. Additional services such as hosting and other platform services can be additive products that we procure. We would like to have the telco industry as an infrastructure partner who opens the door to business for us and acts as a partner in the house, someone we can join to provide innovative solutions to customers.

Working in the Digital Age: Not an Easy but a Thrilling One for Organizations, Leaders and Employees

Marc Wagner, Frank Heil, Laura Hellweg, and Dennis Schmedt

1   Introduction

We live in exponential times—readers come across this much-quoted statement in the relevant trade press on an almost daily basis. The digital revolution is creating almost inconceivable volumes of data, with computing power reaching dizzying heights, so much so that computers will outperform the human brain by 2045, according to Ray Kurzweil. Self-learning, reproducing artificial intelligence and robots are just some of the fruits this development is bearing. And all of this raises the question of how work will be structured in future.

Will there be a humanization of work in which we are finally liberated from activities that do not create value or give meaning to life? Will the rapid elimination of existing job profiles be offset by entirely new activities which are still beyond the realms of imagination? Or do we face an age in which we are thrown into major social conflict as the middle class disintegrates, resulting in crisis and unrest? These are just a few of the questions that emerge at a macro level in relation to the future structure of work.

For enterprises these developments mean that the notion of work has to be re-examined in its entirety. Market entry barriers are almost nonexistent for digital goods. International distribution takes no time at all—opening up virtually endless opportunities. Market foreclosure is possible only at a political level (see China or USA). Many large enterprises and conglomerates in Germany have been overwhelmed by this development. While in the past millennium, they had systematically focused on creating large and efficient organizations which deliver

M. Wagner
Detecon International GmbH, Cologne, Germany
e-mail: marc.wagner@detecon.com

F. Heil · L. Hellweg (✉) · D. Schmedt
Deutsche Telekom AG, Bonn, Germany
e-mail: heilf@telekom.de; laura.hellweg@telekom.de; dennis.schmedt@telekom.de

© Springer International Publishing AG, part of Springer Nature 2019
P. Krüssel (ed.), *Future Telco*, Management for Professionals,
https://doi.org/10.1007/978-3-319-77724-5_36

economies of scale through their sheer size and hence dominated markets, "big is beautiful" no longer applies in the digital age. It rather seems they are weighed down by organizational ballast and "belly fat," which restricts flexibility and agility, and prevents them from responding to market changes. Taleb in particular in his book "Antifragility" set out in detail that classic structures tailored to efficiency make enterprises susceptible to interferences created by all types of change. The systematic eradication of each and every process redundancy suddenly constitutes an absolute innovation killer.

This does definitely not mean that successful companies should in future live in exuberance and waste their resources. Quite the opposite, in fact. They rather need to create structures which fine-tune efficiency and incrementally enhance existing business (exploit), and at the same time enable the company to deliver disruptive innovations which relentlessly challenge existing business models (explore). This so-called ambidexterity is regarded in many quarters as the "holy grail" for an organization's ability to survive in the digital age. While a very logical concept in theory, its specific organizational implementation poses major challenges for enterprises. And this applies in particular to the element designed to provide employees and managers with sufficient spare time to tap into their personal creativity potential (source: WEF) and develop new, innovative ideas. An organization that can attract top talents and respond optimally to their particular period of life. Yet how is work structured exactly in this kind of organization? What are the levers for greater flexibility, agility and innovativeness?

2 New Work … The Holy Grail for Agile Organizations?

And it is at this point that New Work comes into play. First there's the question of what actually lies behind New Work. This is less about an abstract definition than about what impact the introduction of New Work may have and which sub-aspects can be used to structure the "meta topic" New Work.

New Work ultimately provides an answer to how work can be structured meaningfully in the digital age. Originally, New Work was introduced by Frithjof Bergmann in response to similar radical change when the first computers came on the scene in the first half of the twentieth century. The predictions back then were similar to what they are today. The jobs of many employees are becoming obsolete. That calls for a new understanding of work and the distribution of work. For Bergmann, the societal effects, a personal sense of meaning and the search for what "is really, really important" to the particular individual were crucial. Technological development therefore is an opportunity to find time for the really important topics and to live a balanced life. We have already touched on these societal discussions in the past and they are currently becoming increasingly relevant, in part through discussions surrounding universal basic income and alternative social systems. However, in the following, we would like to focus on the organizational implications and the definition and structuring of the concept of New Work.

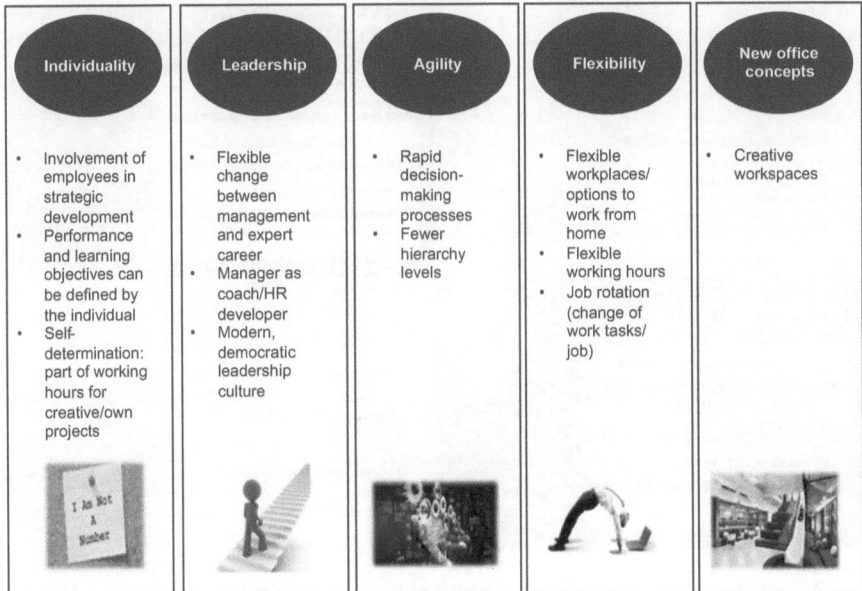

Fig. 1 New Work topics

Based on a Management Association analysis of the success factors for businesses in the twenty-first century (see Fig. 1), we have divided the concept of New Work into the following topics:

In a subsequent analysis (see Fig. 2) we have also found that the tools available have a significant impact on key performance indicators, such as revenue development, employer attractiveness, employee satisfaction and employee turnover (for a complete/detailed presentation cf. the study: "Ich war noch niemals in New Work" ["I've never been to New Work"]).

In the following chapters, we explore some of the wide facets of New Work (digital mindset, flexible working, agile organization and working differently) and illustrate these in the context of Deutsche Telekom AG.

3   Digital Mindset

It is generally accepted that the telco industry is currently undergoing a major process of transformation. In very general terms, telcos are engaged in a transition from being primarily operators of physical infrastructure and networks designed for the effective delivery of analogue voice and packet data services, to being providers of cloud-based (distributed software, IT and virtualized) infrastructure, platforms and digital services.

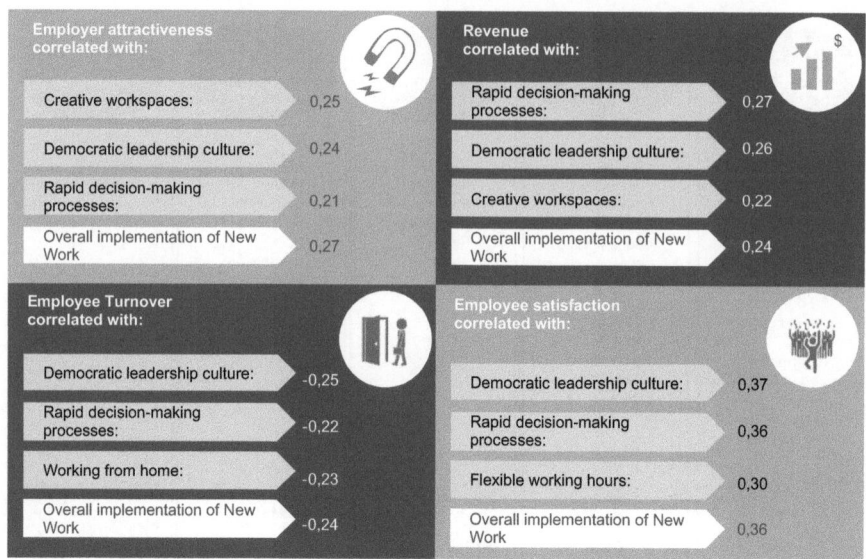

Fig. 2 New Work—Correlation between tools and key performance indicators. Source: Detecon study on the future of the world of work and work organizations (New Work). Figures = Level of implementation $p < 0.05$ (error probability $< 5\%$)

This "digital transformation" refers to how companies are reinventing themselves for the digital world. Businesses have always had to disrupt themselves, but now, to revolutionize their industries, they must infuse their business with technology—essentially becoming technology companies themselves.

The resulting change demand occurs in numerous aspects of transformation: Technology, processes, business models, organizations and culture. The following chart underlines the meaning and impact of culture and skills as the greatest barriers to transformation.

Changing an entire corporate culture will not be done overnight and requires multiple angles and a long-scale program. New Work hence is a major contributor in bringing a new way of working and thus a new culture of collaboration into an organization. In all this, the core and focal point in transformation is the most personal and individual one—the digital mindset.

The following Fig. 3 illustrates one approach to nourish a "more" digital mindset on three levels: basic, intermediate and advanced.

Basic A solid fundament basis for a digital understanding is the knowledge and skill how to use and apply the most common digital tools and instruments. In a corporate environment, it is essential for both managers and employees to communicate and collaborate on a common base without barriers, either of mental or technical kind. What does that mean? Example: Looking at most recently available communication

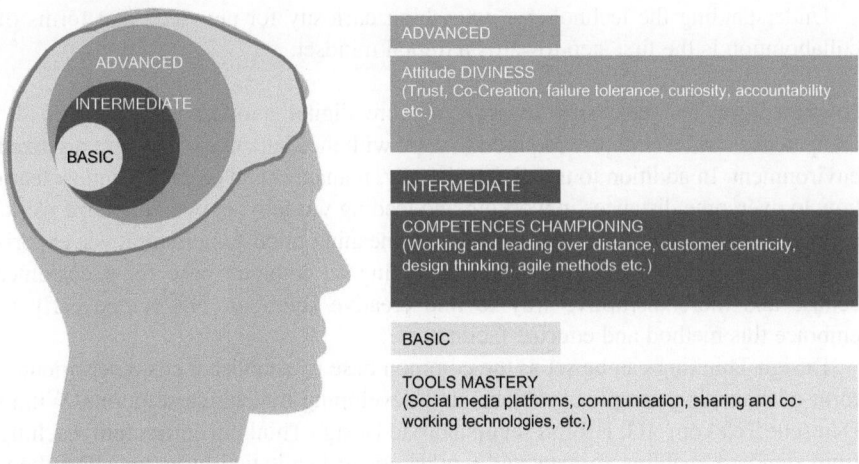

Fig. 3 Digital mindset

and co-work platforms (i.e. Slack) shows, that a new way of collaboration (building channels instead of teams, chat instead of emails, etc.) arises and could become a true alternative to the classical forms of communication and collaboration we know (emails, telephone calls, meetings). Another example: Within Deutsche Telekom's internal social media platform the most viewed and emotional discussed threat was about the internal approval of a globally leading messenger platform. The vital discussion (550 comments, 380k views) perfectly shows the great interest but also concerns in availability, usage and purpose of new forms of communication.

For individuals, the consequence is that in addition to the question of how they are supposed to achieve their objectives, another question about the right technology/tool they should apply for this follows.

This is a call for both, the HR and the IT department, to give orientation, raise the awareness and provide enabling formats and trainings to raise the digital literacy of the entire workforce.

Rooted in the corporate program digital@work within Deutsche Telekom AG, HR & IT investigated the basic user needs first, to understand the real demand on digital tools, collaboration and communication. In parallel, IT started to interlink existing communication systems to one seamless collaboration platform as common base for collaboration. Afterwards HR set up a virtual guide (Digital Guide) to find the right tools and instruments for the required use-cases. For senior managers, HR also set up the reverse mentoring program, with the purpose that digital-native employees help out managers to cope with digital tools, social media platforms and new technologies.

Understanding the technology, nourishing curiosity for new tech and forms of collaboration is the first step towards a digital mindset.

Intermediate The next step to grow a more digital mindset is to develop the competences and techniques required to cope with the challenges of a more digitized environment. In addition to the digital literacy, managers and employees must learn how to overcome distances in working and leading virtually. How to organize work, solve complex problems and understand the meaning of customer/employee experience. Taking the example of Design Thinking as common base for a customer centric and more disruptive way to find creative solutions, HR started early to embrace this method and educate facilitators.

Design Thinking can be set as the common base to establish a customer oriented form of managing complex problems and developing innovative solutions. Within Deutsche Telekom AG, HR has set up a broad Design Thinking curriculum reaching from standard enabling formats and workshops to facilitation programs. The theoretical knowledge is put into practice in several Design Thinking sprints with close connect to strategic business challenges.

Digital leadership: One of the most important attributes of a digital leader is a flexible management style that is adaptable in specific situations and the area of conflict between stability and innovation. The company must focus on its core business. Therefore, a traditional leader should try to avoid significant risks. In contrast to that, it is also important to explore new business to keep succeeding commercially and to form the digital future actively. Managing this ambidextrous environment is only possible if leaders become more innovative and willing to change. Consequently, the leader of the future has to fulfill a hybrid function to balance and support both kind of characteristics.

Advanced Competences and basic digital literacy are preconditions to develop a more digital mindset but the basic attitude towards change, transformation and digitization is crucial (see Fig. 4). It is about the attitude towards change and the ability to apply new ways of thinking, acting and also leading.

What Is Attitude?
We have an attitude towards people, objects and situations. Our attitude influences our behavior, thus how we are thinking, feeling and acting. Our attitude is defined by our identity—motives, values and preferences—and reflects our inner mindset. Attitude is the way how we think and feel about something. It is our personal perception of life and also defines which things we perceive, how we perceive them and how we interpret experiences

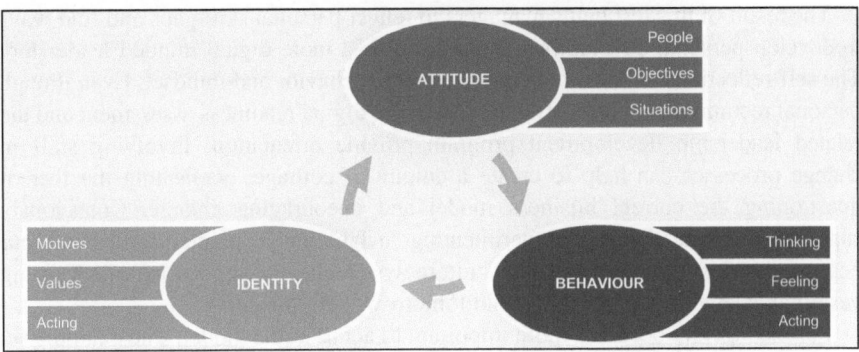

Fig. 4 Attitude

As opposed to the previous two fields of digital minds (tool mastery and competences), attitude can be hardly taught or imposed, copied and pasted. A company cannot impose the right attitude, but it can open a forum and allow a broader involvement on this concern. As one example within Deutsche Telekom AG, the corporate happiness program was piloted to begin diving into the concept of the individual pursuit of happiness and its impact on personal well-being and success. A very simple approach was tested: A day without concerns would positively influence the way we interact with each other and collaborate more effectively.

Another example is the individual reflection of one's personal digital abilities. In close collaboration with Viadrina University, Frankfurt-Oder, Deutsche Telekom's HR department developed a digital readiness assessment to test managers on digital readiness and attitude (see Fig. 5).

Fig. 5 Digital readiness assessment

The result of this test helps managers to reflect personal skill gaps and find ways to develop personal development areas towards a more digital minded leadership. The self-reflection also helps to calibrate skills, behavior and mindset. Even though personal recommendations are not included, the digital readiness assessment and the related leadership development program provide orientation. Involving staff in change processes can help to create a culture of courage, eradicating the fear of questioning the current business model and encouraging change. Consciously supporting critical thinkers, experimenting in labs and appreciating error as our best teacher, as a productive error culture systematically applies learnings, using yesterday's small errors to help avoid tomorrow's big mistakes.

For managers it becomes more important to act as a role model. The example set by leaders is mission-critical.

In Summary, one can see that a digital mindset comprises more than just technological skills. But it is much more than just the extract which we have discussed in this chapter. Becoming a digital minded laborer is more than rethinking your attitude, considering the changes happening around and always staying curios and attentive to the tremendous changes that happen around.

For a corporate culture starting to implement the New Work, it is essential that you gather individuals around you, who are conscious about their digital mind and know that they need to unlearn and remain curious and hungry for the unknown.

4   Flexible Working

As illustrated in the earlier sections, agility and flexibility are key pillars of a future-proof organization in the digital age and cornerstones of New Work concepts. A flexible work structure—in terms of time, location and working practices—can be described easily from a conceptual and theoretical perspective. Flexibilization, however, poses major challenges for large enterprises and organizations in particular as it represents a significant cultural shift for many of them. This is especially the case when existing management structures are disbanded and status symbols (such as private office, title, parking space, affiliation to an exclusive circle) appear to be threatened. One tried-and-trusted concept at the heart of various (successful) New Work rollouts is activity-based working or flexible working (see Fig. 6).

The flexible working approach focuses on the activities performed in the business (rather than status-related or hierarchical considerations). A comprehensive analysis of employees' activities is normally used to reduce complexity and set up various activity clusters (e.g. creative, communicative, concentrated, relaxed) or derive activities through roles and use cases (see Fig. 7).

This gives rise to specific requirements regarding the working environment ("places"), the relationship between employees and managers as well as the organizational structure ("people") and supporting technology ("tools"). The goal is to create a work structure that is optimal for the particular activity clusters or roles of employees and managers. The flexible working approach also involves maximum flexibility regarding the choice of working hours and the place of work

The task determines the environment and type of the employee's work
From "presence and control culture" to "results culture."
Is the key – for each activity there is an optimum working environment.
The working environment can be tailored flexibly to these requirements.
The result counts
Maximum flexibility
Task of hierarchy
Control is good, trust is better
Activity Based Working
Flexible Working
At its heart is the desire to create an optimum working environment for the particular activity of employees (type, location, nature).
The hierarchical status and associated privileges take a back seat.
Trust and mutual respect are the basic pillars of a modern working and leadership culture

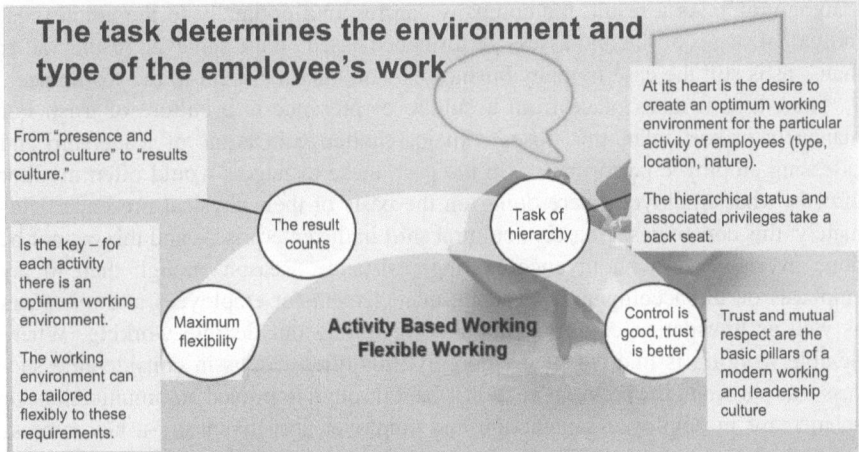

Fig. 6 Activity-based working

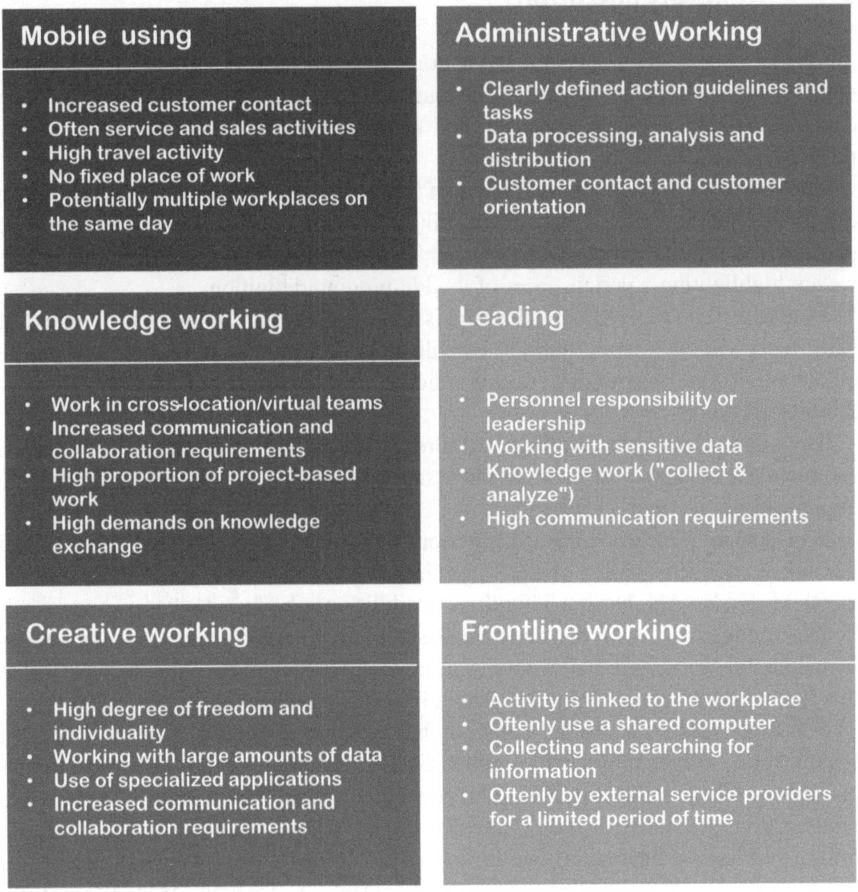

Mobile using
• Increased customer contact
• Often service and sales activities
• High travel activity
• No fixed place of work
• Potentially multiple workplaces on the same day
Administrative Working
• Clearly defined action guidelines and tasks
• Data processing, analysis and distribution
• Customer contact and customer orientation
Knowledge working
• Work in cross-location/virtual teams
• Increased communication and collaboration requirements
• High proportion of project-based work
• High demands on knowledge exchange
Leading
• Personnel responsibility or leadership
• Working with sensitive data
• Knowledge work ("collect & analyze")
• High communication requirements
Creative working
• High degree of freedom and individuality
• Working with large amounts of data
• Use of specialized applications
• Increased communication and collaboration requirements
Frontline working
• Activity is linked to the workplace
• Oftenly use a shared computer
• Collecting and searching for information
• Oftenly by external service providers for a limited period of time

Fig. 7 Employee workflow

("autonomy"). As a result, the company can respond optimally to the employees' individual stages of life and assess performance based on the achieved results rather than—as is still the case in many businesses—on the time spent in the workplace.

Thus, a shift takes place from a culture of presence to a culture of trust. For managers in particular, this poses a major challenge in terms of leadership and assessing employee performance. In the past, these managers would often monitor directly what employees were doing on the basis of their physical presence. Ultimately, this constitutes a massive cultural shift in the enterprise—and this cannot be done overnight or without encountering resistance. Reason enough then to lay emphasis on the accompanying change management for employees and managers as well as transparent communication when rolling out flexible working. Where flexible working is implemented smoothly, this often results in considerable success—as shown in the previous section—be it through improved communications or an increase in employee satisfaction and employer attractiveness—a key success factor for business competitiveness in the digital age.

5   Agile Organization

"Will that work?" they ask in a DAX company like Deutsche Telekom AG. And other questions follow: "What is the purpose of an agile organization?" "What distinguishes this kind of organizational form?" "What does the notion of agility mean in the context of organization?"

The following section answers a few of these questions, in particular through own experience and knowledge sharing with other businesses. Nonetheless, it should be stressed that "agility" in particular is used in many different ways, and in this article we are highlighting a certain facet of a wide-ranging definition.

Deutsche Telekom AG considers the issue of agile organization in the context of New Work via two distinct pillars: Establishing the framing conditions for flexible organizations and promoting cross-functional collaboration supported by digitally thinking managers.

The issue of manager enabling was already touched upon in a previous chapter. As such, this chapter focuses on the issue of framework conditions for flexible organizations.

At Deutsche Telekom AG a distinction is drawn between the terms agility and flexibility. The term agility is used in conjunction with working methods, such as SCRUM, KANBAN, Design Thinking; flexibility can be found wherever organizational structures are involved. The business not only sees more and more advantages in making the organizational structures more flexible; it even becomes very clear that this type of organization is indispensable in certain business segments.

For this reason, Deutsche Telekom has defined a concept for the "flexible organization," which applies throughout the Group and is based on various internal and external benchmark findings. The concept's overarching assumption: Flexible organizations are no silver bullet. Flexible organizations can, however, be used as a bridge that helps a business move towards a new type of work with flat hierarchies and self-organized teams. Consequently, we at Deutsche Telekom AG look carefully at the situation before implementing flexible organizational structures: What influencing factors affect the particular unit and does a changed organizational structure help deliver the desired objectives?

Basically, Deutsche Telekom AG believes the following framework provides a meaningful structure: 70% of the units, the stable core, are organized along classic pyramid lines, 20% of all units are in an agile environment and are generally organized in hybrid structures that enable them to work flexibly as the situation dictates. The innovation areas make up 10% of the organization and are structured purely alternatively, e.g., in the form of Holacracy (see Fig. 8).

Benchmarking underlines the diversity of the flexibilization options, but also the need for flexible organizational structures.

The options at Deutsche Telekom AG range from the classic pyramid structure including line and matrix organization with a focus on efficiency and processes (e.g., service centers), to hybrid structures such as project organizations (e.g., product management units), to alternative organizations based on Holacracy or swarm intelligence, for which there are still no officially implemented examples. The pros and cons of the different organizational forms are obvious—on the one hand, the classic pyramid structure provides planning reliability, but on the other hand, it is fairly rigid. The hybrid structure allows greater complexity, but complicates, among other things, interaction with interfaces.

Adaptability is a key notion when it comes to alternative organizational structures, however to the detriment of centralized control (see Fig. 9).

During the implementation of flexible organizational structures, we have quickly discovered that the concept must be as adaptable as the customer requirements themselves. Nonetheless, the concept includes cornerstones which must be applied whenever organizational structures are implemented:

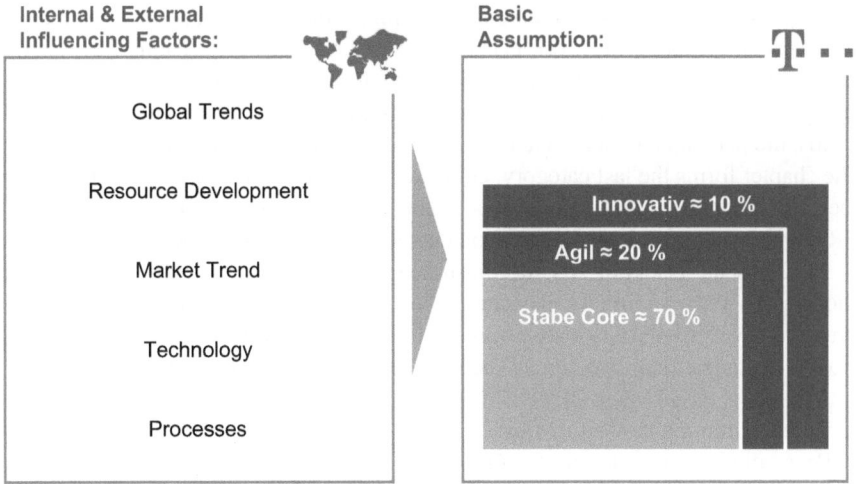

Fig. 8 Share of company units accounted for by different forms of organization

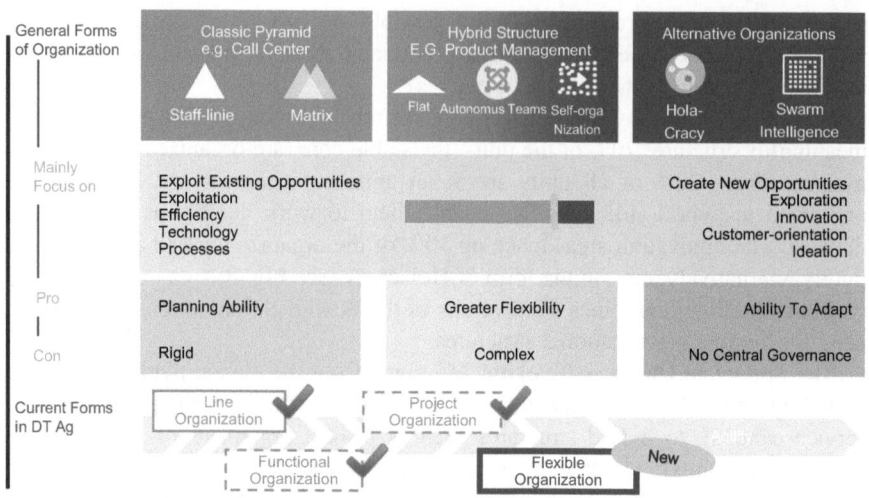

Fig. 9 Range of organizational structures

By default a line organization is preferred. A flexible organization can always be rolled out if the focus is on continuous project staffing, the unit does not work operatively, and no division is running the risk of being split or reorganized. The flexible organization is then implemented as part of a reorganization project (see Fig. 10).

One possible implementation of the flexible organization involves working in a structure based on tribes, squads and chapters. Tribes are comparable with today's business units, which integrate different thematically connected squads and bring these together under a single roof in each case. The head of a tribe coordinates the various squads. Squads are, in turn, comparable with business teams. Normally these teams work autonomously, on an interdisciplinary basis, have nine team members, and are able to define the work packages and take business decisions themselves. Each squad has end-to-end responsibility for the particular business requirement. The product owner prioritizes a squad's activities, controls the backlog and the to-dos, the person is not a classic manager and therefore does not give orders either. The chapter forms the last category, equivalent to a group of experts who share their specialist know-how and best practices. The chapter leader assumes the task of an HR developer, which lays the focus on developing its chapter's skills (see Fig. 11).

At Deutsche Telekom AG the illustrated concept for the flexible organization allows us to reach an initial milestone on the way to forward-looking organizational structures. Nevertheless, we are aware of the challenges associated with rolling out flexible organizational structures. The challenges can be split into two categories:

1. Roles within the new structures
2. Possibility to integrate into the Group structure

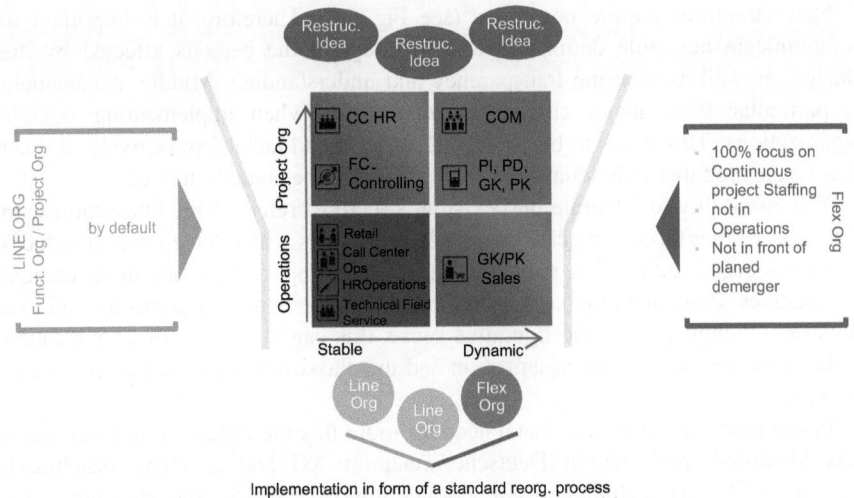

Implementation in form of a standard reorg. process
existing general frameworks for reorganizations apply[2]

[1] First phase of implementation in Germany → International rollout t.b. followed after successful start in Germany
[2] e.g.: job hierarchy, grading, spans & layers, collective agreements (adaptation possible → business case necessary)

Fig. 10 Criteria for a flexible organization

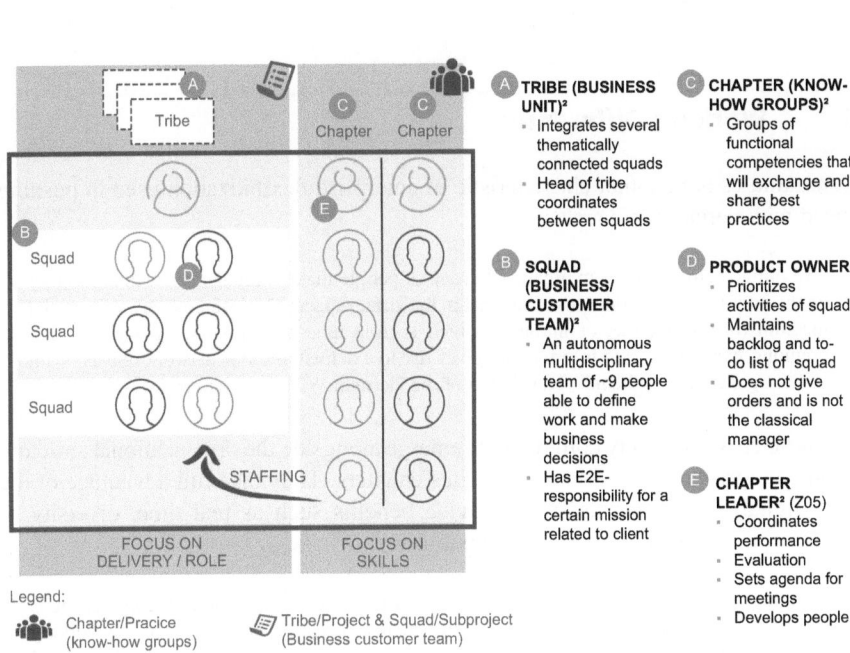

A TRIBE (BUSINESS UNIT)[2]
- Integrates several thematically connected squads
- Head of tribe coordinates between squads

B SQUAD (BUSINESS/ CUSTOMER TEAM)[2]
- An autonomous multidisciplinary team of ~9 people able to define work and make business decisions
- Has E2E-responsibility for a certain mission related to client

C CHAPTER (KNOW-HOW GROUPS)[2]
- Groups of functional competencies that will exchange and share best practices

D PRODUCT OWNER
- Prioritizes activities of squad
- Maintains backlog and to-do list of squad
- Does not give orders and is not the classical manager

E CHAPTER LEADER[2] (Z05)
- Coordinates performance
- Evaluation
- Sets agenda for meetings
- Develops people

Legend:

Chapter/Pracice (know-how groups)

Tribe/Project & Squad/Subproject (Business customer team)

[1] Source: McKinsey
[2] Spans & layers, average of 1:8 per organizational unit according to S&L rule set for MCF

Fig. 11 Structures and roles within the flexible organization

New structures require new roles (see Fig. 11). Therefore, it is important to communicate new role definitions and to integrate the persons affected by the change, as well as ensuring transparency and understanding. Middle management in particular faces major changes in their roles when implementing flexible organizations. This needs to be addressed early on and tackled proactively. It must also be ensured that individuals involved are or will be suitably trained.

The possibility to integrate into existing structures refers to the interaction with the adjacent interfaces. Especially organizational units which have many interfaces with units organized in a line structure must ensure proper integration in the context of processes. Particular emphasis should be paid to the handover points as well as a consistent "language." In the figurative sense, this can also mean that "translation tools" between the flexible organization and the classically organized units have to be built.

In summary, it can be said that a need for more flexible organizational structures was identified, both within Deutsche Telekom AG and at many benchmark enterprises. An overarching concept and a definition of which units are suitable for flexibilization help deliver a structured rollout. However, this does not automatically eradicate the challenges associated with flexible organizational structures. These challenges must be analyzed closely and will need to be resolved in the long term if, as a company, we aim to successfully combine various organizational structure models under a single roof.

6   Working Differently

Communities are another characteristic of enterprise flexibilization used in pushing ahead with certain topic areas.

> "An online community is an organized group of people that communicates with each other on the internet and, in some cases, interact in the virtual space. The technical basis of a social medium (on the internet or intranet), which is used as a platform for mutually sharing opinions, experiences and information, gives rise to a definable social network of users with content generated by them." (Translation of the German Wikipedia article, 2017)

These cross-company communities emerge alongside the organizational structure of an enterprise described in the previous chapters. They take full advantage of the new digital media and, as such, provide benefits such as real time, diversity of opinions, individuality, agility and flexibility.

> "Employee networks are the organizational form of the future—they outlast reorganizations and form the backbone of the organization. Employees with a high degree of networking become important internal influencers within these thematic networks. Internal communications must structure the interaction with employees together with these influencers to remain relevant and credible." (W. Ebner, Head of Social Media Business Program, Telekom Deutschland, 2017)

Most communities are in principle organized on a democratic basis. Communities must be set up, maintained and supported, with members also asked to play their part. Depending on the target group, the functions are agreed and tailored to the users' interests. In this respect, feedback, inquiries and ideas from users are encouraged since such interaction helps increase attractiveness and acceptance. Communities develop successfully above all when their driving force is not an enterprise's marketing idea, but they grow organically from the community's wishes.

Here and there an increasing trend toward hierarchization and fixed institutions can be seen. In the ideal scenario the community imposes its own rules. Even jurisdiction-style, parliamentary or police-style institutions were introduced—mostly at the users' request. In this respect, a development toward rigid rules and "laws" is also discernible, or at least the desire for such a development. Legal concepts such as "inadmissible," "unlawful" or even "defendant" are increasingly used in the discussions.

At Deutsche Telekom AG there are a host of communities which cannot all be mentioned here. The YAM ("you and me" social intranet) has existed for several years and has grown into a large community over this period: More than 100,000 registered users that interact in over 20,000 groups. Even if this number includes a few groups under the heading "I'll set up my own test group," many communities have already developed that foster a lively exchange of ideas.

Committed colleagues maintain and provide support for these communities. In doing so, they often face limitations, particularly with regards to HR resources. In this area too, the successful implementation of business objectives depends on the magic triangle of time, resources and quality.

A core idea of the enterprise, namely being virtual, agile, collaborating across units and internationally within the Group, becomes reality in these YAM communities.

One Example: A Community—#TheGuides The Guides are convinced that they can contribute toward the future-oriented development of Deutsche Telekom's company culture and working practices using the new dynamic options offered by YAM. They actively promote an exchange of knowledge and information to overcome the challenges of the future. The basic principles of this community serve as an example; they aim to support the company's employees to use the YAM for their aims and purposes.

1. **Being a guide is a frame of mind—the willingness for personal development**
   As digital natives, the Guides are fired up about the idea of working differently. They are at home in the future work world.
2. **The activity of a Guide is voluntary and takes place within the framework of their personal possibilities**
   There is no set task description and also no skill set, but the Guides convey a consistent, up-to-date knowledge base by using a "toolkit."

3. **The Guides are there for each other**

They support each other and learn from each other. The Guides utilize the opportunities provided by the company for networking and knowledge transfer (in coherence with the YAM) to improve their skills on an ongoing basis.

4. **The Guides enthuse others by embracing working differently**

The guides motivate interested colleagues for working differently by highlighting the advantages and simplification potential compared with the familiar working models. In this respect, open feedback from the Guides is seen as an important component in extending skills and knowledge.

5. **The Guides share knowledge and provide support to promote working differently**

The Guides share their superior knowledge on working differently with all colleagues who are not yet familiar with Future Work working practices. Irrespective of function or unit affiliation, the Guides help any colleague. The Guides thus accelerate knowledge transfer in the company.

"As "digital" grows," analog "becomes all the more important. Through increasingly digitized communications and distributed digital working in networks, analog formats are becoming ever more important—they're becoming the lubricant of collaboration. Elements such as participatory all-hands meetings, theme-specific open spaces and open regular meetings help create collaborative offerings for employees that provide meaning and value." (W. Ebner, Head of Social Media Business Program, Telekom Deutschland, 2017)

7   Conclusion

In summary, it is evident that a great deal of work is being done at Deutsche Telekom AG on the key areas in the context of New Work. Issues such as leadership/mindset, flexible working, agile organization and communities have been developed and pushed forward over a long period in the framework of the DIGITAL@WORK program.

Today, the further development of these and other topics is no longer happening solely as part of the program, but rather in the aforementioned organizations or communities.

Apart from the general euphoria, there are also downsides in this respect. Challenges are emerging—especially with regard to culture—particularly at the interfaces between the worlds of analog/digital, green/blue, exploit/explore, etc.

But the company is rising to these challenges and continues to develop further in all its facets—for the sake of a future-proof working environment and work culture in the digital age.

About the Authors

Bettina Arnegger is a Global Automotive Sales Manager at NVIDIA, working in the field of Autonomous Driving. Prior to that, she has been employed with ZF, where she held positions in Corporate Market Development and Car Powertrain Technology. After graduating from the executive Master program "Mobility Innovations" at Zeppelin University, she recently entered a PhD program at the Chair of Univ.-Prof. Dr. Wolfgang H. Schulz for Mobility, Trade, and Logistics.

Clemens A. Aumann, Managing Consultant at Detecon International, contributes since 1992 to the rise and evolution of telecommunications. Combining visionary sight with the ability to deliver on site, he strives for end user value-add and pushes for the (re-)invention of the classic telecommunications business model.

Jörg Borowski, Managing Partner at Detecon International, has been advising clients in the telecommunications and ICT industry for more than 20 years. His main field of expertise revolves around the optimization of the interface between technology and business. In addition to network economy issues, he examines future technologies in networks and their impact on process and organization structures.

Edgar B. Cardozo Larrea, Managing Consultant at Detecon International, has more than 16 years of experience in telecommunication markets. During his career, he gained profound expertise in regulatory matters and wholesale strategies, with a focus on network economics, financial modeling, and cost modeling for mobile and fixed operators as well as for regulators in Europe, Asia Pacific, the Middle East, and Latin America.

Christian Dietze, Partner at Detecon Consulting in Abu Dhabi, has more than 12 years of consulting experience in the international telecommunications industry. He has held various leading positions in the telecommunications industry. In TM Forum he has a key role in the development of eTOM and in the elaboration of Digital Maturity Models for Communication Service Providers.

Ulrike Eberhard, Managing Partner at Detecon International and responsible for Detecon in Latin America, has more than 20 years of experience in

© Springer International Publishing AG, part of Springer Nature 2019
P. Krüssel (ed.), *Future Telco*, Management for Professionals,
https://doi.org/10.1007/978-3-319-77724-5

telecommunications consulting and management. She worked with the C-Level of mobile and integrated operators in EMEA and America. Her areas of expertise are corporate strategy, finance, innovation and product development, regulations, and wholesale.

Tillmann Eckstein is Senior Consultant at Detecon International. He has more than 25 years of experience in telecommunications, especially in design, planning, regulation, and rollout of radio systems. His focus is still on mobile radio systems with advanced features as massive MIMO aiming for a holistic view starting from the backhaul/fronthaul connection up to EMF issues.

Claus Essmann is Managing Consultant at Detecon and has more than 18 years of experience in international IT and telecommunication projects with operational and strategical background. His focus is on consultancy and implementation of end user device, M2M/IoT, and Smart City service strategies.

Eva Faust is Director Corporate Development at Telekom Deutschland GmbH. She has cross-segment experience in strategy with a focus on transformation and on market and strategy development.

Lutz Fritzsche, Managing Consultant at Detecon International, has more than 25 years of experience in consultancy especially for planning and optimization of telecommunications networks. Moreover, he actively participates in the advanced development and sale of Detecon's own planning tool NETWORKS and is the contact consultant for numerous clients using the program.

Michael Gerke, Managing Consultant at Detecon International, founded in 2016 the Robotics Community and is responsible for the RPA/AI advisory business. He has more than 20 years of experience working for Global Telco and IT companies focusing on process and service improvements.

Carsten Glohr, Managing Partner at Detecon International, is Member of the Management Board and Practice Leader "Public and ICT Service Provider." He has more than 20 years of experience in consultancy for ICT service providers. His focus is on Digitization, ICT strategies, ICT restructuring, and ICT efficiency programs.

Friederike Göbel, Business Analyst at Detecon International, studied business administration at the WWU Münster and the University of Cologne with a focus on marketing. During her studies, she gained experience in strategy consulting and in the telco industry. Since joining Detecon, she has been working on strategic and digital transformation projects for large companies in the telco and travel industry.

Christoph Goertz is Managing Consultant with a focus on communication technology. He is an expert in mobile networks and solutions such as Edge Computing,

5G, LTE Advanced, IoT, IMS/VoLTE, and WebRTC and provides consultancy in several innovation topics based on his broad technology background.

Jan Grineisen is Senior Consultant at Detecon International and with the strategy, innovation, marketing, and sales practice. He has more than 6 years of experience in consulting for international telecommunication operators with a focus on commercial strategy, customer experience management, and omni-channel strategy.

Joachim Hauk, Managing Consultant at Detecon International, is Member of the Global Industry Practice Board Carrier/Core Telco. He has more than 18 years of experience in consultancy for telcos, IT service providers, and financial industry customers. His topics of expertise are customer centricity, customer experience management, omni-channel design, and marketing strategies for telcos.

Frank Heil is Vice President Organizational Development in the global competence center HR Management at Deutsche Telekom AG. He has more than 17 years of experience in consultancy, strategic and restructuring projects in the telco and high-tech sector. He focuses on strategy and organizational development balancing reorganization projects, organizational guidelines, and design of future and market-oriented organization concepts in the telco industry.

Thomas Heilen is Head of Product Management Consumer Telekom Deutschland GmbH. His understanding of telco business is based on more than two decades in product development for mobile and fixed network business. He strongly believes in the right to play for incumbents like DT in the digitalization game as a next level for the industry.

Jan Philipp Heinemann, Senior Consultant at Detecon (Schweiz), is member of the Financial Services industry team. He has worked on various projects in the telecommunications, public sector, and financial services industries. His focus is on digital strategy including new technological trends like Blockchain or AI and the assessment of their economic, social, and political implications.

Laura Hellweg is a business psychologist and works as a senior expert in the field of organizational development within Deutsche Telekom group. With a focus on future-proof organizations, she develops concepts for new organizational structures. Due to her experience gained in the in-house consulting department at Deutsche Telekom, she has a well-rounded understanding of the DT group and a solid background in project management.

Arnulf Heuermann, Managing Partner at Detecon International, is leading the global competence center for wholesale and regulation. He has more than 30 years of experience in consultancy for regulatory authorities, international investors, and telecommunications operators. His focus is on regulatory topics, corporate and marketing strategies, and corporate finance.

Chin-Gi Hong, Managing Consultant at Detecon International, has more than 10 years of consulting experience in telecommunications. As an eTOM- and ITIL-certified process expert, he has managed large transformation projects for clients all over the world, as well as successfully led Detecon's Global Knowledge Community "Digital Business Process Management."

Riem Jalajel, Consultant at Detecon International, is experienced in complex transformation programs as well as digitalization and innovation projects. Her special interest and expertise lie in partnership strategies, partner management, and the relation between go-to-market success and the integration of start-ups and OTT players within established corporations.

Birinder Singh Khurana, Senior Consultant at Detecon International, has extensive experience in both corporate transformation programs and software development for telecommunication companies. His field of expertise revolves around strategic transformation projects across different business units and business functions for telecommunication companies.

Krzysztof Korzunowicz is Managing Consultant at Detecon International. For the last ten years, he has worked for two major European telecom operator groups in multiple different areas including procurement/vendor relations, network technology, and software/automated test dev. Lately, he has been involved in new telco production methods and, even more importantly, culture change. He is a proponent of Scrum and Scaled Agile concepts.

Peter Krah is Managing Consultant at Detecon International. He has more than 20 years of international experience in the telecommunications industry for operators, vendors, and investors. His focus is on strategic and technical topics, especially in the fields of planning, rollout, modernization, and optimization on the basis of current and future radio technologies.

Christian Krämer, Managing Consultant at Detecon International, is a generalist in strategy, organization, and sales/marketing with more than 40 managed projects and 17 years' consulting experience. As an expert for digital transformation, he drives the German All-IP transformation and promotes partnerships/partner ecosystems by leading Detecon's Partnering Excellence community.

Sascha Krpanic is a Senior Management Consultant at Detecon International since 2013 dealing with multinational clients in the telecommunication and insurance industry. He specializes in corporate and commercial strategies, market intelligence, as well as market positioning.

Peter Krüssel is Head of the Global Industry Practice Carrier/Core Telco and Member of the Management Board at Detecon International. He has more than 20 years of experience in consultancy for telecommunications operators. He focuses on business development and the interrelationship between regulation, technology, innovation, and marketing in the telco industry.

Andreas Lischka is the Director Strategy at Telekom Deutschland GmbH. He has many years of experience in the telecommunications industry working in the fields of technology, product management, marketing, and strategy.

Thorsten Lotz is Senior Consultant at Detecon Asia-Pacific with a focus on future network infrastructures and telco transformation. He has around 10 years of experience in telecommunications with various positions in R&D and consulting at leading vendors and operators in Germany, the Middle East, North America, and South-East Asia.

Gregory Lukowski, Consultant at Detecon International, has experience in the realm of start-ups, consulting various early-stage start-ups especially in the areas of business, go-to-market, and pricing strategy. Since he is with Detecon, his focus has been on innovation management, venture building, design thinking, and business model innovation.

Martin Lundborg, Managing Consultant at Detecon International, has worked as consultant and manager for telecommunications regulations and wholesale service the last 17 years. Based in Germany and Austria, he has managed projects for operators, regulatory authorities, and industry organizations on five different continents.

Patrick Ma is Senior Consultant at Detecon International. He has successfully delivered many managed service projects across the world, for both network operators and service providers. In the previous employment, he was responsible for network interconnection, contract management, regulatory issues, and industry-wide number portability and operations code of practice.

Andreas Mark has been working in portfolio management of stocks at Union Investment since February 1998. He manages European stock orders for institutional investors and is in charge of the analysis of European and North American telecommunications companies. From 1988 to 1998, Mark worked as senior portfolio manager at DZ Capital Management, where he was responsible for the management of European equity funds and mixed portfolios.

Björn Menden is Managing Partner at Detecon International and global practice leader Core Functions and Optimization. He has extensive experience with clients in EMEA. Björn has been responsible for restructuring major players in telecommunications and other industries. He supports clients in increasing their competitiveness and implementing sustainable strategies.

Nikolai Nölle, Consultant at Detecon International, advises in marketing, CRM, and innovation projects with proven expertise in the field of value-added services, M2M, IoT, and big data. His focus is on customer orientation and commercialization within the (further) development of products and use cases and in relationship management, customer experience management, and communications.

Carolin Obernolte works as consultant within the telecommunications sector at Detecon International. She has international project experience in the areas of strategy, marketing, and innovation and also supported multiple transformation projects.

Andreas Penkert, Managing Consultant at Detecon International, has more than 18 years of experience in project management and consultancy. He is expert for customer centricity, digitization, and omni-channel strategy. He managed lots of projects in different industries and is author of several publications concerning customer topics.

Hans-Peter Petry has been a Managing Partner and Member of the Executive Board at Detecon International until 2011 and after his retirement still works part-time as a Senior Advisor. He has more than 35 years of experience in telecommunication technologies. Currently, he is also President of the German Center of Satellite Communications on a voluntary basis.

Mario Pieper studied business and economics at the University of Münster. He started his career in 1999 at the strategy consultancy of Accenture. In 2002, he moved on to Deutsche Telekom. After different positions within the strategic corporate development and the operational management of Telekom Germany, Mario Pieper became Cofounder of a start-up in the smart home environment. Since 2014, Mario Pieper is the Chief Digital Officer of BSH.

Andreas Rauch, Consultant at Detecon International, has 4 years of experience as an academic researcher in the field of innovation management and 2 years as a consultant. For his PhD, he investigated success factors of the integration of suppliers and sales in new product development processes. So far, his focus as a consultant has been on IoT, digital twin technologies, and developing and implementing digital strategies.

Giulia Rehme is a Consultant at Detecon International focusing on strategy and innovation in the telecommunications industry. Her current activities focus on the analysis of future scenarios for the global telecommunications market and the development of strategies to compete against OTT players.

Lothar Reith, Senior Consultant at Detecon International, is a recognized expert in the area of network transformation with deep knowledge in ICT, based on over 40 years of experience in digitization (30 years in telco industry) in various roles at

Nortel and over the last 12 years in consulting for national and international operators and enterprises in the areas of fixed, mobile, data, voice, and IT.

Stefan Rinkel is Vice President of Network Strategy at Telekom Deutschland. After his studies of electrical and telecommunication engineering at the University of Karlsruhe, he joined Deutsche Telekom Group in 1990. He has over 20 years of experience in national and international telecommunication market strategy with a special focus on technology strategy.

Steffen Roos, Partner at Detecon International, helps different clients in different industries to handle the challenges of digital transformation especially in the area of technology innovation, customer touchpoints, and social and agile collaboration.

Thomas Rosendahl, Project lead at Deutsche Telekom, is a thought leader for the next-generation Internet and the responsible manager for the project IMAGO which aims at building the virtual representation layer that powers the next-generation Internet. With more than 20 years' business experience and a strong product development and marketing background, he has driven multiple innovations and market introduction.

Saher Salem, Technology Consultant at Detecon International, has more than 5 years of extensive experience in mobile networks and satellite communications. He conducted various projects at Deutsche Telekom as well as different government entities in the Middle East and North Africa (MENA) region.

Carolina Schiefer, Senior Consultant at Detecon (Schweiz), has more than 5 years of consulting experience focusing on business development and digital transformation for various industries. Moreover, she works on innovation management and innovation scouting topics in order to elaborate digitization potential.

Christian Schilling, Management Consultant at Detecon International, has more than 4 years of experience in consulting for international clients with a focus on digital strategy, innovation, and the Internet of Things in particular.

Dennis Schmedt is project manager and digital operator in Deutsche Telekom's new founded department HR digital and innovation. Embracing both profound management know-how from various management positions in HR, latest as chief of staff HR, and deep understanding of digital technologies and tools is his equipment for working and leading in the digital age.

Manfred Schmitz, Managing Partner at Detecon International GmbH, provides 20 + years of experience in telecommunication business, having worked for suppliers, tier-1 carriers, and top strategy consulting firms. He is Detecon's thought leader for CAPEX and technology strategy and planning, operational business models, and efficiency topics, e.g., managed Services, sharing, and automation including RPA.

Dominik Schneider is Business Analyst at Detecon International. He holds a degree in business informatics and has more than 4 years of experience in the telecommunication industry. His focus is on big data value creation as well as on IoT business models. He has gained experience in various projects on these topics and has also written several publications.

Falk Schröder, Managing Partner at Detecon International, is since January, 1,2018, Member of the Management Board and Practice Leader Networked Infrastructure. He has more than 20 years of experience in international consultancy for telecommunication operators. His focus is on latest technology developments like 5G, low latency, and its impact for the digitization of industries.

Wolfgang H. Schulz is an economist and Dean of the Corporate Management and Economics Department at Zeppelin University. He is founder and CEO of the Institute for Economic Research and Consulting. The research focus is on mobility as a service, cooperative, connected and automated mobility, and electric mobility. The CONVERGE-project was an empirical use case for the Institutional Role Model. Related projects are SIM-TD, eCo-FEV, DRIVE C2X, and STILLE.

Mathias Schweigel Managing Consultant at Detecon International GmbH, planning and optimizing telecommunications networks since 1998. He is supporting network operators worldwide, mainly by using Detecon's planning software NetWorks. He is product manager of NetWorks Mobile Forecast for network capacity monitoring and NetWorks TCM for strategic network planning.

Wolfgang Specht began his professional career as a strategy consultant in the telecommunications and media industry in 1996, when he began working for a subsidiary of Deutsche Telekom. He became a sector specialist and changed over to the sell side in 2000. He has more than 15 years of experience as an equity analyst in the sectors telecommunications, media, and Internet while working at DZ Bank, Oppenheim Research, WestLB, and Bankhaus Lampe.

Julia Steffens is a consultant at Detecon International. Her main areas of expertise lie in developing company strategies for omni-channel management, customer experience, marketing, sales, and service with regard to digitalization. In her current project at a major telecommunication company, she is developing a digital marketing concept for after-sales campaigns.

Markus Steingröver is Managing Partner at Detecon International in the Practice Carrier/Core Telco. He has more than 20 years of experience in consultancy in international telecommunication markets serving operators and regulators. His focus areas are wholesale and regulation as well as the evolving digital regulatory framework.

Christoph Stummer is Managing Consultant at Detecon International. He has more than 20 years of experience in the field of end-to-end process management consulting, process automation, and transformation management in several markets, especially in the market of telecommunication provider.

Thomas Vits is Managing Consultant with more than 15 years in the mobile industry focusing on devices and communication technology. He held various R&D positions in consumer electronics companies where he was involved in the planning, development, and launch of embedded products (e.g., mobile phones) and services, including IMS/VoLTE.

Marc Wagner is Managing Partner and Practice Leader "New Work, Ecosystems and Company Building" at Detecon International. He has more than 20 years of experience in the field of digital transformation. His mission is to unleash the creative potential of employees, to enable organizations' human resource management to succeed in the digital age, and to enhance their innovative capability.

Daniel Weber, Consultant at Detecon (Schweiz), has several years of experience, especially in the telecommunications sector. His focus is on digital strategies and particularly the Internet of Things (IoT).

Horst Wieker has a chair for telecommunications and is head of research for traffic telematics at the University of Applied Sciences of the Saarland. His research focus is on the system design of communication networks and on traffic telematics. In 2000, he founded the EuroTec Solutions GmbH as an AT Institute of the HTWsaar. He is active in many research projects, e.g., Prevent Willwarn, CVIS, AKTIV, SIM-TD, Urban, CONVERGE, and 5GNetMobil.

Frank Wisselink is managing consultant at Detecon International. Frank has 20 years of experience in consulting, managing large organizations and leading key programs within ICT, energy, and high-tech and consumer electronics. He focuses on big data/artificial intelligence value creation, (Narrowband) Internet of Things, smart cities, and the associated organizational transformation.

Michaela Wolfering-Zoerner is a member of the division Corporate Development at Telekom Deutschland GmbH. She has many years of experience in the fields of strategy and marketing in the telecommunications industry.

Rong Zhao, Managing Consultant at Detecon International, is the leader of 5G team. He has more than 18 years of experience in international ICT markets. His focus is on the strategic planning, implementation, and techno-economics of fixed and mobile networks. He is currently the chair for Deployment and Operations Committee at FTTH Council Europe.

Nikolay Zhelev, Managing Consultant at Detecon International, based in the headquarters in Cologne. He has more than 14 years of experience in telecommunications engineering, project management, and consulting. His focus is on technology strategy, network architecture and design for mobile networks, as well as techno-economic analysis for mobile operators and regulators.